Mathematical Models in Ecology

Mathematical Models in Ecology

The 12th Symposium of
The British Ecological Society
Grange-over-Sands, Lancashire
23–26 March 1971

edited by
J. N. R. Jeffers F.I.S.

The Nature Conservancy,
Merlewood Research Station,
Grange-over-Sands,
Lancashire.

Blackwell Scientific Publications
Oxford London Edinburgh Melbourne

Osney Mead, Oxford
3 Nottingham Street, London, W1
9 Forrest Road, Edinburgh
P.O. Box 9, North Balwyn, Victoria, Australia

ISBN 0 632 08740 4

First published 1972

Distributed in the USA by
F.A.Davis Company, 1915 Arch Street,
Philadelphia, Pennsylvania

Printed and bound in
Great Britain by
W & J Mackay Limited, Chatham, Kent

Contents

Preface

There has been an increasing interest in the use of mathematical models in ecology, as revealed by the papers submitted to the journals of the British Ecological Society, and by national and international ecological research projects which have been started during the last few years. Several symposia on the use of mathematical models have been held, notably the symposium held at Yale, in 1968, and the meeting arranged by the Agricultural Research Council at the Grassland Research Institute in 1969. The British Ecological Society asked me to act as Convener for a symposium on the use of mathematical models in ecology, with the aim of providing a meeting point between the various individuals actively concerned in the development and application of mathematical models in this field, and, hopefully, to provide the initiative for a continuing dialogue between mathematicians and ecologists.

The symposium was held at the Grand Hotel, Grange-over-Sands, close to the Nature Conservancy's Merlewood Research Station, which, together with the Biometrics Section in London, is the main focus for the introduction of mathematical models in the Nature Conservancy's research. The staff of the Research Station were, therefore, able to participate fully in the symposium, and to help with its organization. In particular, the Society is indebted to Mr A.J.P.Gore, who took charge of much of the organization of speakers, the visual aids, and the arrangements for the actual presentations, and to Mrs P.A.Ward, who made all the arrangements for accommodation, bookings, and the collection of payments. Mrs Ward has also been primarily responsible for the processing of the manuscripts and figures for this volume. Several members of the staff acted as rapporteurs for the discussions, including Dr O.W.Heal, Mr P.J.A.Howard, Dr Helen E.Jones and Mr D.K.Lindley, and the Society's grateful thanks go to them for undertaking this difficult task.

The papers collected together in this volume represent, in a very real way, the current state of the art of using mathematical models in ecology and related sciences. For this reason, every effort has been made to publish the symposium proceedings as soon as possible after the symposium. In a field of research which is developing rapidly, and for which new applications are being found at an increasing rate, papers which are topical at the time of presentation quickly become dated. It is hoped, nevertheless, that these proceedings will prove valuable in collecting together, in a more permanent form, the wide range of experience which made the symposium itself both stimulating and exciting.

J.N.R.Jeffers

The challenge of modern mathematics to the ecologist

J.N.R. JEFFERS *Merlewood Research Station, Grange-over-Sands*

Summary

Modern developments in mathematics and in computer technology have made available new techniques and new conceptual models to the working scientist. Ecology, being essentially concerned with the complex interaction of living organisms with their environment, has the greatest need for such developments, but can ecologists and mathematicians bridge the gap between their disciplines?

1. Which challenge?

In contributing the opening paper to this symposium, it falls to me to try to set the scene for the symposium as a whole. The title of the symposium is 'Mathematical models in ecology', and I have chosen to develop this theme in relation to the developments which have taken place in mathematics, and particularly in mathematical statistics, and in electronics, and more especially in the use of analogue and digital computers.

Essentially, I see the present state of the art of mathematical modelling in ecology as a challenge of modern mathematics to the ecologist. It is, however, essential to know which challenge we are talking about, because there was an earlier challenge which many ecologists (and others) have never taken up. This earlier challenge came from the work of R.A.Fisher at Rothamsted, and was developed by him and by such eminent statisticians as Yates, Finney, Wishart, and Bartlett, to name only a few from this country alone. The development of an adequate theoretical basis for small-sample statistics has led to efficient methods for the design of experiments and surveys, and for the valid analysis of the data collected by such methods. The theory demonstrated unequivocally the need for replication and randomization in experiments and surveys, and has stressed the dependence of valid inferences on the role of fair samples from defined populations. The power of the methods developed has enabled vast amounts of information to be derived from relatively small samples, but the really important lesson to be learnt is that no amount of data, or analysis, can compensate for the confounding of experimental factors with environmental influences, and unless experiments are planned to include interactions between separate factors, no information can be obtained about such interactions. There are many other important facets of experimental and survey design and analysis which are derivatives of the first challenge that the mathematicians gave to scientists from the early 1920s onwards, but those that I have mentioned will suffice to make my main point, which is that the end results of scientific research are inextricably bound up with the methods of sampling that are employed. If the research worker is not himself competent to assess the implications of the methods he uses, he is wise to seek the advice of a qualified statistician before embarking on a particular programme of scientific research.

Now all this may seem patently obvious to everyone attending this symposium. Nevertheless, a very great deal of research is still undertaken by biologists and others where even the most elementary principles of the design of experiments and surveys are ignored, and, what is worse, this research is backed by money which is drawn from the tax-payer or from the shareholders of industrial and commercial companies. As a statistician, I still have to spend far too much of my time trying to find ways of making use of data for which the methods of collection were inefficient, if not invalid—although not within my own research station I hasten to add. Papers are still being published in reputable journals where the methods of data collection and statistical analysis are deplorable if not impossible, and it is even worse that, with few exceptions, these papers have been read and approved by referees. We still encounter the post-graduate degree student in despair, because he has spent two and a half years collecting his data, and he now finds that even his supervisor has very little idea about how these data can be analysed or presented, or even whether the method of data collection is relevant to the problem he is undertaking. I still have to endure committee meetings at which sometimes distinguished members make statements about the unimportance of mere methodology by comparison with objectives, as if it were possible to separate the ends from the means in scientific research.

My first point is, therefore, that many ecologists, among others, still have not responded to the first revolution in scientific method, and to the first challenge posed by mathematics. My own private estimate, based on experience of consulting in university, industrial and government research, is that at least 50 per cent of biological research is wasted by bad experimental and survey design and by inefficient methods of analysis. Furthermore, the lack of knowledge of basic scientific method applied to numerical information is worst at the more senior levels of research, which is perhaps to be expected, as young biologists are nowadays usually subjected to some training in statistics, if not mathematics. As the planning of research depends mostly on the senior staff in a university or research organization, improvements in the efficiency, or indeed the validity, of much of our research can only be expected to take place fairly slowly.

But time is not on our side. Despite the fact that many biologists have not yet responded to the first challenge—if you agree with my thesis so far—we are facing them with a new challenge. Yates has called this the second revolution in statistics (Yates 1966)—a revolution that was born of the development of the computer, but has since been overtaken by the development of mathematics itself, and by the development of the role of mathematics in conceptual models. It is this challenge on which I intend to concentrate in the remainder of my introduction to this symposium.

2. The nature of the challenge

Having said that the second revolution in statistics was born of the electronic computer, it will be as well to clarify the precise role of the computer in the initiation of the new wave of applications of mathematical techniques. This role may, perhaps, be summarized as follows:

First, it is obvious that the electronic computer, digital or analogue, greatly increases the speed of computation, and by so doing makes possible more complex computations than were possible when the only aids to computation were hand or electric desk-calculating machines. Second, and even more important, however, is the fact that the methods of programming electronic computers which are essential for their use have made available precise descriptions of methods for the interpretation of data and for the calculation of results. So precise are these descriptions that it is even possible for someone who does not understand the computations to carry them out on his own data, and while he may use a model for the wrong purposes, he has little or no excuse for carring it out incorrectly. Third, the use of electronic computers has made it possible to store vast quantities of data in readily accessible forms, and to ensure that these data can be used by many different scientists.

Freed from the need to carry through computations in laborious ways on desk calculating machines, or, even worse, with pencil and paper, mathematicians have been able to develop new ways of looking at problems. The mathematician's approach to both theoretical and practical problems has always been to construct a model of the essential elements of the problems. He differs from scientists of other disciplines in that his models are abstract models of the real situation.

Other scientists, including biologists and ecologists, also use models, but they are more likely to use analogies of physical models of direct cause and effect relationships. The abstract nature of the mathematician's model is, however, its essential feature. The abstraction helps to avoid the identification of essential elements of the model with non-relevant connotations, so often a feature of models which rely on words or on physical analogues. Furthermore, the concentration of attention on the abstract elements helps the mathematician to define models with the smallest possible number of elements, and so to retain parsimony of description as a key principle in the selection of alternative hypotheses or models. The replacement of entities by abstractions also enables the mathematician to employ a calculus of reasoning, using symbols in which the only criterion is the logic of the deductive reasoning. The resulting simplification greatly aids the presentation of the logical argument, although most non-mathematicians, to whom the symbols and the grammar of deductive reasoning are unfamiliar, would probably regard this form of presentation as more confusing than helpful!

The early mathematical models of physical and biological situations were necessarily fairly simple and straightforward. If they were to be applied to practical problems they had to be simple enough to allow the estimation of the basic parameters with such computing aids as already existed. Basic techniques, such as analysis of variance, multiple regression analysis, and factor analysis are dependent on mathematical models which are largely constrained by requirements of linearity, orthogonality, and additivity, and, if valid tests of significance are to be performed, of normality. These constraints originated in the need to define mathematical models which are sufficiently simple for the computing to be feasible with primitive calculating machines. Most of the experimental designs in regular use, for example, are constrained by the need for orthogonality, which simplifies the estimation of the parameters of the underlying mathematical model.

The release of mathematical models from these computational constraints has also released the models from constraints in their formulation. In experimental design, the ability to determine the effects of non-orthogonal factors has enabled experiments to be tailored to the available experimental material, rather than the trimming of the material to meet the design constraints. Provided that the experimenter is prepared to be explicit about the hypotheses he wishes to test, he can now design experiments which have maximum efficiency for the available experimental material (Pearce 1963). Many biological and physical problems can only be expressed with difficulty in terms of linear functions, and the usual expedients of polynomial expressions do not often give satisfactory representations of, for example, biological growth functions. The release from constraints of linearity and additivity enable models requiring non-linear estimation to be employed, often exploiting the power of the electronic digital computer to perform iterative computations easily and rapidly. While the appeal to the Normal distribution still provides the basis of much of the testing of statistical significance, the consequences of non-normality in estimating procedures can now be readily tested by Monte Carlo techniques, and a recent text on statistical methods in engineering (Hahn & Shapiro 1967) demonstrates how important many of the non-Normal distributions have become in the formulation of models of the physical sciences.

But it is not only in the field of statistical mathematics that the release from computational constraints has stimulated the development of the mathematical techniques themselves. Perhaps one of the most important developments has been in the integration of sets of differential equations, so opening up a new wave of interest in deterministic models for which the original parameters are rates of change. Several of the papers at this symposium will be concerned with the exploitation of this kind of model. Another interesting, and much under-exploited range of models is derived from the fast-developing theory of networks. The critical path network is only one of the many applications of this theory in everyday practice. Associated with the subject of operations research, the theory of optimization of functions within constraints has opened up a whole field of conceptual models. Even the relatively simple problem of finding the optimum of a linear function, with linear constraints, required considerable computing power for worthwhile applications. The development of an adequate theory for non-linear functions, with linear or non-linear constraints, has had to wait for recent increases in computational power, but now offers considerable scope as mathematical models in biological research. The theory of games, with Nature as an opponent, may seem somewhat far-fetched as a mathematical model for biological research, but there are many situations where decisions which have an uncertain outcome have to be made, and where this theory might be applied as a first approximation—its development had to wait for the computational power of the modern computer, and for the precision of expression of the computer algorithm. Finally, I should mention the development of an information theory which is capable of handling not only attributes, and discontinuous variables, but also multi-state variables where there is no constraint on the linearity of the states. Numerical taxonomy (Sokal & Sneath 1963) is only one of the applications of this theory, which has wider implications than the classification of plants, animals, habitats, and ecosystems.

3. The implications of the challenge

The vastly increased power of mathematical models and their increased range, resulting from the development of computers and the consequent development of mathematical theory, has three important implications for research workers. First, the conceptual models for scientific research have now been extended far beyond the simple cause and effect mechanisms which are implicit in physical models. Notions such as 'feed-back' have already gained acceptance in everyday speech without gaining similar acceptance in conceptual models for scientific research, and there are many similar concepts which extend the research potential of individuals and the research organization of which they are members where such ideas are allowed to influence the conceptual models which underlie their research.

Second, the increased flexibility inherent in the increased range of conceptual models places even greater emphasis on the formulation of hypotheses to be tested in scientific work. Even where hypotheses are expressed in purely verbal terms, their careful formulation is an essential part of the whole research process, and one which should precede the collection of data and their analysis. This has always been true, but formulation of these hypotheses has become of paramount importance now that the range of conceptual models has been so greatly extended. As scientists, we sometimes seem to be preoccupied with data collection and analysis when we should be more concerned with the definition of the hypotheses to be tested as expressed in the conceptual models which we will choose to use. The choice is no longer limited to a small range of relatively simple models, and we will need to match the form and the complexity of our models to the objectives of our research.

The third implication is, however, perhaps the most important. If objectives and models can now be matched more exactly, it becomes vitally important that those who set the objectives of scientific research should be aware of the models which are available. The chief danger lies in the setting of objectives which are too limited in scope and in practical value because of false judgements about the level of complexity that can be achieved with current models. 'Simplicity' of experimental design is still regarded as a virtue by research workers, even when the simplicity referred to means the failure to investigate interactions between factors. So, too, the failure to allow for non-additivity or feed-back in basic mathematical models by a spurious appeal to simplicity may reduce an otherwise acceptable model to a dangerous and unreliable generalization. In operational research terminology, the result is 'sub-optimization' in the value obtained from research, which may be both expensive and time-consuming. The practical problems which need to be solved in our technological society have to be defined clearly and exactly, but once these problems have been defined, we still have to choose the objectives of our research, and choose these in relation to conceptual models which are now available to us. Nowhere is this need more important than in the environmental sciences, of which ecology is perhaps the most dependent upon adequate models.

The need is for the use of mathematical models in strategic rather than in tactical thinking, for the models to be integrated into the whole strategy of the way in which ecological research is planned, organized, and executed, and not merely relegated to the design and analysis of individual experiments and surveys. The developments in mathematical models which I have outlined are vital in the definition of a new philosophy of ecological research, and cannot be relegated to 'mere methodology'.

4. How can the challenge be met?

It is my contention that neither the mathematician nor the ecologist can meet the challenge alone. It is not sufficient for the mathematician to pose the challenge and then wait for the ecologist to take it up—both are, after all, interested in the outcome, even if the mathematician can always retreat to some quite imaginary world defined solely by his axioms. In the solution of the practical problems of ecology upon which the future of our world depends, the mathematician is unlikely to have sufficient knowledge of ecology to guide the selection of appropriate mathematical models, so that his world may become imaginary against his wishes. Similarly, the ecologist is unlikely to have the necessary mathematical knowledge to enable him to select appropriate models for his research.

It has become almost trite to suggest that the solution must lie in better communication between the mathematician and the ecologist, but the statement is no less true for being frequently made. As the problems that we tackle become larger, and beyond the scope of a single research worker and his assistant—the unit still perpetuated by the way in which research is organized in many universities—it has become necessary to undertake research by project teams. It would seem relatively simple to ensure that such teams had an adequate balance of mathematicians and ecologists as well as the other disciplines, e.g. chemists, physicists, meteorologists, etc., but the current shortage of mathematicians and statisticians has made it difficult to ensure an adequate balance. Even where a mathematician is available, the differences in language, approach, and even mental imagery, may limit severely the degree of rapport which can be established within the project team.

One interesting, but disturbing, consequence of this shortage of mathematicians to work in ecological project teams has been the development by ecologists and others of their own mathematical methods. As a result, we have the situation that the same method is 'rediscovered' a large number of times independently, and, all too often, that method has been discovered and rejected for some important reason several years previously in the mathematical literature. It is true, of course, that the description of the method in the literature was probably incomprehensible to anyone except a mathematician! One area where this mushroom growth of methods which are only apparently new has taken place is in numerical taxonomy, where each practitioner seems more concerned to develop a new technique than to apply already developed methods to practical problems.

In the belief that one should remove the beam in one's own eye before insisting on the removal of the specks from other people's, I am convinced that much of the blame for inefficient research lies with research directors. Research strategy is essentially the business of research directors, and is one part of the director's function that cannot be delegated to the research scientists themselves. He must see, for example, that there is adequate provision for advice on mathematical and statistical problems, and that the advice is taken. There are many problems in the organization of statistical consultancy within a research organization (Sprent 1970) and the solution of these problems depends very largely on the attitude and competence of the research director. There must also be adequate provision of efficient and sufficient equipment for computing and data capture, and, again, this will only be forthcoming if the research director is prepared to see that the equipment is properly budgeted for, and that effective management systems are adopted in its use and supervision. There is little doubt, for example, that the widely reported failure of many computer systems in business and industry is due to lack of direct interest by senior management in their application.

But, by far the most important of the duties of the research director is his constant evaluation of the methodology as well as the aims of the research carried out by the scientists under his control. Only if the objectives of the research are properly matched to the methods used in the collection, analysis and interpretation of the data can there be any hope of an efficient research programme. Where the conceptual models that are used in the research do not take full advantage of methods that have been developed, the wastage of the resources of the organization are as severe as in any form of lack of financial control. Many organizations insist on the latter without paying any regard at all to the former, and, to a large extent, the only person who can readily influence the adoption of adequate conceptual models is the research director.

5. The role of mathematical models in future research

Having outlined the nature of the challenge which I suggest is posed by mathematics to the ecologist, it may be as well to consider briefly the role of mathematical models in future research. While many ecologists accept, perhaps reluctantly, that some mathematics must be involved in ecological research, they often see mathematical calculations as occupying a relatively unimportant place in the analysis and interpretation of results. There is a frequent confusion between mathematics and arithmetic, a confusion that is helped by the fact that many mathematicians are not particularly good at arithmetic, and that there are many good arithmeticians who understand relatively little about mathematics.

I have suggested that the most important impact of mathematics in ecology arises from the adoption of mathematical models as the conceptual models which underlie ecological research and management. One of the most important roles of these models is that of simulators of ecosystems, in much the same way that it is possible to simulate the control and flight of an airliner, or the sending of a man to the surface of the moon. If it were possible to construct a mathematical model of a complete ecosystem, then it would be possible to test out various ideas about the management or manipulation of that ecosystem in anticipation of the practical application of those ideas. Such a simulator requires considerable knowledge of the basic working of the ecosystem, but the expression of the available knowledge in the form of a model enables it to be used practically at the earliest possible time. One of the urgent problems is then to determine which practical problems require the simulation of the total ecosystem, and which can be solved by the simulation of only part of the ecosystem. The more precise definition of the practical problem, in its turn, has the effect of improving the efficiency of the research, but, clearly, can only be achieved if the managers of ecosystems are capable of formulating their problems precisely. While total ecosystem models have been proposed in many of the concepts which underlie the investigations of the International Biological Programme, it seems likely that partial models will be more efficient than models of the total ecosystem, and likely that the former can be achieved more readily than the latter.

Mathematical models are also important in the formulation of new theories about ecosystems and ecology. It perhaps has to be emphasized that there are no mathematical methods for the formulation of hypotheses—although many research workers seem to cherish the illusion that that is what statistical methods are for! The manipulation of relationships expressed in mathematical models does, however, enable the relationships to be explored, and various hypotheses about the ways in which the relationships behave to be formulated. It is still, of course, necessary to test these hypotheses by experiments in which new data are collected—and here is a new trap for the unwary—but much time can be saved in the early stages of hypothesis formulation by the exploration of those hypotheses through mathematical models. Similarly, mathematical models can be used readily to investigate phenomena from the viewpoint of existing theories, by the integration of disparate theories into a single working hypothesis, for example. Such models may quickly reveal inadequacies in the current theory and indicate gaps where new theory is required. So, too, the extraction and processing of experimental results for compilation in a form suitable for mathematical abstraction may indicate the necessary parametrization in conceptual models.

6. The contribution of this symposium

Several symposia have been arranged with the purpose of bringing ecologists and mathematicians into communication. It is usually reported, however, that the meeting ended with the participants split into two groups, the mathematicians talking mathematics among themselves, and the ecologists talking ecology among themselves. It is our hope that this symposium will achieve a dialogue between the mathematicians and the ecologists who are present, and, perhaps, provide the initiative for the continuation of this dialogue over the coming years. To this end, the symposium has been based on a limited number of invited papers, chosen to promote the maximum discussion.

The symposium will begin with a paper on the general philosophy of modelling in the applied sciences, as it is important that the logical basis of our use of models is clearly understood. The following papers will give examples of particular types of mathematical models and their application, covering as wide a range as possible. Then, we have a series of papers which will examine the tools which can help us to use models in ecology, both hardware and software, so as to gain some wider understanding of the available techniques.

On the Friday morning, we are trying an experiment, well aware that this experiment may fail and leave us with an embarrassing lack of discussion. I intend on this morning, but not before, to pose the symposium a practical problem in ecology, and ask for a solution through the use of a mathematical model. Nobody except myself and the chairman for that discussion knows what the problem is, as we want the discussion to be spontaneous. We hope, in this way, to give some insight into the nature of the modelling process, and to promote discussion of the alternative strategies through which real ecological problems can be tackled.

Finally, on the last afternoon of the symposium, we will have an assessment of the symposium from four different points of view. First, what are the implications of techniques and examples for the research director? Second, what roles do modelling strategies play in agricultural research, as an example of just one field of applied ecology? Third, how does the statistician see the development of the use of mathematical models in ecology? Fourth, how does the ecologist view the changes which are introduced into his science by the impact of mathematical models on research? Above all, the symposium will have done its work if it promotes discussion, private or public, and if that discussion bridges the gap between mathematicians and ecologists.

References

HAHN G.J. & SHAPIRO S.S. (1967) *Statistical methods in engineering*. Wiley, London.

PEARCE S.C. (1963) The use and classification of non-orthogonal designs. *J. Roy. Stat. Soc.* (A) **126** (3) 353–77.

SOKAL R.R. & SNEATH P.H.A. (1963) *Principles of numerical taxonomy*. Freeman, San Francisco.

SPRENT P. (1970) Some problems of statistical consultancy. *J. Roy. Stat. Soc.* (A) **133** (2) 139–64.

YATES F. (1966) Computers: the second revolution in statistics. *Biometrics* **22** (2) 233–51.

Some philosophical aspects of mathematical modelling in empirical science with special reference to ecology

J.G.SKELLAM *The Nature Conservancy, London*

Summary

Attention is drawn to some of the philosophical issues which arise when attempts are made to discuss the functions, structure and application of mathematical models in empirical science.

Introduction

It is noteworthy that this symposium, organized in the name of the British Ecological Society, is not devoted to making statements about living organisms and their surroundings, the subject matter of ecology proper, but is more directly concerned with conceptual matters—the functions, structure, and application of mathematical models.

It is fitting that this should be so, for a science considered in isolation from philosophy is like an organism studied in isolation from its environment. At least Sir Arthur Tansley (1939) must have thought so when he pointed out at length that classificatory concepts are schematic fictions and that failure to recognize their philosophical status leads to barren controversy.

For a short while we shall be encroaching on an area of intellectual endeavour, which for more than two millennia has engaged the attention of many gifted minds. It is a treacherous area, cluttered up with numerous discarded attempts to establish empirical knowledge on firm foundations.

Fortunately for ecology, the basic philosophical issues underlying model-making are much the same for all natural sciences. Otherwise, ecologists, in their bold endeavours to consume and integrate knowledge from virtually every other science, would only be adding further complexity to their already bewildering task.

This essay is not written in the spirit of critical philosophy where attempts to define the undefinable exhaust the efforts of a life-time. Words will therefore be used in a rough, intuitive manner, usually in a broad sense much as in Greek thought. Only in this way is it possible to contemplate the subject as a whole, for without generalization there is no schematization.

Modelling

The term 'modelling' is used in so many confusing ways in various fields of human interest that it is not always obvious which is the model and which is the mimic. The Ancient Greeks seemed to regard the general theoretical conception as the model rather than the particular material exemplifications for which classical art is renowned.

With the Renaissance we find Leonardo da Vinci explicitly proclaiming with characteristic genius that *nature* is *his* model. He certainly makes this evident in his magnificent drawings and paintings and in those extraordinary mechanical constructions with which he seemed obsessed.

Life, however, is too short to be spent in semantic quibbles; it is enough if we are able to understand one another. Biometricians usually manage somehow to do this, though some speak of fitting the model to the data and others of fitting the data to the model.

It is evident from the general confusion that some of the issues are subtle and that our thoughts readily oscillate between the original and the copy. Indeed, in those areas of science where modelling has been most successful, the activity is carried out where possible in both directions. Not only is the physical world modelled by mathematical constructions, but, equally, mathematical schemes are modelled by physical constructions. When this reverse modelling combines the advantages of conceptual simplicity with the demand for empirical verification, we encounter the most powerful instrument known for advancing empirical knowledge—the designed experiment. The traditional reluctance of ecologists to engage in highly simplified controlled experiments comparable to those in physical science leads to a superabundance of speculative models of doubtful reliability.

At least one analogy has been employed already, and many more are to come. No apology is offered, for careful analysis of all thought processes soon reveals that analogy plays a dominant role. We are always substituting one thing for another, using symbols for ideas, and signs for symbols. The words 'as if' are implicit almost everywhere (Vaihinger 1935).

The application of mathematics to science bears some analogy to pictorial expression. The pen sketches out fine lines or the brush on damp paper yields all gradations of light and shade, but in all cases much colour and detail are lost. Only those who know how to relate what is seen to what gets drawn, know how to relate what has been drawn to what could have been seen. The uncertainties involved are always two-fold.

Though this is hardly the occasion to engage in linguistic analysis, it may nevertheless be important to point out to those hardened realists who regard mathematical modelling in ecology as inadmissible fantasy that every time they employ natural language they are engaging in modelling activities of extraordinary subtlety. For example, the Roman numerals I, II, III, V, X and the Arabic numerals 1, 3, 5 are easily traced to pictures of fingers and hands and relate the general notion of number to the familiar operation of counting. The Chinese symbol for 'not' is 不, originally a line on top of the symbol for a plant. By suggesting that growth is obstructed, the symbol conveys by example and analogy a much more general conception of negation comparable to that engendered in the mind of a child when a parent puts out a hand and says 'No!'

The connection between written language and pictorial representation is perhaps most apparent when we reflect on early attempts to convey messages in writing.

As a mechanism or medium for evoking memories, images and combinations of abstract ideas, symbolism is undoubtedly a great aid to thought. It can however fossilize the intellect and even deceive. Too great a reliance on symbolism leads to formalism, but formalism as such is mere technology whereas thinking is an art.

Empirical science

In order to discuss the role of modelling in science, some conception of science itself is required, and most definitions are inadequate. The following provisional attempt is perhaps sufficient for present purposes. Empirical Science is a human intellectual activity (pursued both individually and collectively) with the aim of developing a coherent body of reliable knowledge and rational understanding through the systematization of sense-experience (both contingent and contrived).

The word 'reliable' implies that science provides a basis for further action, and the word 'human' indicates both its ethic and its limitations. The social aspects are important because, as has often been said, by standing on the shoulders of giants we can see a little further. Cohesion and rationality are vital on account of the need for intelligibility and the efficiency which ensues. The word 'developing' indicates both the growth of knowledge and its structural advance.

The world pictures adopted by the more mature sciences have all evolved from humble origins by modification and occasional metamorphosis. The philosophy of common-sense comes to us naturally and serves us well. Those who regard the intellect of man as part of his equipment for survival might even say that it comes to us naturally because it serves us well. All men most of the time, and most men all of the time, are common-sense realists. Nevertheless, the common-sense world presents far too many absurdities to be fully acceptable on its face value to a highly rational being. Philosophers have discussed the problem for ages and psychological studies on perception, especially in infants, have thrown much light on the subject. The general opinion which emerges is a conception of man as a world-builder and the commonsense world as a truly remarkable creation, moulded by his intellect out of the stream of raw sense-experience. Without man's innate capacity for synthesis, and the facility with which he resolves paradox by imaginative constructions, science, as we find it today, would not be possible.

Cohesion

To illustrate cohesion, consider that valuable type of representational model we call a map. Maps, whether mathematical projections or not, are usually so closely allied to experience and to photographs that the ideas they generate cohere quite readily with familiar ideas and previous knowledge. We say that we *read* a map just as we say we *read* a book, and thereby recognize intuitively that maps are a form of language. The language is indeed a powerful one because it conveys so much information so quickly and so directly. Condensation is the hall-mark of a good model.

The distortion which arises from an attempt to represent something which is large as if it were small, or what is rounded as if it were flat is inherent in the basic materials. That which arises from emphasis, like the exaggerated width of a road, is intentional. To those who do not understand the medium, both the structural defects and the conventions, maps like other models can be deceptive and even dangerous.

Cohesion is not the same as adhesion. Of the various models which purport to relate to ecology it is worthwhile differentiating where possible between the indigenous and the exotic. Indigenous models are natural developments of prevailing ecological conceptions, for example, models of populations, of dispersal, of productive processes and of ecosystem dynamics. For these, coherence is virtually ensured, and, if they actually work in practice, they become incorporated into the content of established ecological thought. On the other hand, exotic models are introductions from a wide area of highly generalized thinking in which logical concepts prevail and ecological ideas play no active part. This category includes the models which underly sampling schemes, the design of factorial experiments, classificatory devices, multivariate analysis and time-series routines. Such models as these are often valuable adjuncts to the conduct of research, even when their applicability is only an act of faith or a judgment based on scanty internal evidence. Nevertheless, the conclusions reached are necessarily expressed in terms of the concepts of the general purpose model employed, and, having no valid interpretation outside methodology, tend only to adhere to the surface of ecological knowledge without enriching our deeper understanding.

For example, the interactions between treatments in the statistical sense are often highly dependent on the choice of metric. The phenomena being studied are themselves unaffected. The former therefore tell us nothing *directly* about the latter. They are parts of a mathematical decomposition (Fig. 1).

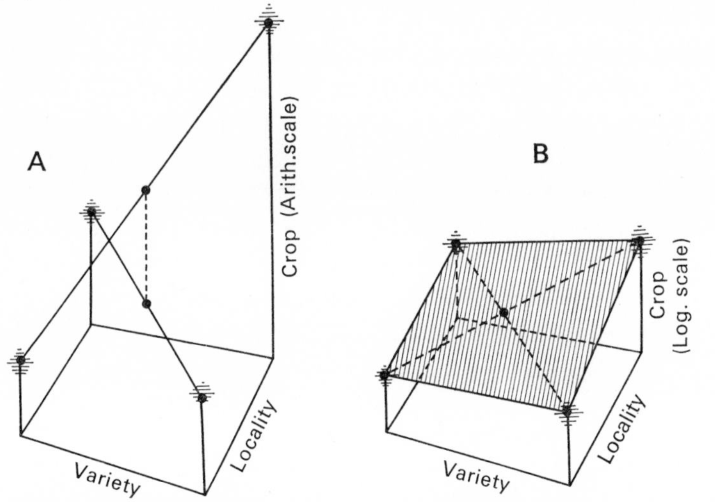

Figure 1. To illustrate the dependence of statistical interactions on the scale of measurement adopted. Both diagrams refer to the same replicated experiment. In case A the statistical significance of the interaction between the treatments (variety and locality) is extremely high whilst in case B there is no interaction at all.

Meaning

Every conceptual advance incurs the penalty of reinterpretation. For example, when at an early age we learn to subsume the various so-called views of a mountain under the concept of a solid geometrical mountain of fixed shape extended in 3-dimensional physical space, we also acquire at the same time some appreciation of perspective, and this provides the link. The word 'mountain' is however retained both for the collection of views given in experience and also for the structure conceived in imagery, with the result that the word undergoes a shift of meaning. Again when the microscope opens up for us a new world of experience, a surface which was formerly described as smooth may now be described as rough. These shifts of meaning are rarely written down explicitly because they normally cause no trouble to those who have served their apprenticeship. At times, however, they are quite formidable, as anyone who has ventured into relativity theory can testify. Those conventional ecologists, who feel tempted to use mathematical models proposed to them, are advised to be on their guard, especially when terms like 'scale of pattern', 'degrees of similarity', and 'trend surface' seep into the language, because model makers dwell at rarefied conceptual levels and are prone to be carried away, particularly by their own constructions.

One way of examining ethereal concepts is to employ the fiction of the simple case, a well known class of models which includes all utopian and desert island fantasies as well as experimental designs.

Consider the term 'random pattern' in the context of the spatial arrangement of seedlings in an exposed lake-bed. The earth is not elastic, but in imagination we can deform it as if it were. Unlike familiar patterns, the random distribution remains random when uniformly stretched or compressed in any direction, as illustrated in Fig. 2. This mathematical curiosity suggests

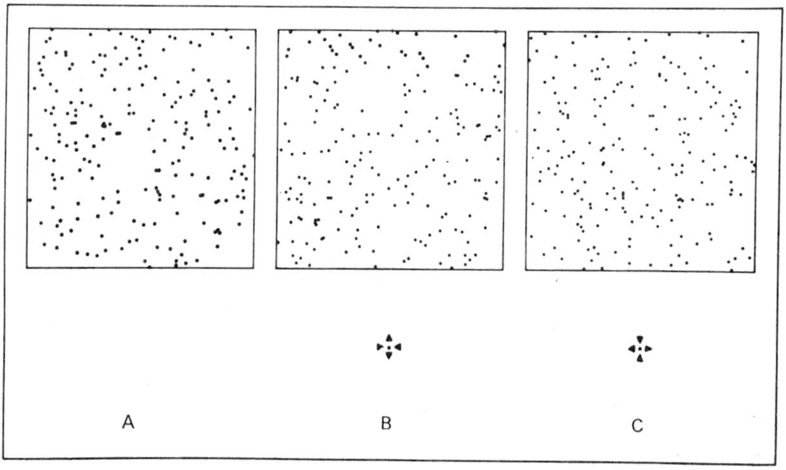

Figure 2. A, B, C are derived from realizations of the same Poisson process. A has its original shape. B was originally twice as long and half as high, whilst C was originally twice as high and half as long. B and C have been uniformly deformed to make them square, and differ from each other by a factor of 16.

that it may be more profitable to think of pattern as being absent in the random case rather than to think that we are dealing with a particular kind of pattern.

The scale of pattern in the sense of Greig-Smith (1952) is determined indirectly from the output of a formal hierarchical analysis of variance. It is apparent in the clear-cut fictional case shown in Fig. 3 that the concept is somewhat ambiguous and does not match our intuitive ideas. This shift of meaning tends to diminish the practical value of the hierarchical procedure as a generator of fruitful working hypotheses.

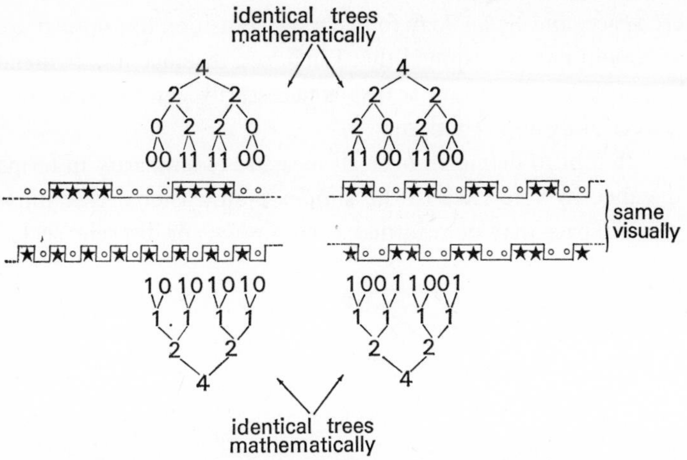

Figure 3. To show some extremely simple linear patterns and the ambiguities which may arise when attempts are made to evaluate the scale of pattern by hierarchical analysis.

As every ecologist knows, seral succession is often revealed on the ground, just as the course of tissue differentiation is often apparent in a single structure (Fig. 4). Cases like these afford natural encouragement to us to represent

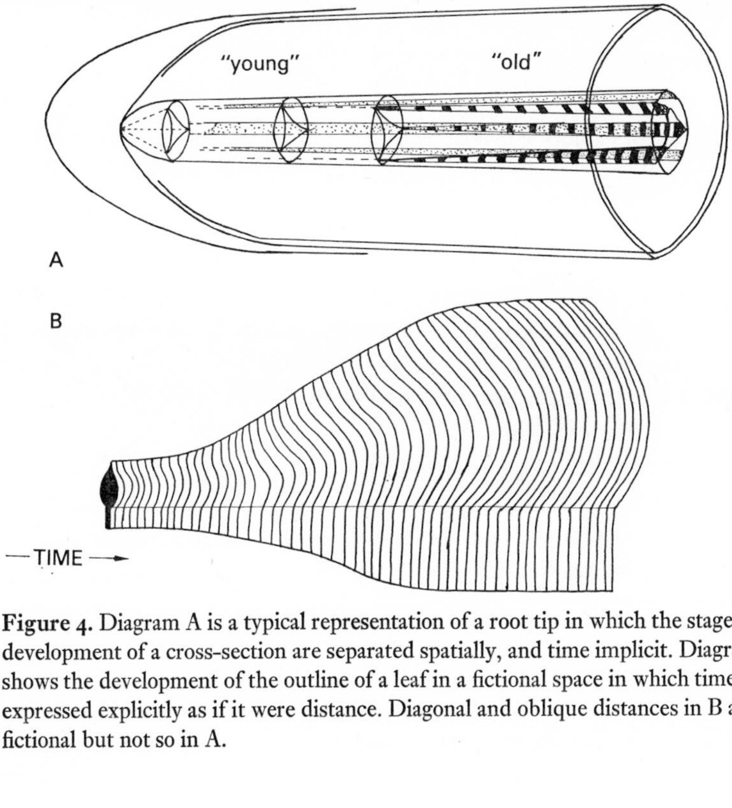

Figure 4. Diagram A is a typical representation of a root tip in which the stages in the development of a cross-section are separated spatially, and time implicit. Diagram B shows the development of the outline of a leaf in a fictional space in which time is expressed explicitly as if it were distance. Diagonal and oblique distances in B are fictional but not so in A.

time as if it were space, and by analogy to make the transition to multivariate thinking. But it should not be assumed that Euclidean distance in a fictional space (lower diagram), no matter how scaled, is necessarily significant scientifically. Physicists certainly don't think so.

Consider the attempt to define an over-all measure of similarity in terms of Euclidean distance by reference to the simple 2-dimensional case illustrated in Fig. 5. The space may be regarded as continuous or discrete, scaled

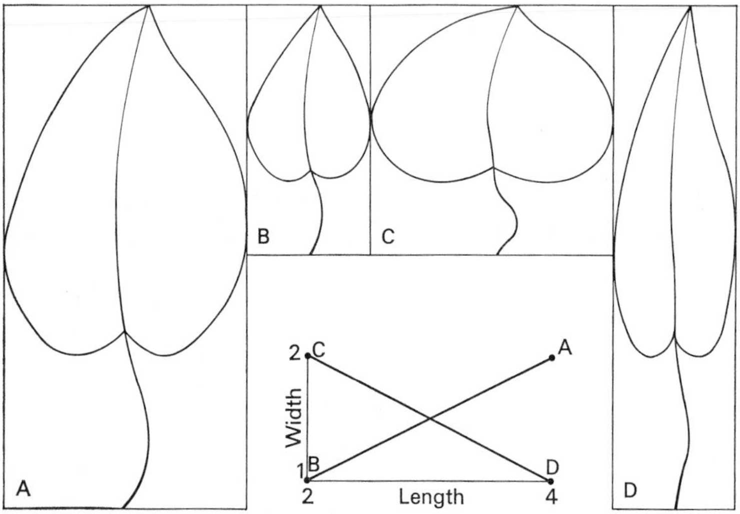

Figure 5. Showing the outlines of four leaves (A, B, C, D) which differ in two respects (length and width) at two levels, and their representation in Euclidean space. A and B, which are normally regarded as very similar, are as widely separated in the fictional 'similarity space' as are C and D, which are normally regarded as being dissimilar.

or unscaled. We can make the example 4-dimensional by introducing meaningful measures of shape and area, and then using matrix algebra, but the moral is still the same, namely, that considerable shifts of meaning are incurred. The concept is bought at the cost of cohesion, and without complete specification is devoid of empirical meaning.

Whereas pure fiction is unrestricted, as in mathematics, the established fictions of science are closely linked to the brute undeniable facts of experience. It is from experience that they derive their meaning, it is in terms of experience that they find expression, and it is by conforming to experience that they demonstrate their value.

Measurement

The need for rules of correspondence is evident whenever we make measurements. Consider, for example, the following quotation from Stamp (1947): 'It has been recently calculated that the coast of England and Wales alone is some 2751 miles in length.' To some, no doubt, this is an informative statement, but for those who are not acquainted with the operational procedure which yields the number 2751 it has no meaning. To sail round the coast is one thing and to run a tape-measure round every boulder is another.

We no longer believe that things have their own true proper names, but we commit a comparable mistake whenever we assume that spatially extended objects necessarily possess an intrinsic mathematical property called 'length', and the error is perpetuated by the structural defects of our language. Strictly speaking, it is not the object but its mathematical model which has the length we assign. Indeed, every measurable property is a relation between a phenomenon and an instrumental procedure, number and counting included.

Because we are free to choose the meaning of such terms as 'length of coastline', there is far more flexibility in modelling than is commonly realized. Indeed physicists are most adept at that variety of intellectual somersault which turns an imperfect empirical relationship into an exact law. The benefits are often considerable, as, for example, in the revision of scales of temperature (Ellis 1966).

The difficulties involved in devising an acceptable practical procedure for measuring the so-called surface area of an avian egg will be apparent to all. Mathematical modelling provides a plausible substitute. The avian egg with all its microscopic roughness is replaced conceptually by a perfectly smooth geometrical shape fitted to the measurable length, breadth and asymmetry. The mathematical area of the mathematical solid is given by calculation. Surfaces of the mathematical family illustrated in Figure 6 appear to fit all shapes of avian egg quite well. Furthermore, this kind of mathematical exercise throws immediate light on the adequacy of the simple index, length $\times$ breadth, which enters into the statistic employed by Ratcliffe (1970) in connection with pesticides and egg breakage in certain bird populations. That this index is correct dimensionally is evident by inspection of the units. The same cannot be said for many other indices occurring in the ecological literature.

Idealization

The art of mathematical modelling is rarely revealed in the literature because the emphasis there is either on the final polished product or on abstract principles. Modelling is an area where that profound and disturbing remark of Lao Tse appears to hold:

'Those who know do not tell; those who tell do not know.'

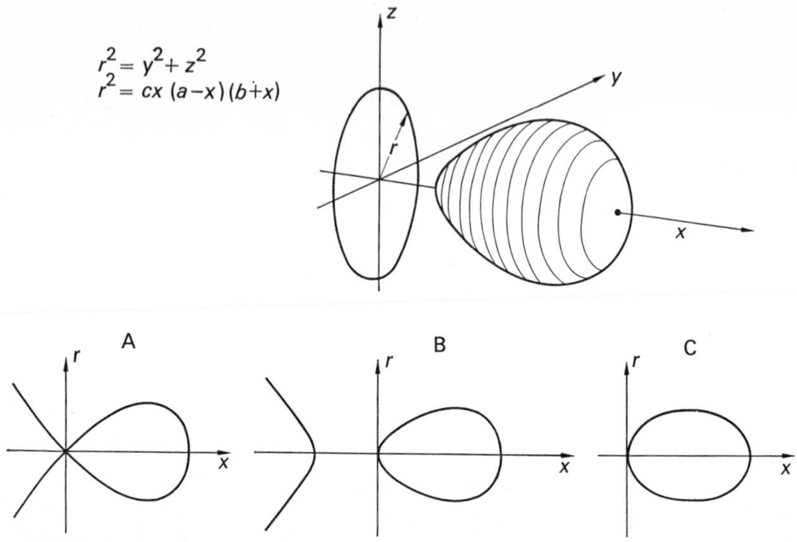

Figure 6. To illustrate the simulation of an avian egg by solids of revolution ($y^2 + z^2 = r^2$) given by the closed loops of curves of the family, $r^2 = c\,x\,(a - x)\,(b + x)$. A and C show the extreme limits of the range of asymmetry whilst B is a typical intermediate case.

As with most skills, we learn by thinking and doing, and thinking again. The forceps of the mind, however, as H.G.Wells once said, are clumsy forceps and pinch the truth a little in taking hold of it.

To illustrate the early steps in cohesive model-making, consider an insulated population of animals in a seasonal environment, and the scientific need to speak meaningfully and rationally about its demographic features and productive processes. The issues are discussed in detail elsewhere (Skellam 1967 a, b) and this is only a brief sketch.

Figure 7 is a representation of the population almost as if it were a rope of

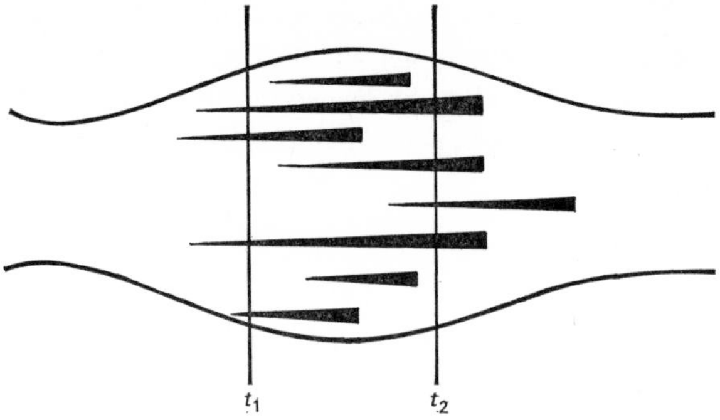

Figure 7. Representation of a population as if it were a rope and the lives of the individuals as fibres of changing form. Only those lives which participate in a short time segment (t_1 to t_2) of the history of the population are shown.

interlaced fibres. Much detail has been abstracted, but the idea that individuals develop and have varying lengths of life is retained. It is immediately apparent that there are difficulties in speaking sensibly, for example, of the average life-span during any short interval of population history, because we are at a loss to know which lives to include and which to exclude.

Figure 8 is an advance achieved by representing age as a separate dimension. This diagram suggests the need for new terms.

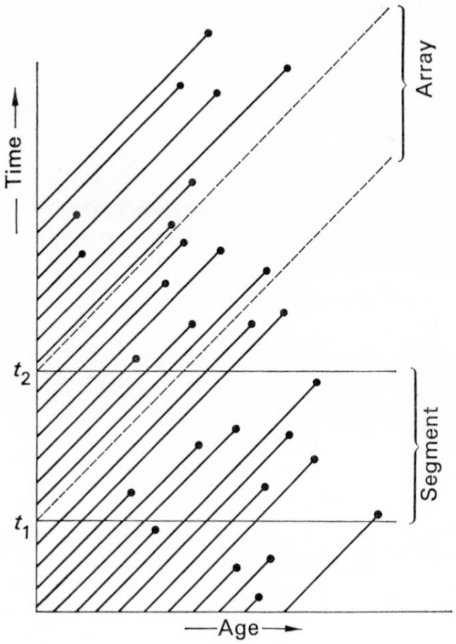

Figure 8. Representation of an insulated population as a set of life-tracks advancing in time and age from birth to death. Horizontal strips refer to segments of population history and oblique arrays to sets of individuals born in the same interval of chronological time.

Figure 9 arises by considering a large population with the lives in each cohort assembled vertically in order of magnitude. Survivorship is exhibited by oblique cuts through the solid figure. The transition from a large population to the probabilistic conception of a small population is straight-forward.

Figure 10 shows how we can start to apply the concepts of the infinitesimal calculus, and how to begin to think abstractly about the outcome of any biological process associated with those fragments of individual life-history which fall in any region of the age-time plane.

Only after these preliminary conceptual advances do we achieve a meaningful representation of productive processes in the conventional symbolism of the integral calculus. Once this has been done, we can then proceed to exploit the logical potentialities inherent in the mathematics and to deduce

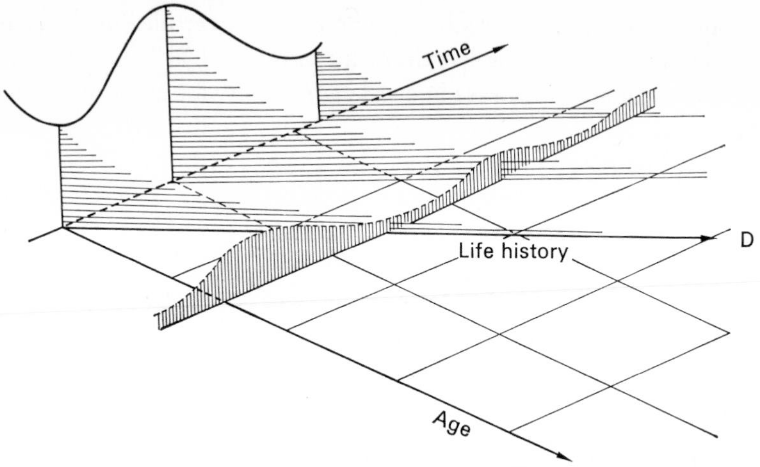

Figure 9. Representation of a large population in which the individual lives are shown as tracks advancing in age and time, and arranged vertically into cohorts in decreasing order of survival so as to produce a 'surface' over the age-time plane.

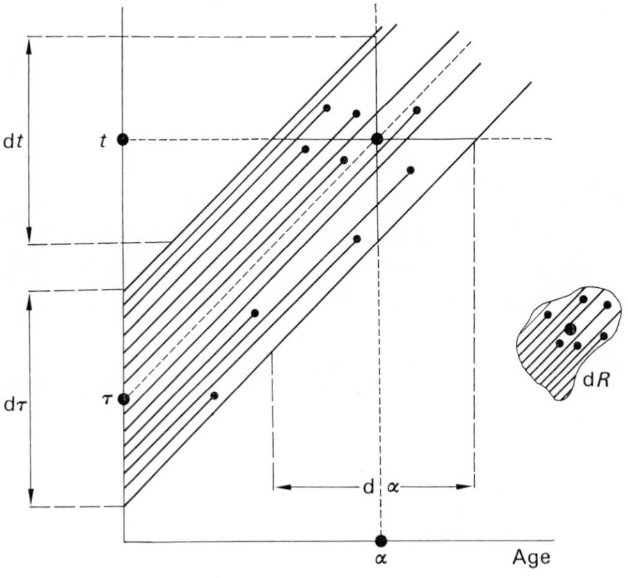

Figure 10. Showing basal part of cohort. In a small region dR of the age–time plane attention is confined to biological activities carried out at virtually the same time by individuals of virtually the same age with virtually the same past history. The elementary intervals, dt, $d\tau$, $d\alpha$, refer respectively to infinitesimally small increments in chronological time (t), instant of birth (τ) and age (α).

useful theorems. In the case of seasonally periodic systems, a simple example of such a theorem would be the equality which exists between the production in a segment and the production in an array when both refer to a whole number of cycles.

Because of coherence, the theorems can all be interpreted in terms of familiar ecological ideas and applied meaningfully to field data, as, for example, in the studies on earthworms by Lakhani and Satchell (1970).

Similarly, anyone who wishes to generate plausible laws of diffusion which might be applicable and intelligible in ecological terms is advised to start at the grass-roots level and think about the activity of the individual organisms and the factors which might induce displacement to a nearby position. The diffusion equation which emerges is not necessarily the same as that for simple Brownian motion or heat conduction (Skellam 1951, 1955).

Applicability

When we say that Ragged Robin has 5 petals and Watercress has 4, as in floristic botany, we are using numbers descriptively much as when we use the colours red and white. It is true of numbers that $5>4$, but few botanists would derive much help or inspiration from the statement that a flower of Ragged Robin has more petals than a flower of Watercress. Similarly, a flower with 5 stamens is not on that account considered to be nearer to one with 6 stamens than one with 10. In themselves the examples are trivial, but the lesson is not. Science demands more from its constructs than mathematical meaning; it also demands relevance. That which matters most in the fruitful application of symbolic forms to real life is purpose. It is idle to ask whether two watches are synchronous, as many statisticians do. The positive question is: 'Can they be regarded as synchronous for the aims we may have in mind?' Without an explicit statement of purpose, models lose their relevance; it becomes difficult to appreciate their finer points, to submit them to critical examination or justify the underlying assumptions—and these, indeed, are often many.

Ordinary numbers have the property that if $a > b$ and if $b > c$ then $a > c$. The birds in a flock do not invariably exhibit the corresponding property that if A pecks B and B pecks C then A pecks C. Though there is no logical objection to the construction of an index of social status, its usefulness as a predictor of pecking behaviour depends on the social structure which prevails.

Similarly, ordinary numbers have the property that $1+1=2$. This fact, however, does not guarantee that 1 litre of liquid added to 1 litre of liquid yields 2 litres of liquid, for in the case when one liquid is water and the other ethyl alcohol, there is a marked shrinkage.

The point is that subjecting numerical quantities to mathematical operations as if they were numbers does not *by itself* yield worthwhile results unless the operational and relational properties of the mathematical system employed are also satisfied tolerably well by their empirical counterparts.

Whether or not we employ an arithmetic mean as in mechanics, or a geometric mean as in growth studies, or a harmonic mean as in Holt's work (1955) on the foraging of the Wood Ant, depends neither on pure logic, nor computational convenience, nor statistical expediency, but primarily on our resolve to ensure that the average we choose leads to something with a meaningful interpretation. Otherwise we lose cohesion, become engulfed in such purely logical fictions as 'the average family' with its 2·4 progeny, and sometimes even get lost in a world of tautology.

Some logical operations exaggerate and others diminish the errors inherent in the initial assumptions. Because of this it is possible to experiment with variants of a model to determine in which ways it is robust and in which ways it is sensitive. Studies of this kind are not only essential in furthering the art of model making and justifying assumptions, but also in throwing light on those features in nature which might possibly be of major importance and those which might be only of minor importance in understanding the phenomena being studied.

Mathematics and science

There is much misunderstanding among ecologists not just of mathematical concepts but of the nature of mathematics itself. The mathematician is essentially a builder of formal languages and an explorer and developer of their inherent potentialities. The world of mathematics is a world of conceivable possibilities, not a world of actuality. Mathematical statements as such are almost void of empirical content. Mathematics can provide us with a framework of thinking but its uninterpreted tautologies are not genuinely informative.

Mathematics is necessarily lucid, not merely because all formal systems are human creations devised in keeping with the ways human beings naturally think, but especially because the mathematician restricts himself deliberately and exclusively to the few clear-cut modes of argument which are recognized as completely convincing and compelling. In compensation for this rigorous discipline his imagination is allowed complete freedom to devise his own systems or develop those of others, provided only that the chain of reasoning does not run into that intolerable situation which stultifies all further thought —the outright contradiction.

The mathematical conception of truth, whereby every well-formed statement that falls short of perfection by the merest conceivable shade is unacceptable, has its origins in Greek thought and has subsequently received the admiration of all. Nevertheless, it is becoming increasingly apparent that it is illusory to suppose that this extreme conception of truth has either meaning or usefulness in any area of thought which is not totally deductive. To use an ecological metaphor, the fields of mathematics and formal logic are its natural habitat and it is not suited to life elsewhere.

From this standpoint it would seem that whenever we encounter statements which appear to assert an absolute truth, whether made by scientists about nature or statisticians about inferences, we should beware of possible misunderstanding and soften them.

To illustrate the scientific attitude to truth, let us take a model which the ecologist might borrow from astronomy in order to compare the amounts of sunlight falling on various sloping surfaces and thereby quantify one component in that complex ecological concept called 'aspect'. The simple heliocentric model of Copernicus proves adequate for the purpose without resort to modern refinements. The model is acceptable in this context.

It is interesting to compare the model with its predecessor, the geocentric model of Ptolemy as improved by the Arabs. Both have their uses; the earlier model reproduces the raw observations more accurately whilst the later one proved more fruitful in developing physical theory. But even a poor model is better than none at all.

Evidently for science there is neither truth nor reality but only the possibility of rationalization and the hope of reliability, and both of these can often be achieved in many ways. Apart from purpose and inspiration little remains but economy to guide our choice.

Simplicity

As every ecologist knows, survival often depends on accepting major advantages at the cost of minor disadvantages. This might even suggest an evolutionary explanation for man's love of simplicity, because the economy of thought that is thereby gained could compensate for the errors incurred. But seeking simplicity with excessive zeal sometimes blinds us to fine distinctions. Indeed, few ecologists would accept, solely on the authority of Aristotle (Physics Bk. 5), that nature is always simple. For this reason there seems to be no deep justification for elevating Ockham's well known advice to the status of an inviolable methodological principle. The maxim is usually quoted as: 'Entia non sunt multiplicanda praeter necessitatem'. Actually his nearest pronouncement seems to have been: 'Nunquam ponenda est pluralitas sine necessitate' (Kneale & Kneale 1962, p. 243). It is ironical that because the rule itself is not necessary, it fails to satisfy its own criterion of acceptability. Ockham's counsel is perhaps better expressed in his statement: 'Frustra fit per plura quod potest fieri per pauciora.'

Ockham in life was the great inconoclast forging a bridge-head from the Middle Ages to the Renaissance. It is more fitting that we honour his spirit than that he be remembered for a few words of practical wisdom handed down to him from the past.

Concluding remarks

Mathematical ecology is moving out of its classical phase carrying with it untold promise for the future, but, as H.A.L.Fisher, the historian, remarks, progress is not a law of nature. Without enlightenment and eternal vigilance on the part of both ecologists and mathematicians there always lurks the danger that mathematical ecology might enter a dark age of barren formalism, fostered by an excessive faith in the magic of mathematics, blind acceptance of methodological dogma and worship of the new electronic gods. It is up to all of us to ensure that this does not happen.

References

BODMER F. (1944) *The Loom of Language*. London.

ELLIS B. (1966) *Basic Concepts of Measurement*. Cambridge.

FISHER H.A.L. (1935) *History of Europe* (Preface). London.

GREIG-SMITH P. (1952) The use of random and contiguous quadrats in the study of the structure of plant communities. *Ann. Bot., Lond.*, N.S. 16, 293–316.

HOLT S.J. (1955) On the foraging activity of the Wood Ant. *J. Anim. Ecol.* 24, 1–34.

KNEALE W. & KNEALE M. (1962) *The Development of Logic*. Oxford.

LAKHANI K.H. & SATCHELL J.E. (1970) Production by *Lumbricus terrestris* (L.). *J. Anim. Ecol.* 39, 473–92.

RATCLIFFE D.A. (1970) Changes attributable to pesticides in egg breakage frequency and egg shell thickness in some British birds. *J. appl. Ecol.* 7, 67–115.

SKELLAM J.G. (1951) Random Dispersal in Theoretical Populations. *Biometrika* 38, 196–218.

SKELLAM J.G. (1955) The Mathematical Approach to Population Dynamics. *The Numbers of Man and Animals* (Ed. by J.B.Cragg and N.W.Pirie), pp. 31–45. Edinburgh & London.

SKELLAM J.G. (1967a) Productive Processes in Animal Populations considered from the Biometrical Standpoint. *Secondary Productivity of Terrestrial Ecosystems* (Ed. by K.Petrusewicz), pp. 59–82. Warsaw & Krakow.

SKELLAM J.G. (1967b) Seasonal Periodicity in Theoretical Population Ecology. *Proc. 5th Berkeley Symposium on Mathematical Statistics and Probability* (Ed. by L.M.Le Cam & J.Neyman), vol. 4, pp. 179–205. Berkeley & Los Angeles.

STAMP L.D. (1947) *Britain's Structure and Scenery* (2nd Edition). London.

TANSLEY A.G. (1939) *The British Islands and their Vegetation* (Preface). Cambridge.

VAIHINGER H. (1935) *The Philosophy of 'As if'* (2nd Edition). London.

Developments in the Leslie Matrix Model

M.B.USHER *Department of Biology, University of York*

Summary

Leslie's model, which predicts the age structure of a population of animals after a unit period of time given both the structure at the present time and a matrix whose elements represent age-specific fecundity and mortality, is reviewed. The model is adapted to studies of the energy flowing through food chains. The 'fecundity' terms in the matrix represent the input of solar energy, and the 'mortality' terms the loss of energy at successive trophic levels. So that such aspects of population ecology as competition can be built into the model, the elements of the matrix are functions of the elements of the vector of energy contents.

Introduction

In 1945 P.H.Leslie put forward a deterministic model which predicted the age structure of a population of female animals, given the age structure at some past time and given the age-specific survival and fecundity rates. A very similar model had, independently, been described by E.G.Lewis in 1942. Their model, in matrix notation, can be written as

$$\mathbf{A}\,\mathbf{a}_t = \mathbf{a}_{t+1} \tag{1}$$

where

$\mathbf{a}_t = \{a_{t,0}, a_{t,1}, a_{t,2}, \ldots, a_{t,n}\}'$ is a column vector representing the population's age structure at time t, and $a_{t,i}$ is the number of females alive in the age group i to $i+1$ at time t;

$\mathbf{a}_{t+1}$ is a column vector, similar to $\mathbf{a}_t$, but representing the age structure at time $t+1$;

$$\mathbf{A} = \begin{bmatrix} f_0 & f_1 & f_2 & \dots & f_{n-1} & f_n \\ p_0 & 0 & 0 & & 0 & 0 \\ 0 & p_1 & 0 & & 0 & 0 \\ \vdots & & & & & \\ 0 & 0 & 0 & & p_{n-1} & 0 \end{bmatrix}$$ is a matrix

describing the transition of the population from one age structure to another over one period of time. The elements f_i ($i = 0, 1, 2, \dots, n; f_i \geqslant 0$) represent the average number of daughters, who will be alive at time $t+1$, born in the interval t to $t+1$ to each female who was in the age group i to $i+1$ at time t. The elements p_i ($i = 0, 1, 2, \dots, n-1; 1 \geqslant p_i > 0$) represent the probability that a female aged between i and $i+1$ at time t will be alive at time $t+1$ in the age group $i+1$ to $i+2$.

Two results immediately follow from this basic model. Firstly, if equation (1) is repeatedly applied, it can be seen that after k periods of time

$$\mathbf{a}_{t+k} = \mathbf{A}^k \, \mathbf{a}_t \qquad (2)$$

Secondly, since the matrix A is square with $n+1$ rows and columns, it follows that there are $n+1$ latent roots and vectors (eigenvalues and eigenvectors) which satisfy the equation

$$\mathbf{A}\,\mathbf{a} = \lambda\,\mathbf{a} \qquad (3)$$

where λ is any latent root and **a** is the latent vector associated with λ.

The matrix A describes the transition probabilities from one time to another, the state of the population at time $t+1$ being dependent upon the state at time t. This situation is reminiscent of the Markov processes of probability theory (see, for example, Feller 1957, chapter 15). If the matrix A were of a Markov type then all its columns would sum to unity. It can be seen that this is not the case since the p_i and f_i elements are independent of each other. Indeed the values of f_i may often exceed unity, particularly if the model is being applied to invertebrate populations where large numbers of offspring are being produced. Markov theory is thus not strictly applicable to these population matrices.

The algebraic properties of matrices such as A have, however, been investigated. The elements of A are composed of the elements f_i and p_i and zeros. Since neither f_i nor p_i can take negative values, A is thus a non-negative matrix. Sykes (1969) shows that the population matrices such as A are irreducible ('unreduced' in the terminology of Brauer, 1962). The Perron-Frobenius theorems for non-negative, irreducible matrices (developed by Brauer, 1957, 1961, 1962) indicate that one of the latent roots of A (denoted by λ_0) has the following four properties:

(i) Corresponding to the latent root λ_0 there exists a latent vector having all its elements non-negative;
(ii) λ_0 is the only latent root of A for which the corresponding latent vector can be chosen with all its elements non-negative;
(iii) λ_0 is not less than the absolute value of any other latent root of A;
(iv) λ_0 is greater than or equal to the smallest row sum of A and less than or equal to the largest row sum. Similar inequalities for the column sums are also true.

Biologically, (i) implies that the Leslie matrix model will always determine a meaningful age structure for the population ('meaningful' in the context of this paper implies that the solution to the model will not indicate negative or imaginary numbers of animals); (ii) states that this age structure is unique since whatever the size of the matrix being used there will be only one biologically meaningful solution; (iii) is a useful property of the model for its numerical solution, since the most simply applied iterative methods of determining a latent root and vector will give the root of largest absolute value (see, for example, Noble 1964); (iv) has less immediate application other than indicating the order of magnitude of λ_0.

Pollard (1966) has developed a stochastic form of Leslie's basic model, giving for each integral point of time the mean and variance of the number of animals in each age group. He illustrates this development of the model by a consideration of the female population of Australia, showing that after 250 periods of time (in this case, one year) the ratio of the numbers in the age groups gives a very close approximation to the dominant latent root, whereas the ratio of the variances and covariances is still slightly different from their expected asymptotic value. Pollard derives a recurrence relation (his equation 26) in which the means, variances and covariances at time $t+1$ are expressed as a linear matrix function of their values at time t. In fact, Pollard is optimistic about the application of the basic model and its variants, since he says '. . . whenever a Leslie deterministic method can be applied to a problem, a branching-process stochastic model can be used instead, and the expectations and quadratic moments easily calculated by a recurrence equation similar in form to the corresponding deterministic iterative equation'. The deterministic iterative equation is that given in equation (1) of this paper.

Application of the model

The data available for the blue whale (*Balaenoptera musculus*) provides an example of the use of a model similar to Leslie's basic model. Usher (in press) discusses the derivation of the matrix elements, based on data provided by Laws (1962) and Ehrenfeld (1970). The model is based on data for the whales in the 1930s before their virtual extinction and the sharp change in their survival versus age curve (Laws 1962, Watt 1968).

The females of the blue whale reach maturity at between four and seven years of age, and they have a gestation period of approximately one year. A single calf is born and is nursed for seven months during which period the female does not become pregnant again. Coupled with the migratory habits of the species this implies that not more than one calf is born to a female every two years. The sex ratio in catches is approximately 1♂: 1♀, and does not vary appreciably with age. Thus, the fecundity terms for a matrix model can be estimated. For ages between 0 and 3 the fecundity terms are zero. For four and five year old whales a value of 0·19 has been estimated as the number of female offspring being produced per female in a two year period. For six and seven year old whales a value of 0·44 has been estimated, and for eight to eleven year old whales the maximum fecundity of 0·5 has been used. For older animals this maximum rate has been reduced to 0·45, since it is known that breeding every second year may not occur for the whole of a whale's life.

The maximum age attained by a blue whale is of the order of 40 years. No direct data concerning the survival rates for the blue whales, in the absence of exploitation, are known. However, estimates can be made from survival curves and from data for the fin whale in the 1930s when this formed only a small proportion of the total whale catch. These data imply that the natural mortality accounted for about 13 per cent of the population per year during the first 10 years of life, and that this average mortality was the same for all the first ten yearly age classes.

Applying the Leslie model, the matrix

$$\mathbf{A} = \begin{bmatrix} 0 & 0 & 0{\cdot}19 & 0{\cdot}44 & 0{\cdot}50 & 0{\cdot}50 & 0{\cdot}45 \\ 0{\cdot}87 & 0 & 0 & 0 & 0 & 0 & 0 \\ 0 & 0{\cdot}87 & 0 & 0 & 0 & 0 & 0 \\ 0 & 0 & 0{\cdot}87 & 0 & 0 & 0 & 0 \\ 0 & 0 & 0 & 0{\cdot}87 & 0 & 0 & 0 \\ 0 & 0 & 0 & 0 & 0{\cdot}87 & 0 & 0 \\ 0 & 0 & 0 & 0 & 0 & 0{\cdot}87 & 0{\cdot}80 \end{bmatrix}$$

where the matrix **A** is designed to operate over a two-year period. The age classes represented by the model are thus 0–1 years, 2–3 years, 4–5 years, 6–7 years, 8–9 years, 10–11 years, and 12 years and older. The element in the lower right corner of the matrix, 0·80, is not a feature of the basic Leslie model. Since the 12+ age group contains many ages of whales up to 40 years, it is obvious that they all cannot die every two years as is assumed by the basic model for the oldest age group. Thus, it has been assumed that 80 per cent of these older animals do not die. This value has been chosen as it gives the mean life expectancy of a 12 year old animal as 7·9 years, implying a total life expectancy of about 20 years for an animal that survives to the age of 12 years.

Following equation (3), and using an iterative procedure to solve the equation, we find that

$$\lambda = 1{\cdot}0986$$
$$a = \{1000, 792, 627, 497, 393, 311, 908\}'$$

The fact that λ is greater than unity implies that the population is capable of increasing. It can simply be shown that λ can be related to the intrinsic rate of natural increase, r, by the equation

$$r = \ln \lambda \tag{4}$$

For the blue whale this indicates an intrinsic rate of natural increase of 0·0940.

The value of λ can also be used to estimate what harvest can be taken. If the population size increases from N to λN over one period of time, then the harvest that can be taken is given by

$$H = 100 \left(\frac{\lambda - 1}{\lambda}\right) \tag{5}$$

where H is expressed as a percentage of the total population.

But how is H affected by aiming the harvest at different parts of the population? This has been investigated by Williamson (1967) who was dealing with the matrix

$$\mathbf{A} = \begin{bmatrix} 0 & 9 & 12 \\ \frac{1}{3} & 0 & 0 \\ 0 & \frac{1}{2} & 0 \end{bmatrix}$$

which has a latent root of 2 and an associated column vector of $\{24, 4, 1\}'$. Using equation (5), the indication is that 50 per cent of the population can be removed every period of time. Williamson shows that a constant population size is maintained if the harvest is spread proportionately over all of the age groups (50 per cent being removed from each age group in this example).

However, if the exploitation is aimed at only one age group then an enhanced exploitation can be taken. Thus, Williamson shows that by removing half of the population solely from the youngest age group the population size still increases (Fig. 1). In fact it was not until he removed 73 per cent of the total population from the youngest age group that he achieved a constant population size.

We can ask why this increased harvest is required to maintain a constant population size. An introduction to the mathematics of the harvesting models is given by Pollard (1966) (Section 3 of his paper). He was particularly concerned with immigration, but the change of the positive signs to minus signs in Pollard's three immigration models makes them useful for an investigation of harvesting. Watt (1968) has discussed the responses of populations to

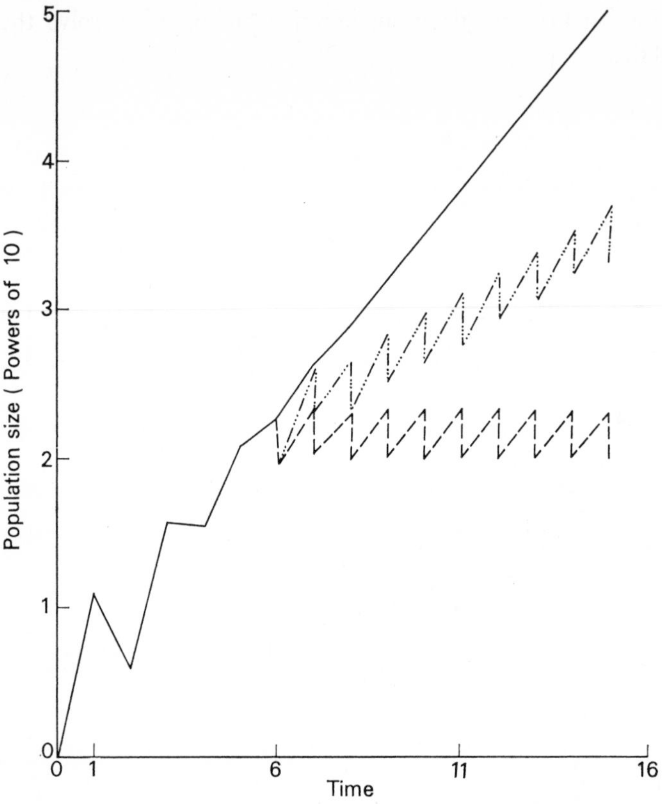

Figure 1. The growth of a population grouped into three age classes.
————————, repeated application of the matrix;
— — — —, as above, but after the sixth period of time 50 per cent of each age class removed;
— ... — ..., as above, but the 50 per cent of the total population removed from the youngest age class.

harvesting, and the results of laboratory studies have indicated that the production of the group that exploitation is aimed at is increased relative to the production of the other groups in the population. Watt (1955) investigated the effects of harvesting populations of the flour beetle *Tribolium confusum*, removing adults, pupae and large larvae. He shows that exploitation increased the numerical productivity of this group relative to the total numerical productivity of the population. Similarly, Slobodkin and Richman (1956), aiming their exploitation at the youngest size class of laboratory populations of the water flea *Daphnia pulicaria*, showed that with exploitation rates of up to 90 per cent of this age group the abundance of the age group increased relative to the abundance of the other age groups. It therefore seems that the matrix model closely follows one of the general features of the production of exploited populations, since Williamson's (1967) model allowed for 73 per cent of the number in the total population to be removed from the youngest class, whilst exploitation aimed over all classes allowed for only 50 per cent of the total population. The production of the youngest age group had therefore been increased.

If the model is to be of practical use in predicting the harvest, then it must be robust and relatively insensitive to small changes in the matrix elements. These elements are derived not as specific quantities but as experimentally determined estimates, for each of which a standard error could be calculated. If these sampling errors were to produce large changes in the predicted harvest, then very little reliance could be placed upon the model and it would have no practical value. It is also important to know just how sensitive the model is to the various elements, since it is worth putting more effort into estimating the elements for which the model is most sensitive. Using Williamson's matrix, where $\lambda = 2$, an increase in all of the fecundity terms of ten per cent gave a value of $\lambda = 2{\cdot}0867$, whilst a similar decrease of ten per cent gave $\lambda = 1{\cdot}9087$. Ten per cent increases and decreases of all of the survival terms gave $\lambda = 2{\cdot}1089$ and $\lambda = 1{\cdot}8865$ respectively. It can therefore be seen that ten per cent changes in the matrix elements would give changes in the latent root of the order of four and a half per cent (fecundity terms) and of five and a half per cent (survival terms). These suggest that the model is robust, and is relatively insensitive to small sampling errors in the matrix elements. It also indicates that the survival terms should be estimated more precisely than the fecundity terms.

It can be seen from the application of the basic matrix model that one is concerned with the increase in size of animal populations. If there is no harvesting the model predicts that the population size continues to increase at a constant rate, namely by a multiplicative factor of λ every period of time. The model thus closely approximates the calculus model describing exponential growth of populations,

$$N_t = N_0 \, e^{rt} \qquad (6)$$

where N_t is the population size at time t, N_0 is the initial population size, e is the base of natural logarithms (2·781828), and r is the intrinsic rate of natural increase. Equation (4) shows how these two models in population dynamics can be related.

The matrix model, although more complicated in that it takes into consideration the age structure of the population, shares many of the over-simplifications of the calculus model. No natural population of animals is subject to continuous growth in population size since some form of regulation operates. It is not my intention to discuss the nature of the regulatory mechanism, but rather to point out that when any form of regulation operates it tends to result either in a sigmoid growth of the population or in oscillations in the size of the population. We shall see in another section of this paper how the matrix model can be adapted to these biologically more meaningful forms of population growth and regulation.

If the results of the calculus and the matrix models can be so easily compared as equation (4) suggests, what are the advantages of the matrix model over the more well known calculus model? I consider that there are four advantages. Firstly, Williamson (1967) has discussed the introduction of students to the concepts of population dynamics, and he has demonstrated the complexity of calculus and the simplicity of matrices. Matrix algebra, although formerly an advanced mathematical topic, is now being introduced to school curricula, and hence the biologists of the future are more likely to understand it. Secondly, matrix processes are assumed to take place in discrete time units, whilst the very nature of calculus implies continuous processes taking place in infinitesimally small time intervals. It would appear then that the matrix approach is more reminiscent of the biological process of birth and death with which we are concerned when modelling. Thirdly, matrices are relatively simply dealt with by numerical rather than algebraic techniques. This is a feature of importance when simulation of populations is required, or for biologists who have only a relatively small working knowledge of mathematics. Finally, the modern computing languages and the implementation of library packages make matrix operations very simple. The Leslie matrix model has, therefore, several advantages over the calculus models in that the computational methods for solution and simulation are more easily dealt with.

Developments of this model

A summary of the developments in the model is shown in Fig. 2, in which I have endeavoured to separate out the various branches of investigation and practical application. Thus, the three boxes at the top of the illustration refer to theoretical developments with which the biologist may not be so concerned, at least in the initial stages of using these models. The eight boxes clustered round the central box relate to variations of the basic model, each corresponding to some application in population dynamics. In all of these cases, one is essentially concerned with the relationship between time and the age or size structure of the population, as well as with any form of stability that might be attained by the population if it were allowed to grow according to the matrix parameters for an indefinitely long period of time, the asymptotic state of the population. The two lower boxes refer to applications of the model that do not directly relate to the effects of age structure and time, and hence are rather separate developments of the basic Leslie model. The development of the

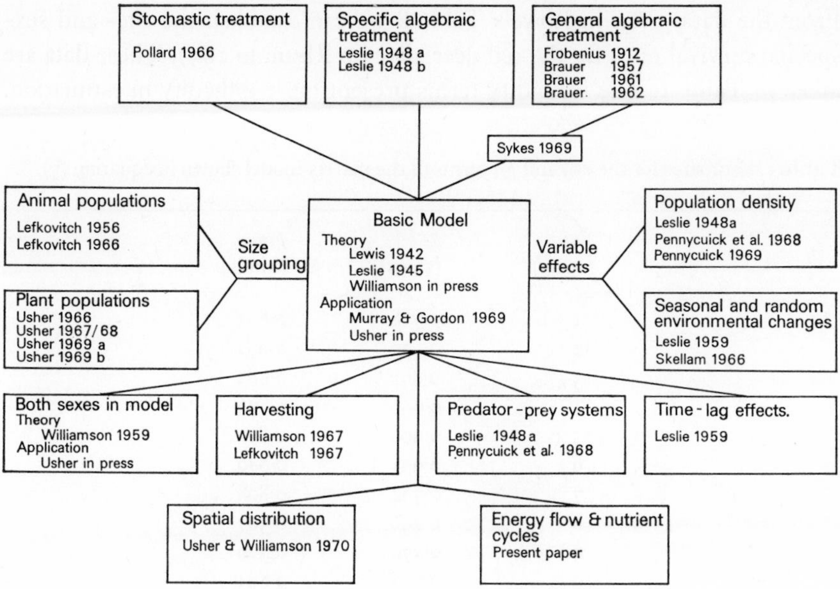

Figure 2. Schematic representation of the developments in the basic matrix model.

model for energy flow and nutrient cycling forms the last part of this paper. A few of the developments in the basic model will, however, be discussed first.

The inclusion of both sexes

The development of the model to include both sexes is due to Williamson (1959). If we consider the simple case where the population is divided into only three age classes, then equation (1) can be written in the form

$$\begin{bmatrix} 0 & f_{m0} & 0 & f_{m1} & 0 & f_{m2} \\ 0 & f_{f0} & 0 & f_{f1} & 0 & f_{f2} \\ p_{m0} & 0 & 0 & 0 & 0 & 0 \\ 0 & p_{f0} & 0 & 0 & 0 & 0 \\ 0 & 0 & p_{m1} & 0 & 0 & 0 \\ 0 & 0 & 0 & p_{f1} & 0 & 0 \end{bmatrix} \begin{bmatrix} n_{t,m0} \\ n_{t,f0} \\ n_{t,m1} \\ n_{t,f1} \\ n_{t,m2} \\ n_{t,f2} \end{bmatrix} = \begin{bmatrix} n_{t+1,m0} \\ n_{t+1,f0} \\ n_{t+1,m1} \\ n_{t+1,f1} \\ n_{t+1,m2} \\ n_{t+1,f2} \end{bmatrix} \quad (7)$$

where f_{mi} and f_{fi} refer to the number of males and females respectively born to a female in the *i*th age class during the period of time; p_{mi} and p_{fi} refer to the probability that a male and a female respectively alive in the *i*th age group will be alive in the $i+1$th age group; and $n_{t,mi}$ and $n_{t,fi}$ refer to the numbers of males and females respectively that are in the *i*th age group at time *t*.

As an example of the application of such a model I will use the data for the red deer (*Cervus elaphus*) on the Island of Rhum given by Lowe (1969). Usher (in press) gives the background to the application of the model to deer.

From the data given in Lowe's Table 8 we can establish the age- and sex-specific survival rates for the red deer alive on Rhum in 1957. These data are given in Table 1. The fecundity terms present more difficulty in estimation.

Table 1. Estimates for the p_{mi} and p_{fi} terms of the matrix model shown in equation (7).

Age (years)	p_{mi} (stags)	p_{fi} (hinds)
1	0·718	0·863
2	0·990	0·902
3	0·990	0·882
4	0·990	0·879
5	0·990	0·862
6	0·991	0·840
7	0·734	0·808
8	0·496	0·507
9	0·370	0·326
10	0·848	0·864
11	0·821	0·824
12	0·781	0·810
13	0·720	0·735
14	0·611	0·680
15	0·364	0·529
16	0	0

However, Lowe's Table 10 shows the proportion of the population that are breeding, and by taking weighted averages for the age groups 2 to 4 years, 5 to 8 years, and 9 years and older from Lowe's Table 15, I have estimated the age-specific birth rates for the production of both male and female offspring. In these calculations it has been assumed that red deer produce only one calf (Southern 1964). The estimates of these fecundity rates are given in Table 2.

The matrix for the deer population is large, having 32 rows and columns. However, in order to show the general form of this matrix, part of **A**, as in equation (7), is

$$\mathbf{A} = \begin{bmatrix} 0 & 0 & 0 & \cdot 202 & 0 & \cdot 419 & \cdots & 0 & \cdot 464 \\ 0 & 0 & 0 & \cdot 214 & 0 & \cdot 444 & & 0 & \cdot 388 \\ \cdot 718 & 0 & 0 & 0 & 0 & 0 & & 0 & 0 \\ 0 & \cdot 863 & 0 & 0 & 0 & 0 & & 0 & 0 \\ 0 & 0 & \cdot 990 & 0 & 0 & 0 & & 0 & 0 \\ \vdots & & & & & & & & \\ 0 & 0 & 0 & 0 & 0 & 0 & & 0 & 0 \end{bmatrix}$$

Table 2. Estimates for the f_{mi} and f_{fi} terms of the matrix model shown in equation (7).

Age (years)	Percentage of population breeding	Percentage of calves that are males	f_{mi} (Births of stags)	f_{fi} (Births of hinds)
1	0	—	0	0
2	41·6	48·6	0·202	0·214
3	86·3		0·419	0·444
4	89·3		0·434	0·459
5	95·1	38·1	0·362	0·589
6	95·2		0·363	0·589
7	93·1		0·355	0·576
8	98·8		0·376	0·612
9	77·5	54·5	0·422	0·353
10	76·5		0·417	0·348
11 and older	85·2		0·464	0·388

Numerical solution of this model for the latent root and latent vector shows that a stable population would have the following characteristics,

$$\lambda = 1{\cdot}1636$$
$$\mathbf{a} = \{1000, 1239, 617, 919, \ldots, 2, 1, 1\}'$$

The vector has been calculated so that there are 1000 stags in the youngest age group (latent vectors are determined exactly by proportions of one element to another, but they can be multiplied by any constant). The full vector for the stable population age structure is shown in Table 3.

Since the data for the matrix were derived for those animals alive on Rhum in 1957, it can be taken as referring to a virtually unexploited population (see Lowe, 1961, 1969 for a history of the Rhum studies). The latent root implies that a harvest of 14·1 per cent can be taken each year (applying equation (5)). This exploitation rate implies that 14·1 per cent would be taken from each age group. The model can, however, be used to predict what percentage of the population could be harvested if the exploitation were to be aimed at only a few of the age groups.

A considerable extension of the basic model for the two-sex situation has been made by Goodman (1969). He applies his models to the structure of the human population in the United States. His model records not the actual population age structure at time t, but rather how many female descendants aged under 5 that a female alive in the ith age group will have. From such studies Goodman is able to predict the eventual age structure for the male and female populations of the United States, exactly analogous to the stable age structure predicted by the basic model.

Table 3. The age structure of a stable population of red deer, predicted by the model given in equations (3) and (7). The structure has been adjusted so that there are 1000 one-year stags.

Age (years)	Stags	Hinds
1	1000	1239
2	617	919
3	525	712
4	447	540
5	380	408
6	323	302
7	275	218
8	174	151
9	74	66
10	24	18
11	17	14
12	12	10
13	8	7
14	5	4
15	3	2
16	1	1

The consideration of size structure

Not all animals can be aged with any degree of certainty, and hence an adaptation of the model is required for other possible groupings of the population. Lefkovitch (1965, 1966, 1967) has been particularly concerned with the insect pests of stored products, and the structure of these populations can most simply be expressed in terms of the four developmental stages—eggs, larvae, pupae and adults. The lengths of time occupied by each of these four stages are in general not equal, and hence one of the premises of the basic model, namely that an animal moves up by exactly one class each period of time, is no longer tenable. Indeed, if one developmental stage is of long duration, animals in that stage at the start of the period of time over which the matrix operates may still be in it at the end of the period of time. If a developmental stage is short, then animals may be able to move up by at least two classes in a period of time.

As an example, consider an insect with the following (average) life history:

egg stage	4 days
larval stage	24 days
pupal stage	5 days
adult stage	no eggs laid during the first 8 days

If we develop a matrix model to operate over a period of one week, then, using terminology similar to that of Lefkovitch (1965), the matrix will take the following form:

$$\mathbf{M} = \begin{bmatrix} 0 & 0 & 0 & m_{a,\,e} \\ m_{e,\,l} & m_{l,\,l} & 0 & m_{a,\,l} \\ 0 & m_{l,\,p} & 0 & 0 \\ 0 & m_{l,\,a} & m_{p,\,a} & m_{a,\,a} \end{bmatrix} \tag{8}$$

where *e* refers to eggs, *l* to larvae, *p* to pupae and *a* to adults. The matrix (8) shows that, for example, a larva in its first 17 days of life will remain a larva ($m_{l,\,l}$), a larva between 18 and 22 days will develop into a pupa ($m_{l,\,p}$), and a 23 or 24 day old larva will pass through the pupal stage to become an adult ($m_{l,\,a}$). Since $m_{l,\,l}$, $m_{l,\,p}$ and $m_{l,\,a}$ are probabilities for the total larval population, then

$$1 - m_{l,\,l} - m_{l,\,p} - m_{l,\,a}$$

is the mortality, being the probability that a larva will die. Lefkovitch (1967) further develops this model in order to investigate the effect of harvesting upon the developmental structure of the population.

Forest trees are generally classified according to their size rather than their age. Usher (1966) developed the basic Leslie model for selection forests; these are forests which contain an uneven age and size structure of the trees and in which there is no definite regeneration phase. In a similar manner to the development for the insect populations described above, trees can either remain in the same size class during the period of time over which the matrix operates, or they can move to a larger class (Usher 1966, 1967/68, 1969a, 1969b has assumed that the period of time over which the matrix operates is sufficiently small for a tree not to move up by more than one size class). The major difficulty for the forest model concerns the analogue to the fecundity terms of the basic model. If we assume a forest with natural regeneration, then when a tree is exploited the gap in the canopy can be utilised either by this natural regeneration or by the surrounding trees enlarging their crowns. Thus, the fecundity terms of the matrix depend not on the population of trees, but on the number of trees that are exploited. If there are n_i trees in size class *i* at time *t*, then at $t+1$ when we have a stable population structure there should be λn_i trees in that class, and $(\lambda-1)n_i$ of these trees will be harvested to reduce the population once more to n_i. The factor $(\lambda-1)$ is therefore required to modify the regeneration terms in the first row of the matrix model. Usher's (1969a) model is

$$\mathbf{Q} = \begin{bmatrix} a_0 + c_0(\lambda-1) & c_1(\lambda-1) & c_2(\lambda-1) & \dots & c_{n-1}(\lambda-1) & c_n(\lambda-a_n) \\ b_0 & a_1 & 0 & & 0 & 0 \\ 0 & b_1 & a_2 & & 0 & 0 \\ \vdots & & & & & \\ 0 & 0 & 0 & & a_{n-1} & 0 \\ 0 & 0 & 0 & & b_{n-1} & a_n \end{bmatrix} \tag{9}$$

where a_i (i=0, 1, . . ., n) is the probability that a tree will remain in the ith size class; b_i (i=0, 1, . . . , $n-1$) is the probability that a tree will move up one size class; c_i (i=0, 1, . . , n) is the number of class 0 trees regenerating in the gap caused by the exploitation of a tree; and, because of forestry conventions during enumerations, $a_i+b_i=1$. It is assumed that no $a_i=1$, with the possible exception of a_n.

Although a matrix that contains functions of its own latent root may seem complex, Usher (1969a, 1969b) has developed a simple and fast method of computer solution for the dominant latent root based on a Newton–Raphson process,

$$\lambda_{n+1}=\lambda_n-f(\lambda_n)/f'(\lambda_n).$$

λ_n is an approximation to the dominant latent root, and λ_{n+1} is a further approximation. $f(\lambda_n)$ and $f'(\lambda_n)$ are given by equations (7) and (9) of Usher (1969a). The model also has an interesting property in that there is a unique solution. Although the matrix **Q** contains functions of its own latent roots, Usher (1969a) has shown that it obeys the Perron–Frobenius theorems for non-negative matrices. There exists only one latent root greater than unity, and associated with this root is a vector of non-negative elements. Thus, a forest manager who applies such a model will be able to determine only one size structure for the forest that is biologically meaningful, and, as this is associated with the latent root of greatest absolute value, this is the structure that maximizes the production of the forest.

An example of the application and solving of this model is provided by the data for a Scots pine (*Pinus sylvestris*) forest in Inverness-shire in Scotland. The data derived from the forest are quoted by Usher (1966). From these data and from Forestry Commission data on the ground area occupied by trees of various sizes (Hummel & Christie 1953), the matrix **Q** is

$$\begin{bmatrix} 0{\cdot}72 & 0 & 0 & 3{\cdot}6(\lambda-1) & 5{\cdot}1(\lambda-1) & 7{\cdot}5\lambda \\ 0{\cdot}28 & 0{\cdot}69 & 0 & 0 & 0 & 0 \\ 0 & 0{\cdot}31 & 0{\cdot}75 & 0 & 0 & 0 \\ 0 & 0 & 0{\cdot}25 & 0{\cdot}77 & 0 & 0 \\ 0 & 0 & 0 & 0{\cdot}23 & 0{\cdot}63 & 0 \\ 0 & 0 & 0 & 0 & 0{\cdot}37 & 0 \end{bmatrix}$$

Solution of this matrix for its latent root and vector by the Newton–Raphson process gives

$$\lambda=1{\cdot}204266$$
$$a=\{1000, 544, 372, 214, 86, 26\}'$$

As the matrix operates over a period of six years, this implies by equation (5) that an exploitation of approximately 17 per cent can be carried out in every six-year period.

Matrix elements as functions

Leslie (1948a) drew attention to the fact that the basic matrix model was similar to the exponential growth equation (equation (6)), whereas populations usually approach asymptotically an upper limit, K, and their growth is approximated by the sigmoid growth equation

$$N = \frac{K}{1 + C\,e^{-rt}} \tag{10}$$

Leslie discusses some of the theoretical basis for adapting the simple model to the sigmoid growth curve, but, with the advent of computers to undertake the calculations that are required, the more common approach is now to simulate the situation. Such a study has been described by Pennycuick *et al.* (1968). However, we must first make another definition. If the population age structure at time t is represented by the vector $\mathbf{a}_t$ as in equation (1), then we can define

$$N(t) = \sum_{i=0}^{n} a_{t,\,i} \tag{11}$$

as being the sum of the elements of the vector $\mathbf{a}_t$, and thus $N(t)$ is the population size at time t.

If the elements of $\mathbf{A}$ are made dependent upon the value of $N(t)$, the rate of growth of the population can either be sigmoid or oscillations can result. As an example, Pennycuick *et al.* (1968) use the matrix

$$\mathbf{A} = \begin{bmatrix}
0 & 0{\cdot}1F & 1{\cdot}2F & 1{\cdot}0F & 0{\cdot}8F & 0{\cdot}6F & 0{\cdot}3F & 0{\cdot}1F & 0 & 0 \\
0{\cdot}3S & 0 & 0 & 0 & 0 & 0 & 0 & 0 & 0 & 0 \\
0 & 0{\cdot}95S & 0 & 0 & 0 & 0 & 0 & 0 & 0 & 0 \\
0 & 0 & 0{\cdot}9S & 0 & 0 & 0 & 0 & 0 & 0 & 0 \\
0 & 0 & 0 & 0{\cdot}8S & 0 & 0 & 0 & 0 & 0 & 0 \\
0 & 0 & 0 & 0 & 0{\cdot}8S & 0 & 0 & 0 & 0 & 0 \\
0 & 0 & 0 & 0 & 0 & 0{\cdot}7S & 0 & 0 & 0 & 0 \\
0 & 0 & 0 & 0 & 0 & 0 & 0{\cdot}65S & 0 & 0 & 0 \\
0 & 0 & 0 & 0 & 0 & 0 & 0 & 0{\cdot}3S & 0 & 0 \\
0 & 0 & 0 & 0 & 0 & 0 & 0 & 0 & 0{\cdot}1S & 0
\end{bmatrix}$$

where $F = \dfrac{15000}{2500 + N(t)}$

and $S = \dfrac{1}{(1 + \exp(N(t)/1389 - 5))}$

The function, F, was chosen so that the fecundity fell off with increasing population size in a more or less exponential manner. The function took a value of 6 when the population size approached zero, and took a value of 1 when the population contained 12500 animals. The survival function, S, is similar to a reflected sigmoid increase function, since the value of S is very close to 1 when the population size is less than 4000, and S approaches zero when the population size exceeds 14000 animals.

Figure 3 shows some results of applying this model. Pennycuick *et al.*

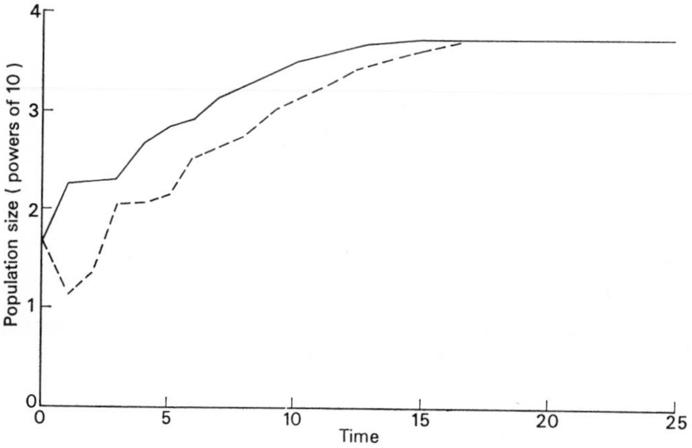

Figure 3. The growth of a population when the matrix elements are functions of the population size. The matrix is that used by Pennycuick *et al.* (1968). The two lines are discussed in the text.

started with a population size of 48 animals, though they do not specify how these were distributed in the age classes. If we assume that there are 6 animals in each of the 8 youngest age classes (i.e. the pre-reproductive and reproductive age groups), then the resultant population growth is shown by the upper graph in Fig. 3. If, however, we assume that the population is composed solely of animals in the youngest age group, then the increase in the population size is shown by the lower graph in Fig. 3. This latter graph shows that the population decreases for a short while before it starts to increase in a sigmoid manner to the asymptotic limit of about 5000.

But, how general a result is this? The answer lies in the generality of the functions that have been used by Pennycuick *et al.* It is known that survival is dependent on the density of the population, particularly the survival of the juveniles (for example the experiments performed on guppies by Silliman & Gutsell, 1958; or the experiments on territorality in the red grouse that show that young can be forced to leave the area if they fail to establish a territory (Jenkins *et al.* 1963, 1967)). It is also known that fecundity is decreased by an increasing population density. The curves shown by Green (1964) for the Collembola *Folsomia candida* are very similar to the sketch of the function given by Pennycuick *et al.* It seems therefore that the functions used by Pennycuick *et al.* are relatively general, and could approximately apply in many circumstances, irrespective of the nature of the regulation of the population or the forces acting to impose an upper limit, K, to the population size.

However, it is useful to simulate some particular cases. In order to do this, we will use the simple matrix that was first discussed by Williamson (1967), and which is illustrated in Fig. 1. The matrix is

$$\mathbf{A} = \begin{bmatrix} 0 & 9 & 12 \\ \frac{1}{3} & 0 & 0 \\ 0 & \frac{1}{2} & 0 \end{bmatrix} \tag{12}$$

If we consider that the effect of crowding is reflected only in the survival of the juveniles, then we can rewrite the matrix as

$$\mathbf{A} = \begin{bmatrix} 0 & 9 & 12 \\ \frac{1}{3}S & 0 & 0 \\ 0 & \frac{1}{2} & 0 \end{bmatrix} \tag{13}$$

where S is a function of the population size. The function that has been used is a generalization of that used by Pennycuick *et al.* (1968), namely,

$$S = \frac{1}{1+e^{N(t)/A-B}}$$

We can select values for the parameters A and B so as to give different shapes to the survival function S. In the simulation work I have taken $AB = 500$. This gives a comparative basis for the different curves, since the value of S is 0·5 for a population size of 500 animals in all cases. Three of these curves are shown in Fig. 4. It will be noticed that the larger the value of B the steeper the survival function becomes in the region of $N(t) = 500$.

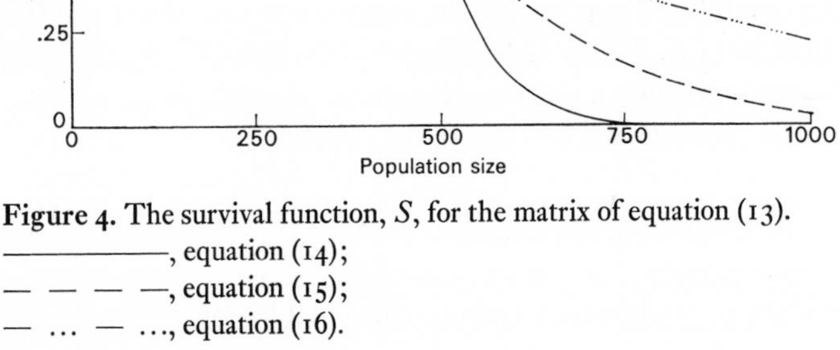

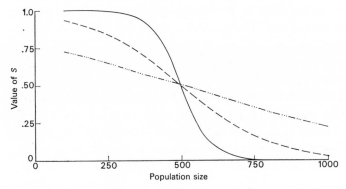

Figure 4. The survival function, S, for the matrix of equation (13).
————, equation (14);
— — — —, equation (15);
— ... — ..., equation (16).

Iterations of the model are shown in Fig. 5. It can be seen that the tendency towards oscillations is proportional to the steepness of S in the vicinity of $N=500$. It is known that, for example, when the matrix elements take just two values, one for when the population size is below a threshold value and the other for a size equal to or above the threshold, the latter value being smaller than the former, the matrix model will yield stable oscillations. If we look at the steepest function in Fig. 4,

$$S = \frac{1}{1+e^{N(t)/50-10}} \tag{14}$$

then it can be seen from Fig. 5 and from computer runs for several hundred

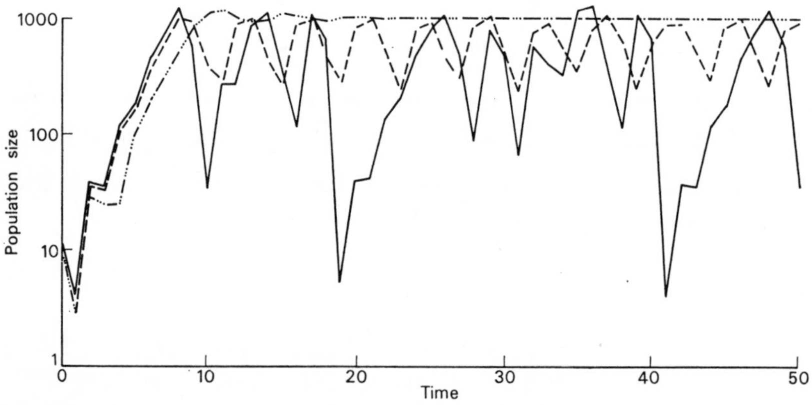

Figure 5. The growth of a population according to the matrix in equation (13). The lines correspond with those in Fig. 4.

periods of time that the model gives a series of stable oscillations. As, however the steepness of the function S is reduced, the fluctuations become less extreme, and become damped. The function

$$S = \frac{1}{1+e^{N(t)/150-3\cdot3333}} \tag{15}$$

in Figs. 4 and 5 yields stable oscillations for at least the first 200 periods of time, although their amplitude is less than the oscillations resulting from the application of equation (14). When the value of A is 350 ($B=1\cdot4286$) the oscillations are damped, yielding a more or less stable population size of 985 animals after about 100 periods of time. Use of the function

$$S = \frac{1}{1+e^{N(t)/400-1\cdot25}} \tag{16}$$

shows that the oscillations are heavily damped, and a stable population size of

1055 animals is achieved after about 30 periods of time. It will be seen from Fig. 5 that the growth of this population during the first ten periods of time very closely approximates a sigmoid growth curve, of the type given by equation (10).

In order to investigate the effects of the population size on fecundity, the matrix (12) can be modified to

$$\mathbf{A} = \begin{bmatrix} 0 & F_1 & F_2 \\ \frac{1}{3} & 0 & 0 \\ 0 & \frac{1}{2} & 0 \end{bmatrix} \tag{17}$$

where F_1 and F_2 are functions relating the fecundity to the population density. In order to preserve generality, these functions are based on expressions similar to those developed by Watt (1960). Thus,

$$F_1 = \frac{2}{a\,N(t)\,t}\left\{1-\exp\left[-a\,N(t)\,t\left(\frac{I_{\min}+\exp\,(d-f\,N(t))}{1+\exp\,(C-b\,N(t))}\right)\right]\right\} \tag{18}$$

$$F_2 = \frac{12\,F_1}{9}$$

Watt's equation, on p. 691 of his paper, adds an A to the right hand side of equation (18). He says (p. 689), 'Sometimes, . . . , the locus of F intercepts the F-axis not at zero, but at some value A.' Investigating the limits of Watt's equation, as

$$N(t) \to \infty \text{ then } F \to A$$

and as

$$N(t) \to 0 \text{ then } F \to A + 2\left\{\frac{I_{\min}+e^d}{1+e^C}\right\}$$

Such limits for F are obviously false, and I prefer to drop the constant A as in equation (18). This then implies that, as the population density becomes infinitely large, the fecundity approaches zero, whilst as the population size tends to zero the fecundity tends to

$$2\left\{\frac{I_{\min}+e^d}{1+e^C}\right\}$$

Three of these functions are shown in Fig. 6, all having in common the parameters $t=1$, $I_{min}=0{\cdot}4$, $d=0{\cdot}1$, $f=-0{\cdot}1$, $C=0{\cdot}1$ and $b=-0{\cdot}05$. The parameter a has been varied. These curves were chosen as Pennycuick *et al.* (1968) suggested that increasing the fecundity elements of the matrix produced an oscillatory situation. The results of simulations for periods of up to 100 periods of time are shown in Fig. 7. It can be seen that in all cases the popula-

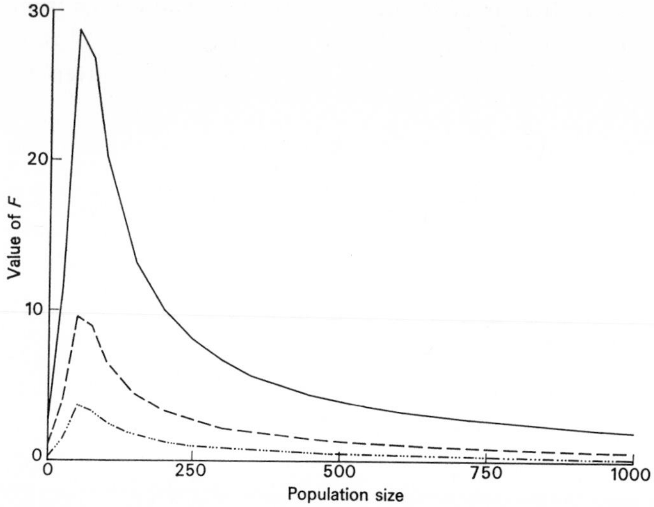

Figure 6. The fecundity function, F_1 of equation (18), used in the matrix of equation (17). The lines relate to changes in the parameter *a*, and are discussed in the text.

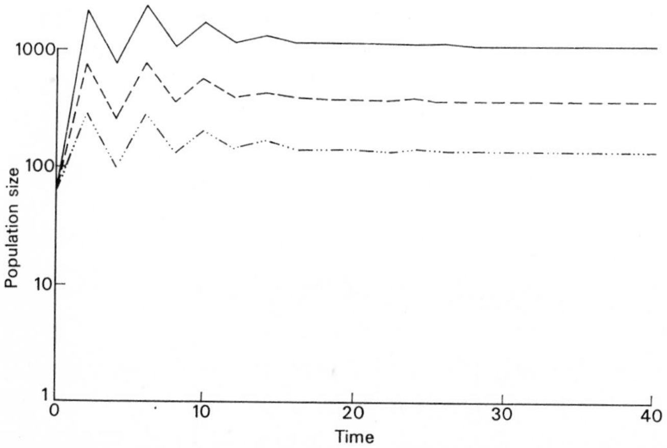

Figure 7. The growth of a population according to the matrix in equation (17). The lines correspond to those in Fig. 6.

tions showed damped oscillations, and that the stable population size is correlated with the general level of fecundity. In no case were stable oscillations produced.

What general conclusions can be gained from these simulation experiments? When the matrix elements are functions of the population size, three sorts of population growth and regulation may result, namely a sigmoid increase, a sigmoid increase with damped oscillations, or stable oscillations (the latter persisting for at least several hundred periods of time). The sensitivity of the functions of the population size would appear to determine which of these methods of population growth and regulation is in operation. When a function changes in value quickly for small changes in the population size anywhere within the range of population sizes likely to be encountered, it would appear that stable oscillations are produced. Such a function can be considered as very sensitive for a part of its range. When a function is insensitive to changes in the population size throughout its range, then the model will demonstrate a sigmoid increase towards the stable population size. Functions with intermediate sensitivities within the range of likely population sizes would appear to have oscillations of small amplitude or to have damped oscillations.

Functional elements in the matrix **A** are not the only cause of oscillations. Studies on predator-prey systems have also shown oscillations, and Leslie's (1959) time-lag model is also oscillatory. The fact that the relation between population size and survival causes oscillations in these simulated experiments does not imply that this is the only cause of oscillations in field populations.

A new model for energy flow and nutrient cycles in ecosystems

The basis of the model

Both Usher (1966) and Goodman (1969) have commented on the fact that the matrix **A** in equation (1) is composed of the sum of two matrices,

$$\mathbf{A} = \begin{bmatrix} f_0 & f_1 & f_2 & \dots & f_n \\ 0 & 0 & 0 & & 0 \\ 0 & 0 & 0 & & 0 \\ \vdots & & & & \\ 0 & 0 & 0 & & 0 \end{bmatrix} + \begin{bmatrix} 0 & 0 & 0 & \dots & 0 \\ p_0 & 0 & 0 & & 0 \\ 0 & p_1 & 0 & & 0 \\ \vdots & & & & \\ 0 & 0 & 0 & & 0 \end{bmatrix}$$

or

$$\mathbf{A} = \mathbf{F} + \mathbf{P} \tag{19}$$

The matrix **F** represents the input of new members to the population, whilst the matrix **P** represents the transition of members between the age groups of the population.

In considering ecosystems, we are usually concerned with such dynamic processes as the cycling of nutrients or the flow of energy. The matrix models can be adapted so as to deal with such processes, since a modified form of matrix **F** would represent the input of energy or nutrients into the ecosystem; a modified form of the matrix **P** would represent the transfer of energy or nutrients within the ecosystem; and the natural compartmentation of the ecosystem into its species composition or into its trophic levels (Phillipson 1966) would facilitate the formation of a column vector describing the distribution of energy or nutrients in the ecosystem at a given time. Losses from the ecosystem are not specified in the matrix model, although they are assumed to be the difference between input and the sum of the output from, and storage in, any one compartment. For ease of developing the model we shall initially assume that the method of compartmentation of the ecosystem is into the trophic levels.

Two examples

The methods of analysis of ecosystems have been considered by Smith (1970). In developing the matrix model, I will use his example for the phosphorus flow in a three compartment system, since this will allow for comparisons to be made. Smith's model is illustrated in Fig. 8, and the parameters are defined by

x_i=the amount of phosphorus in the ith compartment at any specified time;

a_i=the rate of inflow of phosphorus into the ith compartment;

z_i=the rate of outflow of phosphorus from the ith compartment;

f_{ij}=the rate of flow of phosphorus from the ith to the jth compartment.

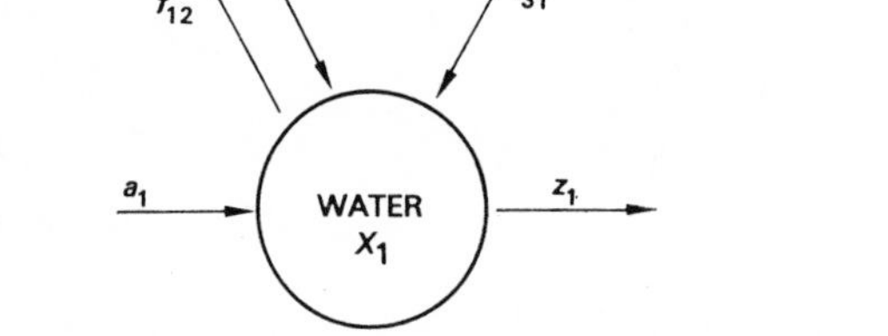

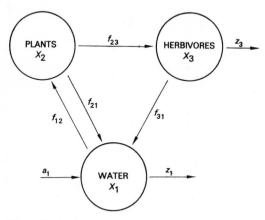

Figure 8. A schematic representation of the phosphorus cycle in a three compartment ecosystem.

In order to develop a matrix model, we need also to define a further set of parameters,

f_{ii}=the proportion of phosphorus that is stored in the ith compartment during a period of time.

The analogous model to that given in equation (1) is

$$\begin{bmatrix} f_{11}+a_1/x_{t,1} & f_{21} & f_{31} \\ f_{12} & f_{22}+a_2/x_{t,2} & f_{32} \\ f_{13} & f_{23} & f_{33}+a_3/x_{t,3} \end{bmatrix} \begin{bmatrix} x_{t,1} \\ x_{t,2} \\ x_{t,3} \end{bmatrix} = \begin{bmatrix} x_{t+1,1} \\ x_{t+1,2} \\ x_{t+1,3} \end{bmatrix} \quad (20)$$

The two vectors in equation (20) denote by $x_{t,i}$ the amount of energy or nutrients in the *i*th compartment at time *t*. As defined above, the matrix elements of the form f_{ij} $(i \neq j)$ denote the transitions between the *i*th and *j*th compartments. The elements in the leading diagonal are composed of two factors, namely the energy or nutrient not transferring between compartments (f_{ii}), and the input into the *i*th compartment, which is, in this initial model, independent of the amount of energy or nutrients already in that compartment, and hence is represented by $a_i/x_{t,i}$.

Having defined the model in these simple terms, it is a logical step, as in the population models, to allow all of the elements in the matrix to become functions of the total standing crop of nutrients or energy, or to be functions of the standing crop in one or a few of the compartments. Smith uses a numerical example for the rates of flow of phosphorus in his hypothetical system. If we use his data, but convert them to a $\frac{1}{4}$-hour period of time, then the numerical constants derived from his functions are shown in Fig. 9. The actual

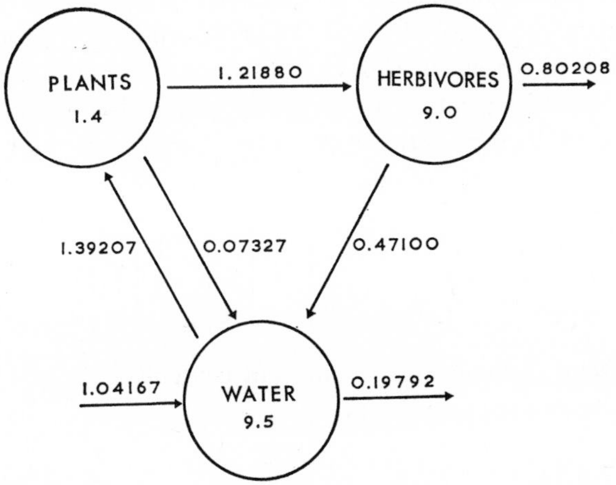

Figure 9. The ecosystem represented in Fig. 8 with numerical values for the parameters.

functions that Smith used, modified for the shorter time period, are given in Table 4. Since the total energy in the *i*th compartment is x_i, then division of the functions in Table 4 by the appropriate x_i expresses the rates of flow on a 'per unit of energy or nutrient' basis. Thus, the matrix in equation (20) becomes

$$\mathbf{A} = \begin{bmatrix} 0{\cdot}94298 & 0{\cdot}05208 & 0{\cdot}05208 \\ 0{\cdot}14584 & 0{\cdot}01042 & 0 \\ 0 & 0{\cdot}93753 & 0{\cdot}85417 \end{bmatrix} \quad (21)$$

using the values of $x_1 = 9{\cdot}5$, $x_2 = 1{\cdot}4$ and $x_3 = 9{\cdot}0$ given by Smith.

Table 4. The functions used in the matrix model in equation (21).

Matrix element	Function describing the process
a_1	constant
f_{12}	$0{\cdot}10417\, x_1 x_2$
f_{21}	$0{\cdot}05208\, x_2$
f_{23}	$0{\cdot}10417\, x_2 x_3$
f_{31}	$0{\cdot}05208\, x_3$
z_1	$0{\cdot}02083\, x_1$
z_3	$0{\cdot}01042\, x_3^2$
f_{11}	$(1 - 0{\cdot}02083 - 0{\cdot}10417\, x_2)\, x_1$
f_{22}	$(1 - 0{\cdot}05208 - 0{\cdot}10417\, x_3)\, x_2$
f_{33}	$(1 - 0{\cdot}05208 - 0{\cdot}01042\, x_3)\, x_3$

The time period was reduced from a day to ¼-hour since this allows the matrix elements to be positive and thus the model to be analogous to those which incorporate probabilities. Investigation of the matrix in equation (21) shows that it has a dominant latent root of 1, and the associated latent vector is $\{9{\cdot}5,\ 1{\cdot}4,\ 9{\cdot}0\}'$, which is in agreement with Smith's results. However, the requirement of positive matrix elements is not necessary, although in such a case the results of the Perron-Frobenius theorems will no longer apply to **A**. Using Smith's actual data for a period of one day, then

$$\begin{bmatrix} 1 - c_1 - x_2 + a_1/x_1 & c_4 & c_6 \\ c_3 x_2 & 1 - c_4 - c_5 x_3 & 0 \\ 0 & c_5 x_3 & 1 - c_6 - c_2 x_3 \end{bmatrix} \begin{bmatrix} x_1 \\ x_2 \\ x_3 \end{bmatrix} = \lambda \begin{bmatrix} x_1 \\ x_2 \\ x_3 \end{bmatrix} \quad (22)$$

would define the stable state of the energy or nutrient content of the ecosystem. Using Smith's values for the parameters,

$$\mathbf{A} = \begin{bmatrix} -4{\cdot}4737 & 5 & 5 \\ 14 & -94 & 0 \\ 0 & 90 & -13 \end{bmatrix} \quad (23)$$

is the matrix that describes the stable state. This matrix has a latent root of 1 and an associated vector of $\{9{\cdot}5,\ 1{\cdot}4,\ 9{\cdot}0\}'$. Unfortunately, it is impossible to use a simple iterative technique to solve for

$$\mathbf{A\,a} = \lambda \mathbf{a}$$

in this particular example. It can be shown that there are an infinite number of solutions, and computer simulations have tended to yield large and negative values of λ.

Why, then, do the usual techniques for the operation of matrix models break down? The reason is based on the functions used by Smith. He says that his functions are based on the simplest available models. For example the functions for f_{12} and f_{23} are based on simple predator-prey equations of the form cx_ix_j. These functions therefore lead to the result that the greater the energy or nutrient content in the ith and jth compartments the greater the flow, without there being any asymptotic upper limit to the amount of flow or the amount that can be stored in the compartment. All of the functions specified by Smith and listed in Table 4 are of this form, since none of them impose any asymptotic limits or maxima on the input, output or flow of phosphorus. The matrix model can, however, be shown to yield the 'correct' result, since one of the infinity of solutions does show the stable ecosystem nutrient structure. The model can be solved easily if we put $\lambda = 1$, and then solve for the values of x_1, x_2 and x_3. However, since the functions used in the matrix are not really representative of a field condition, the matrix model is unstable and is incapable of direct numerical solution.

In order to develop a more comprehensive model, an energy flow situation is illustrated in Fig. 10. The energy input is represented by the function a_1. Blackman (1968) has discussed the manner in which dry matter production is related to the plant standing crop, and he shows this to be a curvilinear relation. The data of Black (1964) indicates a parabolic relation between the

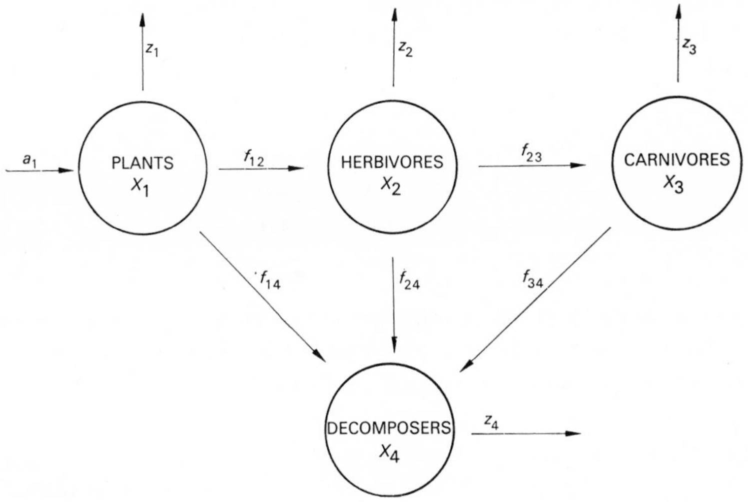

Figure 10. A schematic representation of the energy flow in a four compartment ecosystem.

gross dry matter production and the leaf area index (L.A.I.) of *Trifolium subterraneum*. It can be argued that the L.A.I. gives a measure of the standing crop of energy in the plant population, and it is here assumed that this relation is linear. When the L.A.I. is zero there can be no production, and therefore the parabola passes through the origin and has an equation of the form

$$a_1 = c_1 x_1 - c_2 x_1^2 \tag{24}$$

where c_1 and c_2 are positive constants, and x_1 is the energy content of the plant population.

For the transfer of energy between the trophic levels I have used the model developed by Watt (1959) for the relation between the numbers of attacked and attacking species. Thus, the functions are

$$\left.\begin{aligned} f_{12} &= c_3 x_2 \left(1 - \exp\{-c_4\, x_1\, x_2^{\,(1-c_5)}\}\right) \\ f_{23} &= c_6 x_3 \left(1 - \exp\{-c_7\, x_2\, x_3^{\,(1-c_8)}\}\right) \end{aligned}\right\} \tag{25}$$

where c_3, c_4, c_5, c_6, c_7 and c_8 are positive constants, and x_2 and x_3 are the energy contents of the herbivores and carnivores respectively.

The respiration rates of the plants, herbivores and carnivores, have been assumed to be proportional to their gross production, and are thus defined as

$$\left.\begin{aligned} z_1 &= k_1 a_1 \\ z_2 &= k_2 f_{12} \\ z_3 &= k_3 f_{23} \end{aligned}\right\} \tag{26}$$

where k_1, k_2 and k_3 are positive constants.

The flow of energy into the decomposer food web has been assumed to be proportional to the net production of the trophic level, and is given by the functions

$$\left.\begin{aligned} f_{14} &= k_4(a_1 - z_1) \\ f_{24} &= k_5(f_{12} - z_2) \\ f_{34} &= k_6(f_{23} - z_3) \end{aligned}\right\} \tag{27}$$

where k_4, k_5 and k_6 are positive constants.

The loss of energy from the decomposer food web has been assumed to show a parabolic relationship, being 100 per cent as the energy level approaches zero, increasing to 110 per cent, and then decreasing again. This function allows for all of the energy to be respired or exported, and for the system to be potentially able to deal with more energy than is available to it. However when the energy input increases to such an extent that all the energy cannot be respired, then the function allows for energy to be stored in the decomposer part of the system. The function that has been used is

$$z_4 = 100 + \frac{20}{k_7} I - \frac{10}{k_7^2} I^2 \qquad (28)$$

where $I = f_{14} + f_{24} + f_{34}$ is the input of energy to the decomposer system, and k_7 is a positive constant determining the position of the maximum of the parabola.

The matrix model is thus

$$\mathbf{A} = \begin{bmatrix} a_1/x_1+(x_1-f_{12}-f_{14}-z_1)/x_1 & 0 & 0 & 0 \\ f_{12}/x_1 & (x_2-f_{23}-f_{24}-z_2)/x_2 & 0 & 0 \\ 0 & f_{23}/x_2 & (x_3-f_{34}-z_3)/x_3 & 0 \\ f_{14}/x_1 & f_{24}/x_2 & f_{34}/x_3 & (f_{14}+f_{24}+f_{34})/x_4-z_4/x_4 \end{bmatrix} \qquad (29)$$

The matrix elements have to be divided by the appropriate energy content in order that their values are expressed on a 'per unit of energy' basis. In applying this model for the flow of energy through an ecosystem, the values of the constants are specified in Table 5. An initial approximation to the energy

Table 5. The values of the constants used in deriving the functional elements of the matrix in equation (29).

Constant	Value	Constant	Value
c_1	1·0	k_1	0·6
c_2	0·0025	k_2	0·8
c_3	0·3	k_3	0·7
c_4	0·05	k_4	0·4
c_5	0·9	k_5	0·3
c_6	0·4	k_6	1·0
c_7	0·01	k_7	2·0
c_8	0·9		

distribution in the system was described by the vector $\{100, 10, 1, 20\}'$, and the model iterated on a computer until there was convergence. The process of convergence was extremely slow, since approximately 200 iterations were required to give an accuracy to two places of decimal. These iterations are shown in Fig. 11. It can be seen that the values of the energy contents of the plants, herbivores and carnivores, fluctuated with damped oscillations, eventually taking their final values. Since the model allowed the standing crop of energy in the decomposer system to become zero, it can be seen that this happened for a time before the energy content of the decomposers again

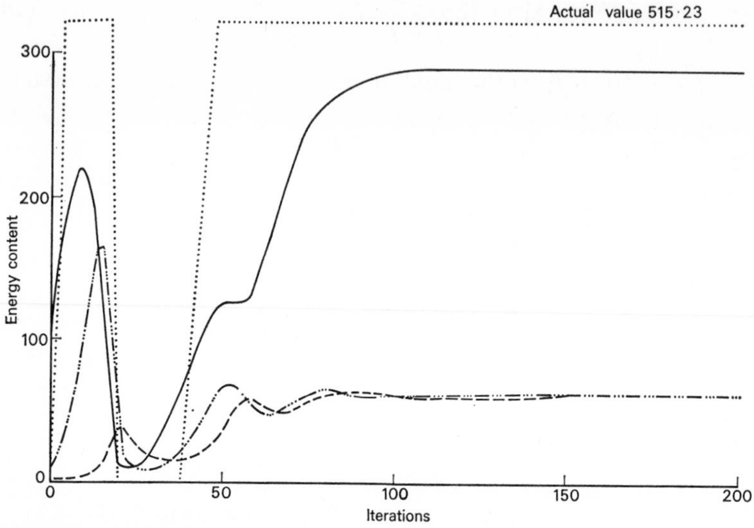

Figure 11. The operation of the matrix equation (29) to the ecosystem depicted in Fig. 10.

increased. The final values of the energy contents derived by this model are

$$\{293{\cdot}19,\quad 62{\cdot}63,\quad 62{\cdot}63,\quad 515{\cdot}23\}'$$

and as this forms a latent vector of the matrix its associated latent root is 1. The matrix A in equation (29), with numerical elements, is

$$\begin{bmatrix} 1{\cdot}0000 & 0 & 0 & 0 \\ 0{\cdot}0641 & 0{\cdot}7000 & 0 & 0 \\ 0 & 0{\cdot}2449 & 0{\cdot}7551 & 0 \\ 0{\cdot}0427 & 0{\cdot}0038 & 0{\cdot}0735 & 0{\cdot}9663 \end{bmatrix}$$

This matrix describes the stable situation for the ecosystem based both on the matrix model and on the models for each of the component processes operating in the ecosystem.

Conclusions

If a new model is developed, there are several criteria that it should possess. Such criteria have been discussed by Watt (1960), who concludes that models should incorporate insight into the biological mechanisms involved as well as being sufficiently general for the model to be applicable in a variety of situations merely by changes in the parameters. I consider that these two features of modelling should be demonstrably apparent when any new model is put forward.

The idea of using matrices in ecosystem modelling is not new, since the use of matrices has been described by Van Dyne (1969a, 1969b). However, this particular model is new since it is not reliant upon the calculus-based models which have been previously used. But does this feature of the matrix model incorporate any biological insight? The matrix is defined as operating over a period of time, and thus the model can take the structure of the ecosystem at time t, and predict the new structures at times $t+1$, $t+2$, $t+3$ etc. There is no way of using the model to predict what the structure would be at time $t+\frac{1}{2}$, or at any non-integer value of time. However, many biological processes occur in discrete periods of time and are not continuous in their operation. Enumerations or determinations of energy or nutrient content are made at specific time intervals, often annually, and hence the nature of the data recording is closely fitted to the characteristics of matrix operations. It would seem, therefore, that the model is at least as biologically meaningful as any models based on calculus.

The generality of the model is also demonstrable. In describing the example for the four compartment ecosystem we have had to specify fifteen different parametric values in Table 5. These were required not basically for the matrix model but for the models of the various component processes of this simplified ecosystem. The elements of the matrix could take fixed values or they could consist of any of the functions that have been developed for the component processes. The matrix model therefore has considerable generality, although the degree of generality is dependent upon the generality of the component models.

In this sense the matrix can be considered as a macro-model. It forms a model framework within which many other models are required. Once all of these smaller models have been specified the matrix approach allows for a simple method of computation of the stable state of the ecosystem, and it allows for simulation experiments to be performed in order to investigate the effects of changes in management practice or environmental variables. Being a macro-model the matrix approach will direct the efforts of research at the areas that are unknown. It incorporates requirements from all branches of ecology. We have seen how one feature of the model is an analysis of the relation between energy standing crop and energy uptake, but in the real world where the compartments are species and not trophic levels we will also require models for the competition between species within a trophic level for the limited resources available to the ecosystem. Such could be the inclusion within this model of the community matrix of Levins (1968). The matrix has been introduced as having elements that are either constant or functions which take a single value, but even more generality can be assumed when the elements of the matrix are themselves matrices. It would therefore be a relatively simple matter to have one of the main diagonal elements as a more usual type of Leslie matrix which would indicate the changes in the structure of the population of one of the species in one of the trophic levels. The matrix model is therefore capable of being built up and developed so that it not only represents the trophic levels in the community, but also the species within each of the trophic levels, as well as the ages or sizes of the individuals of each of the species.

References

BLACK J.N. (1964) An analysis of the potential production of swards of subterranean clover (*Trifolium subterraneum* L.) at Adelaide, South Australia. *J. appl. Ecol.* **1**, 3–18.

BLACKMAN G.E. (1968) The application of the concepts of growth analysis to the assessment of productivity. In *Functioning of terrestrial ecosystems at the primary production level.* Ed. F.E.Eckardt. Liège: UNESCO. pp. 243–59.

BRAUER A. (1957) A new proof of theorems of Perron and Frobenius on non-negative matrices: I. Positive matrices. *Duke math. J.* **24**, 367–78.

BRAUER A. (1961) On the characteristic roots of power positive matrices. *Duke math J.* **28**, 439–46.

BRAUER A. (1962) On the theorems of Perron and Frobenius on non-negative matrices. In *Studies in mathematical analysis and related topics.* Ed. S.Gilbarg *et al.* Stanford: University Press. pp. 48–55.

EHRENFELD D.W. (1970) *Biological conservation.* New York, etc.: Holt, Rinehart & Winston.

FELLER W. (1957) *An introduction to probability theory and its applications.* Vol. 1. New York & London: Wiley.

FROBENIUS G. (1912) Uber Matrizen aus nicht negativen Elementen. *Sber. preuss. Akad. Wiss.* (1912) (26), 456–77.

GOODMAN L.A. (1969) The analysis of population growth when the birth and death rates depend upon several factors. *Biometrics,* **25**, 659–81.

GREEN C.D. (1964) The effect of crowding upon the fecundity of *Folsomia candida* var. *distincta*. *Ent. Exp. & Appl.* **7**, 62–70.

HUMMEL F.C. & CHRISTIE J. (1953) Revised yield tables for conifers in Great Britain. *Forestry Commission For. Rec.* 24.

JENKINS D., WATSON A. & MILLER G.R. (1963). Population studies on red grouse, *Lagopus lagopus scoticus* (Lath.) in North-East Scotland. *J. Anim. Ecol.* **32**, 317–76.

JENKINS D., WATSON A. & MILLER G.R. (1967) Population fluctuations in the red grouse *Lagopus lagopus scoticas*. *J. Anim. Ecol.* **36**, 97–122.

LAWS R.M. (1962) Some effects of whaling on the southern stocks of baleen whales. In *The exploitation of natural animal populations.* Ed. E.D.Le Cren & M.W.Holdgate. Oxford: Blackwell. pp. 242–59.

LEFKOVITCH L.P. (1965) The study of population growth in organisms grouped by stages. *Biometrics* **21**, 1–18.

LEFKOVITCH L.P. (1966) The effects of adult emigration on populations of *Lasioderma serricorne* (F.) (Coleoptera: Anobiidae). *Oikos* **15**, 200–10.

LEFKOVITCH L.P. (1967) A theoretical evaluation of population growth after removing individuals from some age groups. *Bull. Ent. Res.* **57**, 437–45.

LESLIE P.H. (1945) On the use of matrices in certain population mathematics. *Biometrika* 35, 183–212.
LESLIE P.H. (1948a) Some further notes on the use of matrices in population mathematics. *Biometrika* 35, 213–45.
LESLIE P.H. (1948b) On the distribution in time of the births in successive generations. *J. Roy. Stat. Soc. Ser. A*, 111, 44–53.
LESLIE P.H. (1959) The properties of a certain lag type of population growth and the influence of an external random factor on a number of such populations. *Physiological Zool.* 32, 151–9.
LEVINS R. (1968) *Evolution in changing environments. Some theoretical explorations.* Princeton: University Press.
LEWIS E.G. (1942) On the generation and growth of a population. *Sankhya* 6, 93–6.
LOWE V.P.W. (1961) A discussion of the history, present status and future conservation of the red deer (*Cervus elaphus* L.) in Scotland. *Terre Vie* 1, 9–40.
LOWE V.P.W. (1969) Population dynamics of the red deer (*Cervus elaphus* L.) on Rhum. *J. Anim. Ecol.* 38, 425–57.
MURRAY M.D. & GORDON G. (1969) Ecology of lice on sheep, VII. Population dynamics of *Damalinia ovis* (Schrank). *Aust. J. Zool.* 17, 179–86.
NOBLE B. (1964) *Numerical methods I. Iteration, Programming and algebraic equations.* Edinburgh & London: Oliver & Boyd.
PENNYCUICK C.J., COMPTON R.M. & BECKINGHAM L. (1968) A computer model for simulating the growth of a population, or of two interacting populations. *J. Theoret. Biol.* 18, 316–29.
PENNYCUICK L. (1969) A computer model of the Oxford great tit population. *J. Theoret. Biol.* 22, 381–400.
PHILLIPSON J. (1966) *Ecological energetics.* London: Arnold.
POLLARD J.H. (1966) On the use of the direct matrix product in analysing certain stochastic population models. *Biometrika* 53, 397–415.
SILLIMAN R.P. & GUTSELL J.S. (1958) Experimental exploitation of fish populations. *U.S. Fish Wildlife Serv. Fishery Bull.* 58, 215–41.
SKELLAM J.G. (1966) Seasonal periodicity in theoretical population ecology. *Proc. 5th Berkeley Symp.* 4, 179–205.
SLOBODKIN L.B. & RICHMAN S. (1956) The effect of removal of fixed percentages of the newborn on size and variability in populations of *Daphnia pulicaria* (Forbes). Limnol. Oceanogr. 1, 209–37.
SMITH F.E. (1970) Analysis of ecosystems. In *Analysis of temperate forest ecosystems.* Ed. D.E.Reichle. Berlin etc.: Springer-Verlag. pp. 7–18.
SOUTHERN H.N. (1964) *The handbook of British mammals.* Oxford: Blackwell.
SYKES Z.M. (1969) On discrete stable population theory. *Biometrics* 25, 285–93.
USHER M.B. (1966) A matrix approach to the management of renewable resources, with special reference to selection forests. *J. appl. Ecol.* 3, 355–67.
USHER M.B. (1967/68) A structure for selection forests. *Sylva* 47, 6–8.
USHER M.B. (1969a) A matrix model for forest management. *Biometrics* 25, 309–15.
USHER M.B. (1969b) A matrix approach to the management of renewable resources, with special reference to selection forests—two extensions. *J. appl. Ecol.* 6, 347–8.
USHER M.B. (in press) *Biological conservation and management.* London: Chapman & Hall.
USHER M.B. & WILLIAMSON M.H. (1970) A deterministic matrix model for handling the birth, death and migration processes of spatially distributed populations. *Biometrics* 26, 1–12.

VAN DYNE G.M. (1969a) *Grassland management, research, and training viewed in a systems context*. Colorado State Univ., Range Sci. Dept., Sci. Ser. **3**, pp. 1–39.

VAN DYNE G.M. (1969b) Some mathematical models of grassland ecosystems. In *The grassland ecosystem: A preliminary synthesis*. Ed. R.L.Dix & R.G.Beidleman. Colorado State Univ., Range Sci. Dept., Sci. Ser. **2**, pp. 3–26.

WATT K.E.F. (1955) Studies on population productivity. I. Three approaches to the optimum yield problem in populations of *Tribolium confusum*. *Ecol. Monogr.* **25**, 269–90.

WATT K.E.F. (1959) A mathematical model for the effect of densities of attacked and attacking species on the number attacked. *Can. Ent.* **91**, 129–44.

WATT K.E.F. (1960) The effect of population density on fecundity in insects. *Can. Ent.* **92**, 674–95.

WATT K.E.F. (1968) *Ecology and resource management*. New York, etc.: McGraw-Hill.

WILLIAMSON M.H. (1959) Some extensions of the use of matrices in population theory. *Bull. Math. Biophys.* **21**, 13–17.

WILLIAMSON M.H. (1967) Introducing students to the concepts of population dynamics. In *The teaching of ecology*. Ed. J.M.Lambert. Oxford & Edinburgh: Blackwell. pp. 169–75.

WILLIAMSON M.H. (in press) *The analysis of biological populations*. London: Arnold.

Models and analysis of descriptive vegetation data

M.P.AUSTIN *Division of Land Research CSIRO, Canberra, Australia*

Summary

The compatibility of numerical descriptive techniques and current conceptual models of vegetation is considered. The problem of non-linearity in ordination is examined in some detail. The influence of non-linearity and variations in diversity on classification techniques is reviewed. Neither the current methods of numerical description nor the usual conceptual models of vegetation appear satisfactory.

Techniques for the correlation of vegetation with environment are briefly reviewed. It is suggested that functional models of the interaction of environmental variables are required in descriptive analysis, and that regression provides a more flexible technique than other current numerical methods. An example of a regression description using functional environmental scalars is given.

I. Introduction

Numerical techniques are now used extensively in descriptive ecology. They have three main purposes.

(1) Simplification of multivariate data.

(2) Assistance in hypothesis-generation.

(3) Domain definition, that is definition of the vegetation type within which experiments may be carried out, and the probable limits of extrapolation for the experimental results obtained.

The justification for using numerical techniques is twofold.

(1) Reduction in subjectivity; though not objective numerical techniques are repeatable and internally consistent, while the subjective decisions are explicit rather than implicit.

(2) Compatability with computers allows the rapid analysis of very large sets of data. An extended review of the user's viewpoint has been provided by Grieg-Smith (1969), but it is perhaps time we began to assess these techniques in detail.

In particular, three questions may be asked.

(1) Are the assumptions of the numerical descriptive methods compatible with our current conceptual models of vegetation?

(2) How effective are our present methods of analysing descriptive data in providing ecological insight into vegetation environment interactions?

(3) What role can current modelling activities play in this field?

II. Assumptions of numerical descriptive techniques

Two complementary descriptive techniques can be recognized, namely ordination and classification (Anderson 1965, Greig-Smith, Austin & Whitmore 1967, Webb, Tracey, Williams, & Lance 1967) and though the assumptions invoked are often similar it is convenient to treat them separately here.

(*a*) *Ordination*

Two important assumptions can be recognized; first that linear models are appropriate, and second, that species richness and stand biomass can be ignored (Austin & Greig-Smith 1968).

(i) Linearity

This problem has been raised in various forms by several workers (Goodall 1954, Groenewoud 1965, Austin 1968) and more recently Swan (1970) has given a clear demonstration using artificial data. Most ecologists would not expect species to interact or respond to an environmental gradient in a linear manner, though current ordination methods assume this. If one makes the reasonable ecological assumption that species respond to an environmental gradient in such a way as to produce a bell-shaped abundance curve, then the effect of a linear model on the position of stands along such a gradient can be demonstrated by the artificial example taken from Fig. 1. The example is trivial, but the conclusions are applicable to the multivariate case of 'n' species and 'p' environment gradients. The vegetation distance between stand 1 and stand 3 is only slightly affected by the curvature of the species responses compared with the environmental distances. The same is not true for the distance between stand 1 and stand 4, where the distortion is considerable; because of the non-monotonicity of the response curve of species 'b' (see data matrix in Fig. 1), the distance contributed by this species appears to

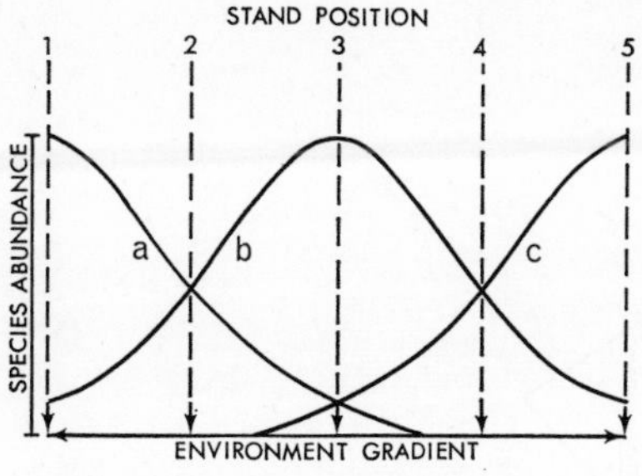

STAND POSITIONS AND SPECIES OPTIMA ALONG AN ARTIFICIAL ENVIRONMENTAL GRADIENT

DATA MATRIX

Species	Stands 1	2	3	4	5
a	100	50	10	-	-
b	10	50	100	50	10
c	-	-	10	50	100

Euclidean vegetation distance

$$D_{jh} = \sqrt{\sum_{i=1}^{3}(X_{ij}-X_{ih})^2}$$

ENVIRONMENTAL DISTANCE BETWEEN STANDS

	1	2	3	4	5
1		1.0	2.0	3.0	4.0
2			1.0	2.0	3.0
3				1.0	2.0
4					1.0

SCALED VEGETATION DISTANCE (D_{jh}/D_{12}) BETWEEN STANDS

	1	2	3	4	5
1		1.0	1.995	1.885	2.209
2			1.012	1.105	1.885
3				1.012	1.995
4					1.0

Figure 1. Artificial example of distortion effect of bell-shaped species abundance curves on linear ordination.

suggest that stand 4 is as close as stand 2 to stand 1. Stand 5, though further from stand 1 than 4 owing to the increase in abundance of species 'c', is not proportionately further as species 'a' is absent in both stands 4 and 5, and species 'b' has the same abundance in 5 as in 1. There are three contributing factors to distortion: curvature of the response, lack of monotonicity, and the joint absences of species. The latter is the major factor, particularly when a gradient is long compared with the breadth of individual species distribution, and the end stands have no species in common; the degree of distortion is influenced by the amount of species overlap (Swan 1970). This type of distortion occurs in all the present methods of ordination used in ecology and can produce a distorted picture of a single gradient (Fig. 2) which will extend beyond three dimensions (Swan 1970, Noy-Meir & Austin 1970).

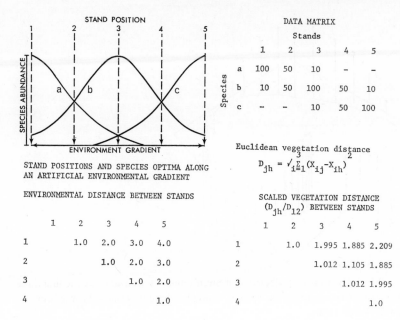

We (Austin & Noy-Meir 1971) have since extended this approach to the case of a two-dimensional environment. Individual species response to two environmental gradients was represented by a surface equivalent to a bivariate normal distribution and the environmental plane defined by a series of such species was sampled by means of a grid of points (stands). If we consider only the first two axes of an ordination of the data derived from these stands, distortion is readily apparent (Fig. 3). The effects of several common standardizations of the data matrix were studied; these sometimes improved the two dimensional projection, e.g. the Bray and Curtis (1957) standardi-

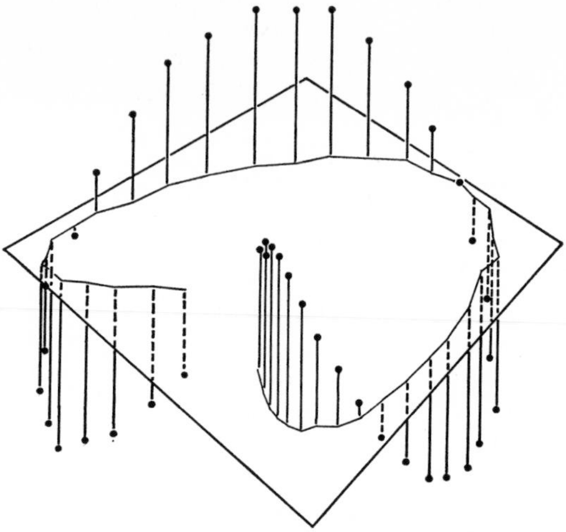

Figure 2. Principal component ordination of Swan's model $\bar{V}$ data showing the first three axes; the original linear gradient has become a complex three-dimensional curve (from Noy-Meir & Austin 1970).

zation (Fig. 4a). The marked lack of success of the Bray and Curtis ordination procedure when studied by Austin and Orloci (1966) may be due to the fact that the data were left unstandardized.

(ii) Species richness and abundance

Some workers have explicitly tried to remove some of the effects of species richness and abundance. The Bray and Curtis double standardization is one of these; species are expressed as a proportion of their maximum observed value to remove the effect of differences of abundance between species and these standardized scores are then expressed as a proportion of the total abundance in the stand to remove abundance differences between stands. It is very much open to question whether or not significant information is being lost in this way.

In the two dimensional grid model (Fig. 3) the corners of the grid are relatively species-poor; in this case, the Bray and Curtis standardization reduces the degree of distortion (Fig. 4a). If the model is modified in such a

Figure 3 (a). Artificial two-dimensional environment. Grid represents sampling points (30 stands) for vegetation data. Species abundances are represented by a series of bell-shaped response surfaces with the maximum value occurring at one of the sampling points (i.e. 30 species, larger figures for species with optimum at stand 15).
Figure 3 (b). Results of ordinating two-dimensional artificial vegetation data which has been sampled by an equally spaced grid of stands.

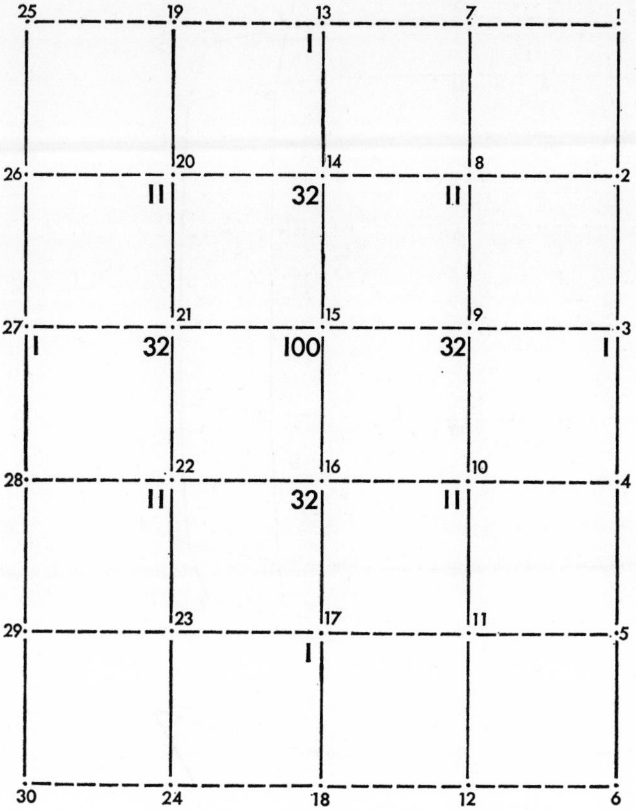

25
19
13
7
1
I
26
20
14
8
2
II
32
II
27
21
15
9
3
I
32
100
32
I
28
22
16
10
4
II
32
II
29
23
17
11
5
I
30
24
18
12
6

(a)

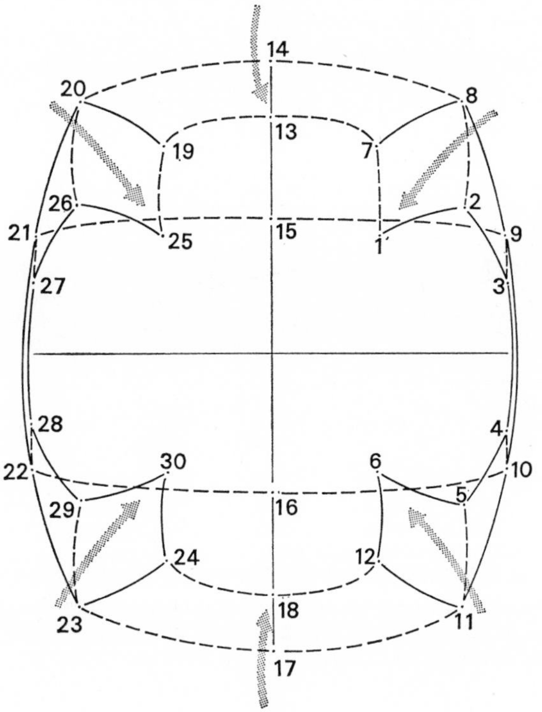

14
20
8
13
19
7
26
2
21
25
15
1
9
27
3
28
4
30
6
22
10
5
29
16
24
12
18
23
11
17

(b)

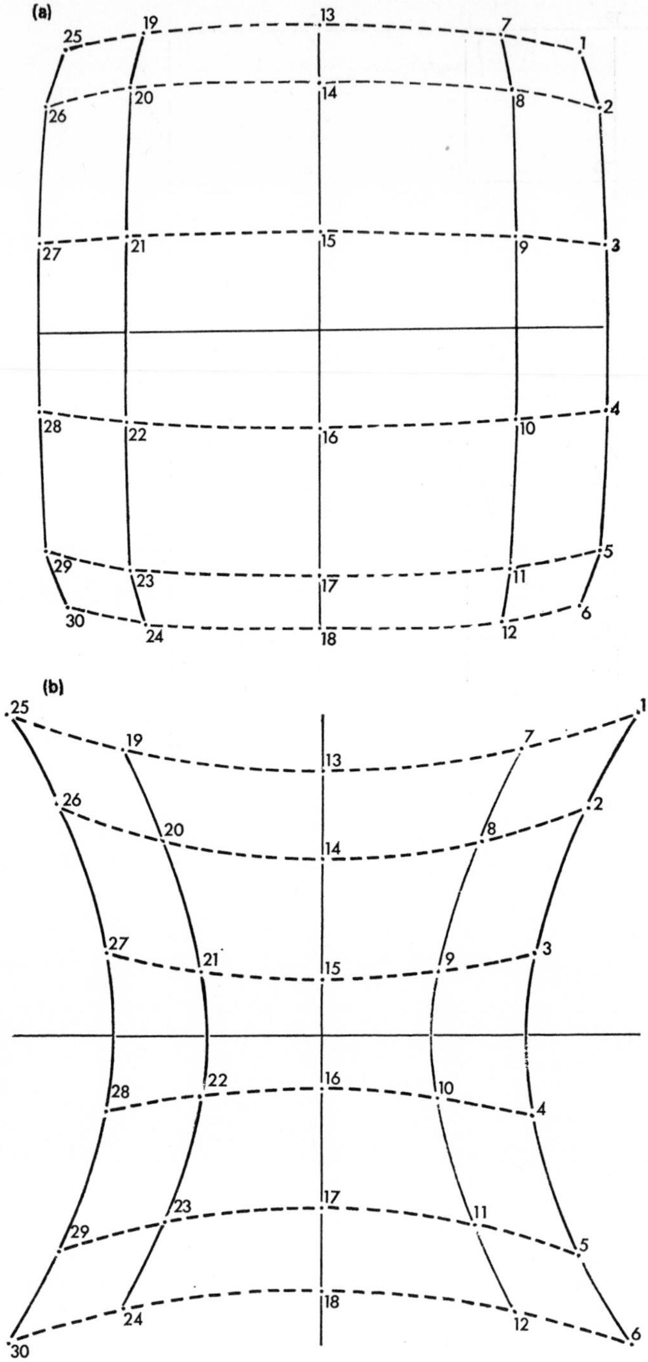

Figure 4. Effects of stand abundance. (a) Results of ordinating artificial data after standardization using Bray & Curtis procedure; closely approximates the original grid. (b) Results of ordinating artificial data modified by reducing differences in stand abundance and applying Bray & Curtis standardization.

way as to reduce this differential effect by adding further species with optima outside the grid, then the Bray and Curtis standardization is less successful (Fig. 4b), as is the case with the simple example given in Table 1. It is un-

Table 1. Effect of Bray & Curtis methods on data from Fig. 1

A: Bray & Curtis standardized data $x = \dfrac{Xi/Xi\text{max}}{\Sigma Xij}$

	1	2	3	4	5
a	91	50	8	—	—
b	9	50	84	50	9
c	—	—	8	50	91

B: Euclidean distance calculated from the above data*

	1	2	3	4	5
1		1·0	1·93	1.92	2·22
2			0.94	1·22	1·92
3				0·94	1·93
4					1·00

C: Distance calculated using Kulczynski's coefficient (c) expressed as l-c*

	1	2	3	4	5
1		1·0	1·30	1·35	1·35
2			1·01	1·06	1·35
3				1·01	1·30
4					1·00

* Scaled $D = D_{jh}/D_{12}$

likely that any simple transformation or standardization will give an intuitively acceptable result in more than a few types of data matrix. The general problem of the parameters of a floristic data matrix and the role of standardizations has been discussed by Austin and Greig-Smith (1968), but the ecological implications of these components of floristic data need further consideration.

It is clear that the linear ordination methods which have been used are not compatible with the non-linear, non-monotonic conceptual model used by ecologists, and some reconsideration of these methods is necessary.

(*b*) *Numerical classification*

(i) Heuristic methods

Classificatory techniques will suffer from exactly the same effects as ordination.

It is important to appreciate, however, that heuristic techniques may be appropriate, depending on the level of ecological understanding or the type of assumptions which the ecologist is prepared to make.

1. *Physiognomic classification.* In the absence of any specific hypotheses, extremely powerful heuristic methods of classification are now available for preliminary sample stratification and hypothesis-generation. Methods such as MULTCLAS can incorporate mixed data, i.e. qualitative, quantitative and multistate data (Lance & Williams 1967). These methods are at present particularly appropriate to studies of the physiognomic or structural variations of vegetation, where ecological concepts regarding the relationships of physiognomic features are poorly developed. Webb, Tracey, Williams, and Lance (1970) have applied such methods to variations in rainforest physiognomy and related it to latitude and altitude. It can also be applied to more local studies of physiognomy (Austin 1970), see Fig. 5. The results can be evaluated subjectively in relation to environmental factors; Fig. 6 shows the relationship of the physiognomically defined vegetation types to rainfall and altitude. However, even apart from the problem of evaluating classifications and combining different types of attributes (Austin 1970), these methods are sensitive to the problems mentioned previously. This is most clearly demonstrated by an ordination of a selected group of stands from the previous example (Fig. 7); the curvilinear distortion is readily apparent. Neither dry sclerophyll forest nor the rainforest transition possess many of the attributes common to the several intermediate forest types. This figure also indicates one advantage of classification—it tends to break such curves into linear segments.

2. *Common divisive techniques.* The development of classificatory techniques for floristic data has been less arbitrary, as concepts about floristic variation are better developed (Williams & Lambert 1959). The most widely-used methods have been qualitative, monothetic and divisive, principally for the practical reasons of ease of computation and applicability to large amounts of data and these methods have proved successful in stratifying stands into relatively homogeneous groups. I have summarized the sensitivities of three of these methods in Table 2a. Association-analysis is markedly sensitive to the presence of rare species (or the absence of very common ones) due to the effect of low marginal or cell totals on the value of χ^2 (or related measures) from the 2×2 contingency table (Goldsmith 1967, Williams pers. comm.); this gives rise to 'chaining'—the splitting off of small groups of stands characterized by rare species. In intrinsic classifications, there is no *a priori* reason for giving weight to rare species in this way.

Divisive information analysis (Lambert & Williams 1966, Lance & Williams 1968) is a more recent technique. This method tends to produce groups equal in size as the method is likely to divide a group on a species

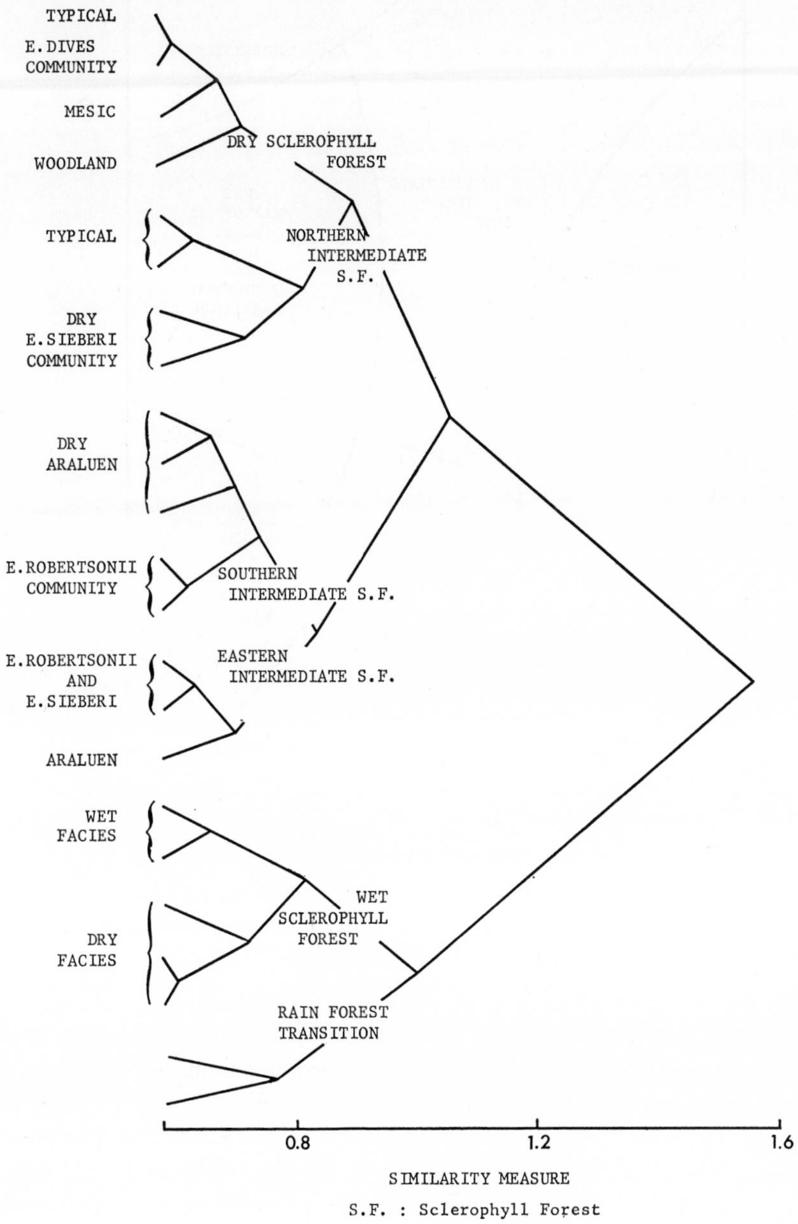

Figure 5. Classification of vegetation based on structural characters using MULTCLAS with Gower coefficient, flexible sorting ($\beta = -0{\cdot}1$), qualitative, numeric and non exclusive multistate types of characters (data from Queanbeyan-Shoalhaven region N.S.W. Australia). From 'An applied ecological example of mixed data classification' in Data Representation, edited by R.S. Anderssen and M.R. Osborne, 1970.

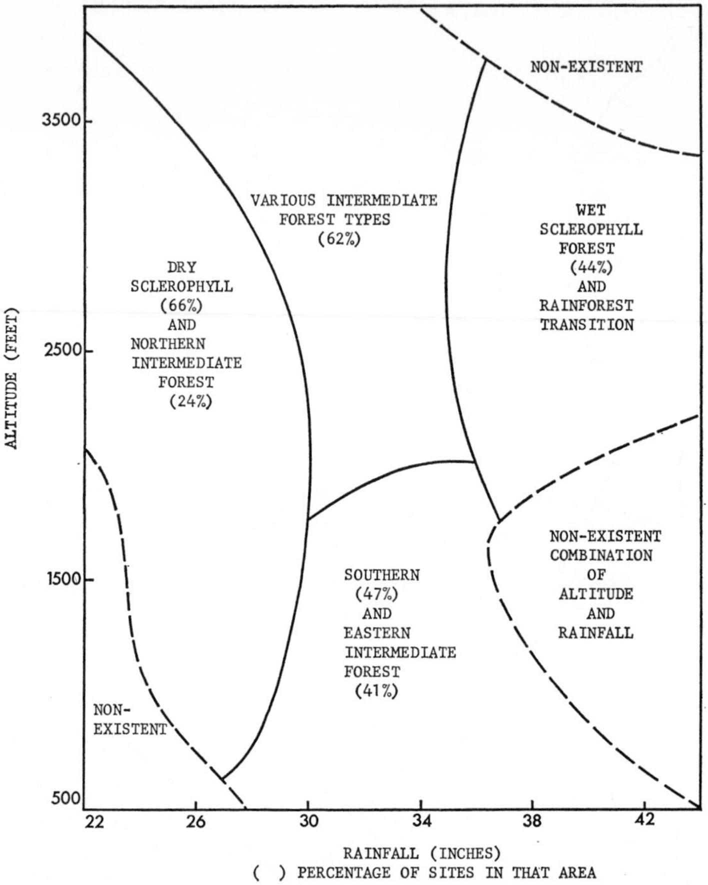

Figure 6. Environmental analysis of structural vegetation types obtained from MULTCLAS. From 'An applied ecological example of mixed data classification' in Data Representation, edited by R. S. Anderssen and M.R. Osborne, 1970.

which occurs in approximately 50 per cent of its stands. Field (1969) has pointed out that the information statistic used is symmetric with respect to zero, i.e. 0 = 0 is as important as 1 = 1 in assessing similarity values (as is χ^2/N), this means that stands with large numbers of zero values in common will appear similar. In effect, stands with few species, i.e. low stand richness, appear similar. A further information statistic has been developed (Dale 1969, Dale, Lance & Albrecht 1971) which is asymmetric to absences; it effectively considers the proportion of the total stand composition contributed by each species but is still therefore sensitive to stand richness.

These numerical methods were developed without consideration of detailed ecological models or concepts of vegetation. It is important to be aware of their implicit assumptions; association-analysis assumes rare species are important; divisive information analysis that important vegetation groups

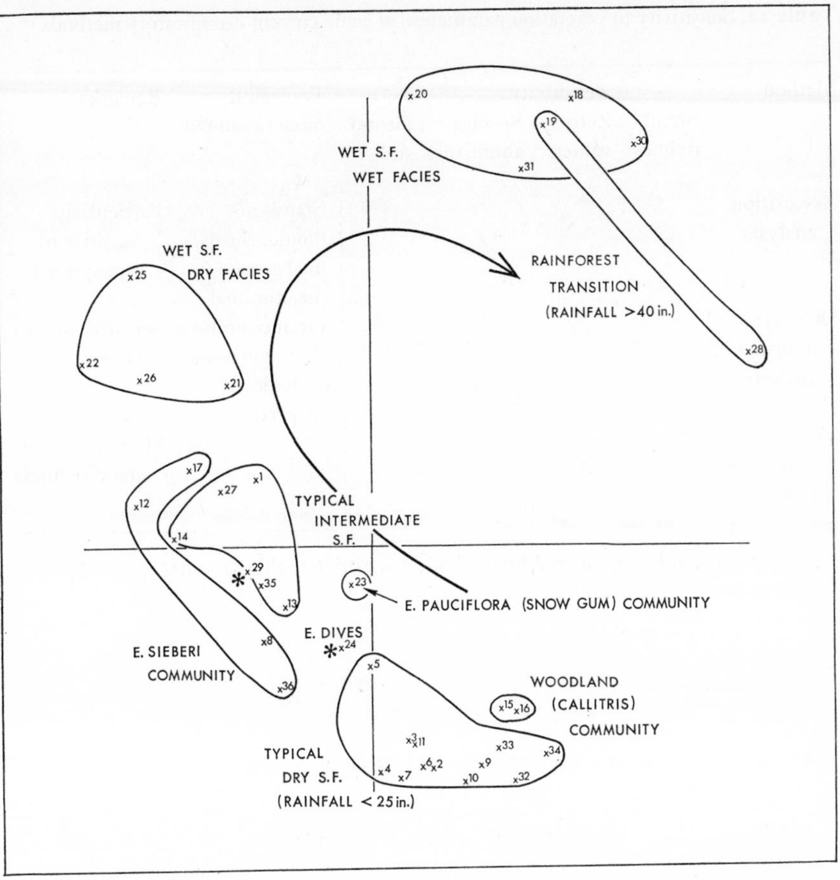

Figure 7. Ordination of selected stands from data used for the physiognomic classification shown in Fig. 5. Gower coefficient and method of ordination used. (E.=*Eucalyptus*, S.F.=Sclerophyll forest).

are likely to be equal in size; while most methods assume that vegetation groups (types) are spheroidal in some abstract hyperspace and in many cases that common absences should indicate similarity.

In discussion, Dale has pointed out that there does not exist a global strategy or method that is equally effective for dealing with the various combinations of richness and abundance which occur in floristic data matrices. A method is needed which will allow us to decide which technique will give the most effective classification for providing ecological insight. To do this, a more general model of vegetation is needed.

(ii) Other models

There are a number of other classification models (Dale pers. comm., Table 2b) which are of interest though they are clearly incomplete. I will mention

Table 2a. Sensitivity to vegetation parameters of some current classificatory methods

Method	Sensitivity: Stand richness	Zero values	Species abundance	Group size	Ecological model assumed	Remarks
Association analysis	*	*	*	(*)	Stands in 'homogeneous' final groups (i.e. minimal variance groups) share common ecological properties	Particularly sensitive to rare species
Divisive information analysis	*	*	(*)	*		Sensitive to zero values
Mutual information analysis	*	✓	(*)	*		Major sensitivity stand richness

✓ parameter explicitly considered by method; * sensitive to this parameter; (*) sensitive to this parameter only indirectly.

Table 2b. Ecological aspects of other recent classificatory methods

Method	Mathematical model	Ecological model	Remarks
Two-parameter model (MacNaughton-Smith 1965, M. Dale pers. com.)	$\frac{\alpha\beta}{1+\alpha\beta}$ α: species 'ability' $1/\beta$: stand 'difficulty'	Groups of stands which fit model, i.e. species arranged in a monotonic gradient of increasing abundance and stands in one of increasing richness, have ecological significance	Assumes monotonicity; diversity characteristics of stands dominate floristic differences
Multiplicative species interaction (Gilbert & Wells, 1966, M. Dale, pers. com.)	$\log n_{ij}=b_i+b_j+c_ic_j$ $(b_i=\frac{1}{2}\log n+\log q_i)$ n_{ij}: no. of co-occurrences q_i: species frequency (as probability) c_i: interaction coefficient	Joint occurrences of species is a result of two species interactions and the average frequency of those species	Competitive coefficients c_j's invarient; method has not been extensively studied

briefly two of these. The two-parameter model of MacNaughton-Smith (1965) can be interpreted as the α's representing the abilities of the species to grow in a particular stand, and the reciprocal of the β's the 'difficulty' of a stand with respect to all species. The technique is to divide the stands into groups on the presence or absence of the species which gives the best fit to the model in each of the groups. In the qualitative case the α's are monotonically proportional to the species frequency within the group, while the β's are similarly related to the stand richness.

A second model of vegetation which has not perhaps received as much attention as it might is the species interaction model of Gilbert and Wells (1966). Dale (pers. comm.) has recently developed this approach as a divisive technique in the same way as the two-parameter method.

The properties of the vegetation on which these two novel approaches concentrate are the diversity of the species and the species competitive interactions—aspects of vegetation not always given full consideration in the analysis of field observations. Our concept of vegetation as either a 'homogeneous plant community' or a 'vegetational continuum', neither of which considers variations of species richness or stand biomass, seems to me to be insufficient for developing numerically expressed models on which to base classificatory methods of any reasonable generality. I do not believe significant advances will be made until better conceptual models of vegetation are developed.

There exist ecological concepts which have not been integrated into vegetation studies. The hypothesis that species richness (diversity) increases from 'poor' to 'rich' habitats is an example. If diversity acts as an indicator of environment in this way, then it should be explicitly incorporated into vegetation descriptions. Slobodkin and Sanders (1969) have expressed the alternative view that species richness increases with the predictability of environmental stress and abundance with availability of resources. These concepts need to be studied in relation to the analysis of vegetation data.

III. Description of vegetation-environment interactions

Once a suitable method of describing vegetation has been adopted, then the correlation between vegetation and environment can be examined. Most of the methods which have been used rely on simple graphical presentation and assessment, and are often based on the assumption that simple (easily measured?) environmental variables can adequately represent the effective environment (Perring 1959, Curtis 1959, Bunce 1968, Gittins 1965, amongst others).

(a) Gradient analysis

This has been one of the most widely used of those methods which explicitly consider the correlation between vegetation and environment (Whittaker 1956 1960 1967, Whittaker & Niering 1965, Waring & Major 1964, Bakuzis 1969).

Several criticisms of gradient analysis can be made.

(i) The gradients are based on stand scores derived from estimates of species behaviour in response to either subjectively estimated environmental variables (Whittaker 1956) or to those variables which are simple to measure (Waring & Major 1964). In effect gradients are derived from species patterns and are used to examine species patterns. This circularity seems likely to hamper efforts at analysing the reasons for the ecological equivalence of sites with different combinations of environmental variables.

(ii) There is no provision in the method for assessing the relative success of one environmental variable (gradient) as compared with another.

(iii) The method of calculating the gradient is linear, and therefore subject to exactly the same distortions as other ordination methods and will also be influenced by species richness, etc.

(b) Environmental scalars

Loucks' use of environmental scalars (1962) in which environmental variables are considered independently of the vegetation, is, in effect, a statement of those environmental components which are thought to contribute to the equivalence of sites. The assumptions involved in combining the simple environmental variables into scalars may be arbitrary, mainly because of a lack of information, but within such a framework the correlation with vegetation can be examined and the separate assumptions involved in the ecological equivalence of sites tested. The problem of incorporating some relative measure of success still remains.

(c) Numerical techniques

Canonical correlation is one method which has been used for correlating vegetation and environmental variables (Austin 1968, Cassie 1969, Cassie & Michael 1968, Barkham & Norris 1970). The apparent success of this method has varied. The mathematical model appears to be unrealistic ecologically, as the method provides orthogonal linear correlations between sets of vegetation and environmental variables. The correlation between separate principal component analyses of the vegetation and environment has also been used (Austin 1968, Barkham & Norris 1970). Visual assessment of the correlation has been preferred for recognizing non-linear correlation, thus no measure of success is available. See Austin (1968), and particularly Austin, Ashton, & Greig-Smith (manuscript) for examples and a full discussion of this approach.

These methods are also based on restrictive linear models. Regression is a less restrictive technique in that curvilinear relationships can be incorporated. Yarranton has recently used this technique in relation to the vegetation of a limestone pavement (Yarranton 1970, Yarranton & Beasleigh 1968 1969, see also Rutter 1955, Greig-Smith 1964, and Austin 1966 1968). There are considerable problems relating to the mathematical assumptions of this method (Box 1954 1966, Draper & Smith 1966), though the proportion of variance accounted for provides a measure of relative success. A significant ecological omission from recent work is also apparent, that is the failure to consider the functional and dynamic relationships between the environmental variables included as 'independent variables' in the regression equation.

IV. Use of models

(*a*) *Functional and dynamic scalars*

Many regression analyses have used simple environmental variables, for example, light intensity, relative humidity and temperature (Yarranton 1970), or simple topographic variables such as aspect and slope (Medin 1960, Stringer & La Roi 1970). Each variable may appear as an independent variable in the regression equation, yet may in fact be part of the same functional relationship. Light, humidity and temperature may all be contributing to the same moisture stress effect on the plants studied by Yarranton and an expression or model of how these variables interact to affect plant moisture stress may lead to simpler, i.e. with few and less complex terms and more successful (in terms of R^2 and interpretability), regression equations. Similarly, variation in solar radiation is a complex function of slope, aspect, time of year and latitude (see for example Lee 1963), and is likely to be the predominant agent in any relationship between vegetation and aspect (Boyko 1947, Loucks 1962, Jenik & Rejmanek 1969). A radiation scalar or index which takes these relationships with aspect into account, together with the effect of sunshine hours, diffuse radiation and obstructions, has recently been developed (Fleming & Austin, unpublished); scalars of this type are functionally related to vegetation and may be expected to provide more interpretable correlations on which to base hypotheses.

Concerning the dynamic aspects of the environment, agronomists appear to have made more progress than ecologists, particularly in relation to soil moisture effects on yield of cereals (Baier 1965; Baier & Robertson 1968, Zahner 1967, Fitzpatrick 1965, Rijtema 1968, McAlpine 1970).

Fitzpatrick and Nix (1969) have shown how effective modelling moisture regimes can be in predicting yield of cereals in Queensland. Using soil moisture storage, rainfall, and other meteorological variables, together with the growth characteristics of the particular crop, they developed iteratively a soil water balance model for Biloela in Queensland (Fig. 8). From this, they were able to derive estimates of moisture stress and show that there was a critical

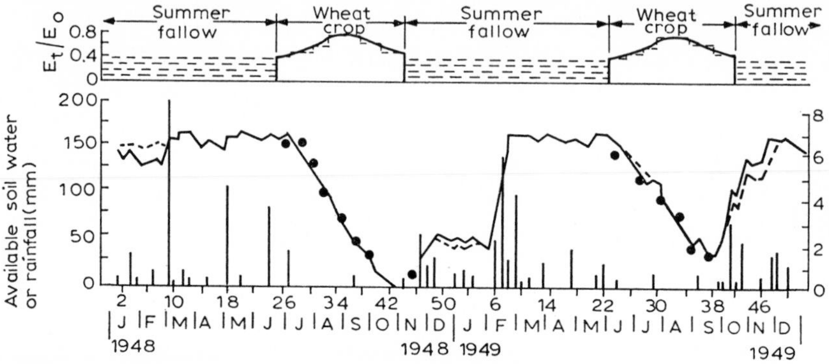

Figure 8. Comparison of estimated and observed soil water changes under a wheat-fallow sequence (1948–49) at Biloela using E_0———, $\bar{E}_0$-----.
Reproduced from Agricultural Meteorology, Volume 6, No. 5, 1969. E.A. Fitzpatrick and H.A. Nix: A model for simulating soil-water regimes in alternating sallow-crop systems (Figure 8, page 317).

period during which yield was dependent on moisture stress (Nix & Fitzpatrick 1969). A stress index was developed from this for data from Biloela and compared with yield data from other areas lacking measurements of the dynamic changes in soil moisture with considerable success (Fig. 9). They

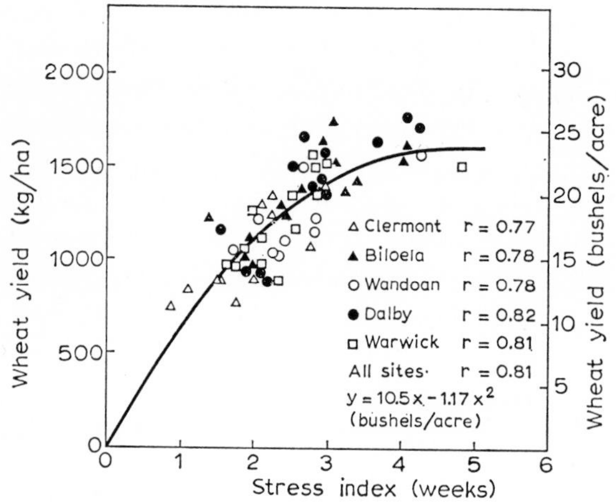

Figure 9. Relationship between wheat yields and computed stress index at five centres in Queensland between latitude 22°S and 20°S based on model developed for Biloela.
Reproduced from Agricultural Meteorology, Volume 6, No. 5, 1969. H.A. Nix and E.A. Fitzpatrick: An index of crop water stress related to wheat and grain sorghum yield (Figure 4, page 332).

have succeeded in integrating the dynamic component of soil moisture status into a moisture stress index (cf. Waring & Major 1964). Zahner and Stage (1966) provide an example of this dynamic water budgeting approach to the growth of forest tree species. Examples applying to natural vegetation appear rare, though see Major (1963) and Mather & Yoshioka (1968).

If adequate analysis of descriptive data is to be achieved, methods which do not assume linearity are necessary. What is needed is a description of the multidimensional response surface of a vegetation variable in relation to environment. Multiple regression appears to offer a possible approach to this. Regression analysis can be used as a descriptive tool rather than a statistical technique for testing hypothesis.

In view of the lack of descriptive regression analyses which incorporate functional independent variables, I shall try to briefly present an example.

(*b*) *Regression with dynamic scalars*

An example of a regression description can be obtained from chalk grassland data previously analysed by principal components ordination (Austin 1966, 1968). In this example *Carex flacca* is the dependent variable, and stepwise regression (Draper & Smith, 1966) is used. The independent variables were *Zerna erecta* performance, percentage soil moisture (by weight), soil pH (in 0·01 M $CaCl_2$), exchangeable calcium and potassium, available phosphate, soil depth, angle of slope, and aspect value. Two seasonal variables were included, i.e. time in days from April 26 and a ranking of the three years over which the data were collected, assessed on the adjusted mean abundance of *Carex* for each year. Various types of equation could be used; in general some simplifying assumptions must be made. In this case, linear and square root terms for each environmental variable were included together with linear and square root interaction terms for each with the major plant competitor *Zerna erecta*.

This is the normal type of regression analysis and the best equation in terms of R^2 is extremely complex (Table 3). The description is not very informative; the seasonal variables indicate that there is an effect, but they do not suggest a possible explanation. Similarly soil variables figure prominently but without consideration of how they might interact. Depending on available knowledge two approaches can be adopted: either modelling the significant process, or simplification when knowledge is limited.

Using the radiation index mentioned previously, and a number of guesstimates regarding the growth and rooting depth of *Carex flacca*, it is possible to utilize the modelling approach of Fitzpatrick and Nix (1969) and Nix and Fitzpatrick (1969) and to derive a moisture stress index. Monthly values of potential evapotranspiration were modified by the radiation index to provide

Table 3. Regression: *Carex flacca* on simple environmental variables plus seasonal factors

b's	Term	'*t*'	Statistics
+102·9	Constant		
+116	Year ranking	5·15	
−0·259	*Zerna* perf. × year ranking	4·33	
+43·6	Sq. root (seasonal time)	2.97	R^2=0·647
+364	Sq. root (avail phosphate)	2·18	adjusted for d.f.
+25·9	Sq. root (soil depth)	2·74	=0·595
−12·3	Soil moisture	4·99	*F* =12·473***
−5·24	Sq. root (exch. calcium)	1·61†	
+0·0264	*Zerna* perf. × soil moisture	4·39	
−16·2	Sq. root (*Zerna* perf. × avail. phosphate)	1·91†	
−2·01	Sq. root (*Zerna* perf. × year ranking)	2·50	

† not significant at 5% level.

estimates for different slopes and a water balance calculated to give estimates of actual evapotranspiration; Nash (1963) adopted a similar approach. Two indices were then derived: the sum of potential evapotranspiration from the beginning of the growing season and the mean ratio of actual to potential evapotranspiration over the period of growth.

The four soil variables which appear in the equation (exchangeable calcium, available phosphate, soil depth and percentage soil moisture) are exactly those variables which were associated with the second environmental component in the previous study of these data (Austin 1968), i.e. they exhibit multicollinearity. It was concluded that this group of variables was part of a factor complex and that no single variable was likely to be the causal factor. In view of this, the first principal component of the correlation matrix between these four variables was calculated (accounts for 51 per cent of variance), and used as an estimate of the factor-complex.

These series of new variables were incorporated into a regression analysis similar to the previous one. The results are presented in Table 4. The adjusted R^2 is identical with the previous best equation. The equation has been re-

Table 4. Regression: *Carex flacca* on environmental dynamic scalars

b's	Term	'*t*'	Statistics
+48·6	Constant		
+91·2	Soil index	9·23	
−0·179	*Zerna* perf. × soil index	7·57	R^2=0·626
−7·38	Aver. moisture stress	5·55	adjusted for d.f.
+98·7	Sq. root (aver. moisture stress)	5·46	=0·595
+1·19	*Zerna* perf. × radiation index	2·89	*F* =20·08***
−40·3	Sq. root (*Zerna* perf. × radiation index)	2·57	

solved into three environmental components together with the competitive effect of *Zerna erecta*. The response surface for *Carex flacca* (Fig. 10) is still complex but the interpretation is much simpler and intuitively appealing: the competitive ability of *Zerna* is greatest under conditions of high soil index, intermediate radiation conditions and high plant moisture status. *Carex flacca* appears to be relatively insensitive to radiation and moisture stress as compared with soil factors and competition.

This description appears more useful and detailed than that given in Austin (1968) where only graphics and principal component analysis (PCA) were used. Regression confers more flexibility than the linear PCA model but it is perhaps the clearer statement of environmental models in this study which was most important.

The assumptions of regression analysis are less restrictive than principal component analysis, but are only slightly less inapplicable to any actual study (Box 1966). More is known about the limitations of regression than PCA and more sophisticated use of the technique than in the example here appears possible and desirable.

Most ecologists who have attempted to set up environmental scalars have assumed, as a first approximation, that four factors are important to vegetation: moisture, nutrients, light and heat. They can be recognized in the regression description; the soil index represents our lack of knowledge regarding nutrients (cf. Loucks 1962, Waring & Major 1964, Fitzpatrick & Nix 1970), and the moisture stress index indicates our relatively greater knowledge of moisture while heat and light are confounded in the radiation index.

Conclusion

Two aspects of the study of vegetation-environment interaction can be recognized, i.e. the direct functional relationship between plant growth and variables such as moisture and nutrient availability, and the effects of indirect environmental variables on this availability. The descriptive ecologist may not need models of the complexity sometimes advocated for studying ecosystems, but he certainly needs some approximate models for the major factors in order to obtain understanding of the ecological equivalence of sites.

At a technical level, it is apparent that the non-linear, non-monotonic species response curve is not satisfactorily considered by current numerical techniques. The bell-shaped curve is an oversimplification, and Fig. 11 provides one example of the possible complexity; this difficulty has yet to be considered.

Numerical methodologists appear to have concentrated on heuristic techniques to the exclusion of ecology. Plant ecologists have not been conspicuous in contributing to recent theoretical developments (see for example

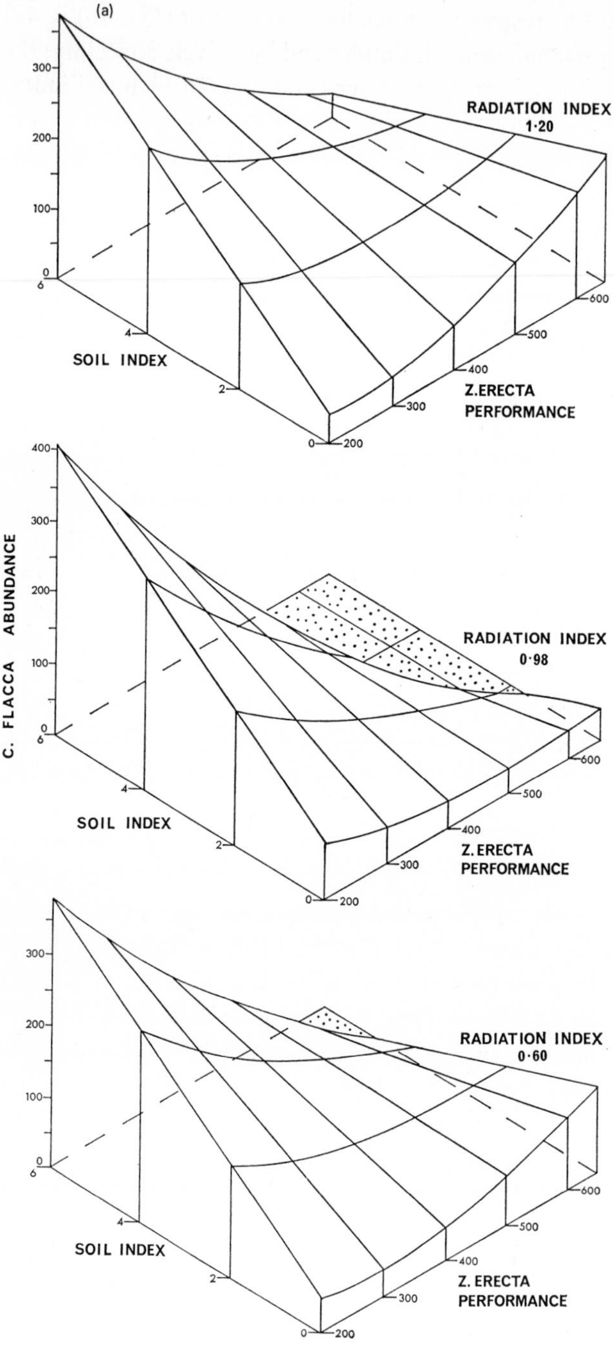
(a)
300
200
100
0
6
4
2
0
SOIL INDEX
RADIATION INDEX
1·20
200
300
400
500
600
Z.ERECTA
PERFORMANCE
C. FLACCA ABUNDANCE
400
RADIATION INDEX
0·98
RADIATION INDEX
0·60

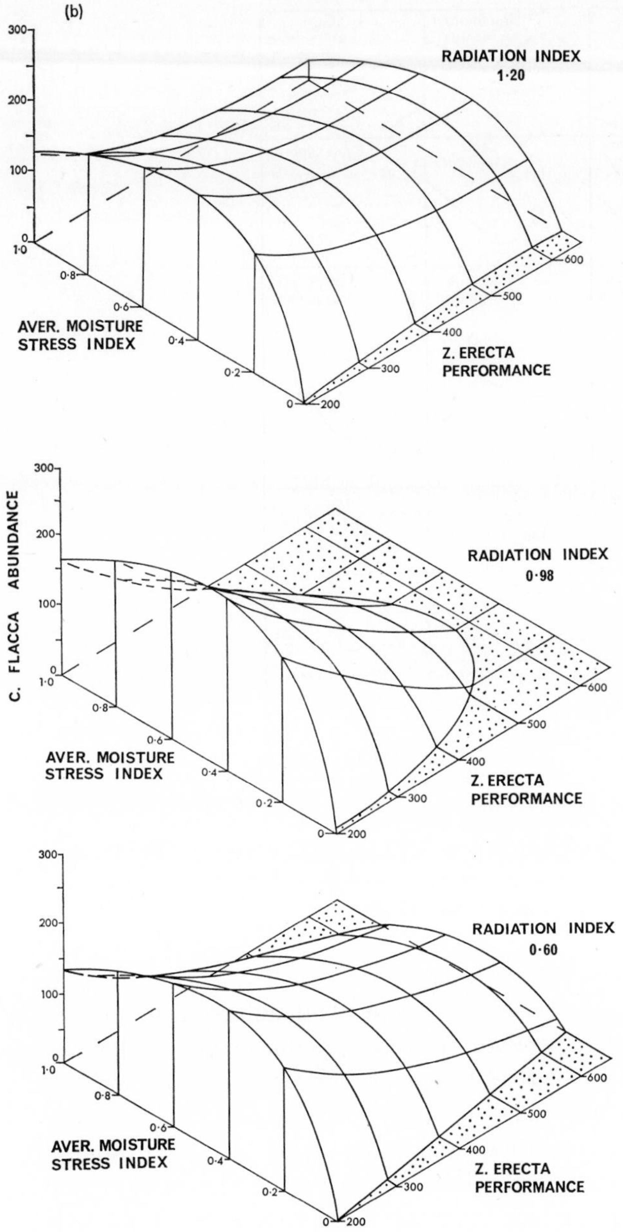

Figure 10. Regression description of *Carex flacca* performance in *Zerna erecta* dominated community. (a) *Carex flacca*'s response to *Z. erecta* performance and soil index under different radiation levels with moisture stress constant (0.50). (b) *C. flacca* response to moisture stress and *Z. erecta* performance under different radiation levels with constant soil index value (3.0).

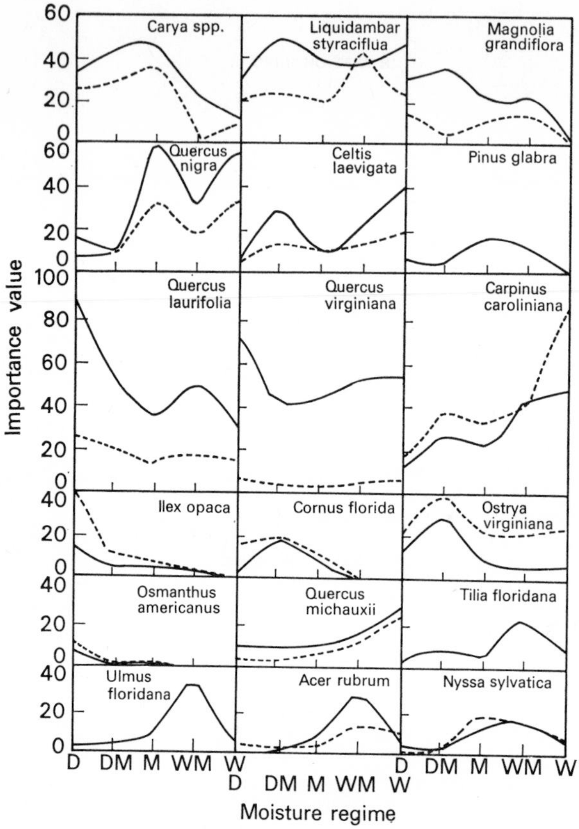

Figure 11. An example of the varied response curves which can occur. Distribution of 18 tree species along a moisture gradient. D, DM, M, WM, and W refer to dry, dry-mesic, mesic, wet-mesic and wet respectively (from Monk 1965).

the papers in the recent Brookhaven symposium on 'Diversity and Stability in Ecosystems'). The consequence of this is that both numerical and conceptual models of vegetation are poorly developed. Experimentalists tend to assume that experiments on two species interaction are relevant to the multispecies interactions of vegetation—an assumption which remains to be proved (McIntosh 1970). Harper (1968) has commented on our inability to study adequately the interactions of vegetation with other trophic levels. If descriptive plant ecologists are to provide information and hypotheses of use to other ecologists, they will need to consider the rest of the ecosystem. Present concepts and methods seem unlikely to achieve this integration.

It is to be hoped that suitable techniques will be developed. Whittaker (1967) has suggested and I quote '. . . the character of species distributions and compositional gradients imply diminishing returns from complex mathematical treatment of vegetation. Advances in theory of vegetation structure and classification have come predominantly from simpler techniques, from the results of which the basis of complex and indirect techniques may be understood. This is by no means to oppose mathematical treatment, but to balance against its interest and value the fact that for some kinds of vegetation research, simpler and more direct methods, carried out with skill and perceptiveness, are more effective.'

Personally I would say progress will come from mathematical treatment of vegetation data which is compatible with and equivalent to our ecological understanding. This is not always the case at present.

Acknowledgements

I wish to acknowledge the co-operation of I.Noy-Meir on the ordination study and the helpful comments, stimulus and discussion of I.Noy-Meir, J.M.Norris, M.B.Dale, B.G.Cook and K.D.Cocks during the preparation of this paper.

References

ANDERSON D.J. (1965) Classification and ordination in vegetation science: controversy over a non-existent problem? *J. Ecol.* 53, 512–26.

AUSTIN M.P. (1966) *A quantitative study of a chalk grassland community*. Ph.D. thesis, University of London.

AUSTIN M.P. (1968) An ordination study of a chalk grassland community. *J. Ecol.* 56, 739–57.

AUSTIN M.P. (1970) An applied ecological example of mixed-data classification. *Data Representation* (Eds. R.S. Anderssen and M.R. Osbourne), 113–17. Queensland University Press.

AUSTIN M.P. & ORLOCI L. (1966) Geometric models in ecology. II. An evaluation of some ordination techniques. *J. Ecol.* 54, 217–27.

AUSTIN M.P. & GREIG-SMITH P. (1968) The application of quantitative methods to vegetation survey. II. Some methodological problems of data from rain forest. *J. Ecol.* 56, 827–44.

AUSTIN M.P. & NOY-MEIR I. (1971) The problem of non-linearity in ordination: experiments with two-gradient models. *J. Ecol.* 59, 763–73.

AUSTIN M.P., ASHTON P.S. & GREIG-SMITH P. (manuscript) The application of quantitative methods to vegetation survey. III. A re-examination of rain forest data from Brunei. To be submitted to *J. Ecol.*

BAIER W. (1965) The interrelationship of meteorological factors, soil moisture, and plant growth (a review). *Intern. J. Biomet.* 9, 5–20.

BAIER W. & ROBERTSON G.W. (1968) The performance of soil moisture estimates as compared with the direct use of climatological data for estimating crop yields. *Agr. Met.* 5, 17–31.

BAKUZIS E.V. (1969) *Forestry viewed in an ecosystem perspective. The ecosystem concept in natural resource management* (Ed. G.M. Van Dyne). Academic Press.

BARKHAM J.P. & NORRIS J.M. (1970) Multivariate procedures in an investigation of vegetation and soil relations of two beech woodlands, Cotswold Hills, England. *Ecology*. **51**, 630–9.

BOX G.E.P. (1954) The exploration and exploitation of response surfaces: Some general considerations and examples. *Biometrics* **10**, 16–60.

BOX G.E.P. (1966) Use and abuse of regression. *Technometrics*. **8**, 625–9.

BOYKO H. (1947) On the role of plants as quantitative climate indicators and the geo-ecological law of distribution. *J. Ecol.* **35**, 138–57.

BRAY J.R. & CURTIS J.T. (1957) An ordination of upland forest communities of southern Wisconsin. *Ecol. Monogr.* **27**, 325–49.

BUNCE R.G.H. (1968) An ecological study of Ysgolion Duon, a mountain cliff in Snowdonia. *J. Ecol.* **56**, 59–75.

CASSIE R.M. (1969) Multivariate analysis in ecology. *N.Z. Ecol. Soc. Proc.* **16**, 53–7.

CASSIE R.M. & MICHAEL A.D. (1968) Fauna and sediments of an intertidal mudflat. *J. exp. mar. Biol. Ecol.* **2**, 1–23.

CURTIS J.T. (1959) *The Vegetation of Wisconsin: an Ordination of Plant Communities.* Madison, Wisconsin.

DALE M.B. (1969) Information analysis of quantitative data. Paper presented to the *International Symposium on Statistical Ecology*, New Haven.

DALE M.B., LANCE G.N. & ALBRECHT L. (1971) Extension of information analysis. *Aust. Computer. J.* **3**, 29–34.

DRAPER N.R. & SMITH H. (1966) *Applied Regression Analysis.* J. Wiley and Sons.

FIELD J.G. (1969) The use of the information statistic in the numerical classification of heterogeneous systems. *J. Ecol.* **57**, 565–9.

FITZPATRICK E.A. (1965) Climate of the Tipperary area. In General report on lands of the Tipperary area, Northern Territory, 1961. *Land Res. Ser.* **13**, 39–52.

FITZPATRICK E.A. & NIX H.A. (1969) A model for simulating soil water regime in alternating fallow-crop systems. *Agr. Met.* **6**, 303–19.

FITZPATRICK E.A. & NIX H.A. (1970) The climatic factor in Australian grassland ecology. *Australian Grasslands* (Ed. R.M. Moore.) Australian National University Press.

GILBERT N. & WELLS T.C.E. (1966) Analysis of quadrat data. *J. Ecol.* **54**, 675–85.

GITTINS R. (1965) Multivariate approaches to a limestone grassland community. I. A stand ordination. *J. Ecol.* **53**, 385–401.

GOLDSMITH F.B. (1967) *Some aspects of the vegetation of sea cliffs.* Ph.D. thesis, University of Wales.

GOODALL D.W. (1954) Objective methods for the classification of vegetation. III. An essay in the use of factor analysis. *Aust. J. Bot.* **2**, 304–24.

GREIG-SMITH P. (1964) *Quantitative Plant Ecology.* 2nd. Ed., London.

GREIG-SMITH P. (1969) Analysis of vegetation data: the user viewpoint. *Intern. Symp. on Statistical Ecology*, New Haven (in press).

GREIG-SMITH P., AUSTIN M.P. & WHITMORE T.C. (1967) Quantitative methods in vegetation survey. I. Association-analysis and principal component ordination of rain forest. *J. Ecol.* **55**, 483–503.

GROENEWOUD H. VAN (1965) Ordination and classification of Swiss and Canadian coniferous forests by various biometric and other methods. *Ber. geobot. Forsch Inst. Rubel.* **36**, 28–102.

HARPER J.L. (1968) The regulation of numbers and mass in plant populations. *Population Biology and Evolution* (Ed. R.C. Lewontin), 139–58. Syracuse Univ. Press, Syracuse, New York.

JENIK J. & REJMANEK M. (1969) Interpretation of direct solar irradiation in ecology. *Arch. Met. Geophys. Bioklim.* Ser. B. **17**, 413–28.

LAMBERT J.M. & WILLIAMS W.T. (1966) Multivariate methods in plant ecology. VI. Comparison of information-analysis and association-analysis. *J. Ecol.* **54**, 635–64.

LANCE G.N. & WILLIAMS W.T. (1967) Mixed data classificatory programs. I. Agglomerative systems. *Aust. Computer J.* **1**, 15–20.

LANCE G.N. & WILLIAMS W.T. (1968) Note on a new information-statistic classificatory program. *Comput. J.* **11**, 195.

LEE R. (1963) Evaluation of solar beam irradiation as a climatic parameter of mountain watersheds. *Hydrology Papers No 2*. Colorado State University, Colorado.

LOUCKS D.L. (1962) Ordinating forest communities by means of environmental scalars and phytosociological indices. *Ecol. Monogr.* **32**, 137–66.

MCALPINE J.R. (1970) Estimating pasture growth periods and droughts from simple water balance models. Proc. XI. International Grassland Congress. 484–7.

MCINTOSH R.P. (1970) Community, competition and adaptation. *Quart. Rev. Biol.* **45**, 259–80.

MACNAUGHTON-SMITH P. (1965) Some statistical and other numerical techniques for classifying individuals. *Home Office Res. Rept. No.* 6, H.M.S.O., London.

MAJOR J. (1963) A climatic index to vascular plant activity. *Ecology*. **44**, 485–98.

MATHER J.R. & YOSHIOKA G.A. (1968) The role of climate in the distribution of vegetation. *Ann. Ass. Am. Geogr.* **58**, 29–41.

MEDIN D.E. (1960) Physical site factors influencing annual production of true mountain mahogany, *Cercocarpus montanus*. *Ecology*, **41**, 454–60.

MONK C.D. (1965) Southern mixed hardwood forest of north central Florida. *Ecol. Monogr.* **35**, 335–54.

NASH A.J. (1963) A method for evaluating the effects of topography on the soil water balance. *For. Sci.* **9**, 413–22.

NIX H.A. & FITZPATRICK E.A. (1969) An index of crop water stress related to wheat and grain sorghum yields. *Agr. Met.* **6**, 321–37.

NOY-MEIR I. & AUSTIN M.P. (1970) Principal component ordination and simulated vegetational data. *Ecology*. **51**, 551–2.

PERRING F. (1959) Topographical gradients of chalk grassland. *J. Ecol.* **47**, 447–81.

RIJTEMA P.E. (1968) On the relation between transpiration, soil physical properties and crop production as a basis for water supply plans. *Institute for Land and Water Management Research Technical Bulletin 58*.

RUTTER A.J. (1955) The composition of wet-heath vegetation in relation to the water table. *J. Ecol.* **43**, 507–43.

SLOBODKIN L.B. & SANDERS H.L. (1969) On the contribution of environmental predictability to species diversity. Diversity and Stability in Ecological Systems. *Brookhaven Symposium in Biology*. **22**, 82–94.

STRINGER P.W. & LA ROI G.H. (1970) The Douglas fir forests of Banff and Jasper National Parks, Canada. *Can. J. Bot.* **48**, 1703–26.

SWAN J.M.A. (1970) An examination of some ordination problems by use of simulated vegetational data. *Ecology*. **51**, 89–102.

WARING R.H. and MAJOR J. (1964) Some vegetation of the California coastal redwood region in relation to gradients of moisture, nutrients, light and temperature. *Ecol. Monogr.* **34**, 167–215.

WEBB L.J., TRACEY J.G., WILLIAMS W.T. and LANCE G.N. (1967) Studies in the numerical analysis of complex rain-forest communities. I. A comparison of methods applicable to site/species data. *J. Ecol.* 55, 171–91.

WEBB L.J., TRACEY J.G., WILLIAMS W.T. and LANCE G.N. (1970) Studies in the numerical analysis of complex rain-forest communities. V. A comparison of the properties of floristic and physiognomic-structural data. *J. Ecol.* 58, 203–32.

WHITTAKER R.H. (1956) Vegetation of the Great Smoky Mountains. *Ecol. Monogr.* 26, 1–80.

WHITTAKER R.H. (1960) Vegetation of the Siskiyou Mountains, Oregon and California. *Ecol. Monogr.* 30, 279–338.

WHITTAKER R.H. (1967) Gradient analysis of vegetation. *Biol. Rev.* 49, 208–64.

WHITTAKER R.H. and NIERING W.A. (1965) Vegetation of the Santa Catalina Mountains, Arizona. II. A gradient analysis of the south slope. *Ecology.* 46, 429–52.

WILLIAMS W.T. and LAMBERT J.M. (1959) Multivariate methods in plant ecology. I. Association analysis in plant communities. *J. Ecol.* 47, 83–101.

YARRANTON G.A. (1970) Towards a mathematical model of limestone pavement vegetation. III. Estimation of the determinants of species frequencies. *Can. J. Bot.* 48, 1387–1404.

YARRANTON G.A. and BEASLEIGH W.J. (1968) Towards a mathematical model of limestone pavement vegetation. I. Vegetation and microtopography. *Can. J. Bot.* 46, 1591–99.

YARRANTON G.A. and BEASLEIGH W.J. (1969) Towards a mathematical model of limestone pavement vegetation. II. Microclimate, surface pH, and microtopography. *Can. J. Bot.* 47, 959–74.

ZAHNER R. (1967) Refinement in empirical functions for realistic soil moisture regimes under forest cover. Proc. Internat. Symp. on Forest Hydrol. Pennsylvania State University. 261–74.

ZAHNER R. and STAGE A.R. (1966) A procedure for calculating daily moisture stress and its utility in regressions of tree growth on weather. *Ecology* 47, 64–75.

Theoretical models for large-scale vegetation survey

J.M.LAMBERT *Botany Department, University of Southampton*

Summary

The most critical phases in the planning and execution of vegetation surveys lies in the selection of suitable sampling methods and of appropriate analytical techniques. Various theoretical models concerned with these problems are discussed in terms of type of information required, economy of effort, and efficiency of operation.

I. Introduction

The main features of the vegetation cover of many parts of the world are still so imperfectly known that a general vegetation survey of such regions remains a primary requirement. Since the overall costs involved will usually prohibit the mounting of more than one large-scale operation for a given region, it is important that the results of the survey should be useful in as many fields as possible. Furthermore, since the cost of even a single large-scale exercise could still prove excessive without the strictest economy in all aspects of the work, the project should be designed from the beginning to provide maximum returns for minimum outlay in manpower, time and money.

The general aim of this paper is to examine a number of theoretical models appropriate to various aspects of this type of work, bearing in mind throughout the kind of information most likely to be required and the general need for economy. The examination covers both data collection and subsequent data analysis, since the two are interdependent. If data of a particular type are needed to fulfil the primary purpose of the survey, then the analytical techniques employed must be compatible with the intrinsic properties of the data; conversely, if a particular method of analysis is envisaged at the outset, then the data must be of a size and conformation amenable to that manipulation. Too often vast quantities of raw data are collected without enough thought in advance as to how, and to what end, and at what cost, the material can be subsequently handled; and the larger the potential data mass, the greater the need to use appropriate models to aid in its collection and digestion.

The provision of an efficient and effective scientific methodology covering all aspects of the work involves examination of an infinite array of possible models employable at every stage. The models under study can range from purely conceptual systems at one end of the scale to full-scale working models at the other. The concept of a model system applies to the whole procedure as well as to its separate parts, and the final production model must therefore be tested for harmonious interaction as well as for individual tolerances of its various components. With large-scale work, involving large-scale capital investment, the use of a faulty or inefficient model at any phase of the work could have disastrous consequences.

A choice between two or more models to serve a particular function involves empirical testing as well as theoretical assessment. It is often not realized, however, that model-testing in a methodological context should be a rigorous procedure conforming to precisely the same rules that apply to any experimental science. There are far too many papers which report the results of tests conducted without controls and with grossly inadequate replication to distinguish the effects of different components of a composite technique. Moreover, an increasing number of pseudo-methodological contributions seem content merely to describe the results of the application of a single method of unknown propensities to a single set of data with unknown properties, with no clear directive as to whether the method or the material is to be regarded as the situation under test.

Our approach at Southampton over the last few years has been to examine existing models for vegetation survey in an attempt to identify the areas where new or better models are most urgently required, and to try to fill a small proportion of the gaps. In addition to purely theoretical considerations, our tests have ranged from drawing-board exercises with simulated situations to empirical tests on genuine vegetational situations of increasing magnitude and complexity. With one or two of the models, we are now approaching the stage when a fairly ambitious test can be applied, and we are currently testing one particular system on two contrasting areas of semi-natural vegetation of reasonably realistic dimensions.

The whole subject of appropriate model procedures for large-scale vegetation survey is so immense that certain limits have had to be set for the purpose of this contribution. In the present context, therefore, the following assumptions have been made:

(i) That the survey is initially concerned solely with vegetational features, either in relation to the immediate vegetational resources of the area or as a form of bio-assay of habitat potential.

(ii) That more, or different, information is required than is obtainable merely from the study of air photographs, so that extensive ground survey of one form or another is involved.

(iii) That the detailed procedure must be as 'objective' (i.e. as definitive) as possible, both to minimize individual bias among the workers involved and to facilitate subsequent interpretation of the results in a variety of fields.

(iv) That the results are required in a spatial framework (e.g. in the form of a vegetational map) to cater for regional interests.

(v) That finance is available for data analysis by computer-based techniques on a scale commensurate with the time and money spent on data collection in the field.

Of these assumptions, perhaps the most controversial is the first, which restricts the scope of the primary recording to the plant cover without a parallel set of detailed habitat observations. The justification for such a phytosociological approach has been given in a previous paper (Lambert & Dale, 1964), and it is on the basis of the theoretical arguments therein that this particular assumption has been made.

II. Data collection

If large-scale work is committed at the outset to comply with more than one subsequent use of the results, a specialized model of high performance but limited versatility must frequently be rejected in favour of a less spectacular general-purpose model. It is important that both the highest and lowest common denominator of anticipated user requirements should be established at the outset, otherwise an unnecessarily restricted or restrictive model may be chosen. Lack of forethought along these lines is often responsible for the use of unsuitable models for data-collection, and any *ad hoc* attempts to adjust the model during the course of the work may have unwelcome repercussions on subsequent data-analysis.

A. The sampling pattern

All sampling models assume at the outset that entities exist whose properties can be represented with varying degrees of confidence by a subset taken from the total population. With vegetation survey, much of the argument as to the most appropriate sampling technique stems from uncertainty as to whether the entities in the population under study should be defined initially on geographical or on vegetational parameters. In the first case, spatial position on the surface of the earth forms the primary criterion for recognition of the entities, and no *a priori* assumption is made about their vegetational properties; in the second, the entities are defined by vegetational criteria, and no assumption is made about their spatial properties.

It is not proposed to rehearse here the various arguments for and against a sampling model based on the field recognition of vegetation 'stands'. In spite of recent papers purporting to show its efficiency (e.g. Moore *et al.* 1970), the model is invalidated in the present context by three fundamental faults. First, it embodies certain prior assumptions about the nature of the vegetation which may or may not be true in the area under study and are therefore better avoided; secondly, it requires experienced workers to operate the model, and hence could be unacceptably expensive in terms of scientific manpower; and thirdly, the model itself is so imperfectly defined that it is difficult for the customer to assess its tolerance limits. The following discussion will deal only with models which are designed to allow a pattern to emerge reasonably free from restrictive preconceptions, which are more economical in the use of highly trained personnel, and which can be defined simply and precisely enough to engender confidence in the results.

With sampling models which are spatially rather than vegetationally defined, the choice then lies between various forms of random and systematic sampling. Theoretically, random sampling has the advantage that it carries no known bias and is statistically the most amenable; against this, there is no provision for equal spread of information in a spatial context. Conversely, a systematic grid provides for equality of spatial information, but also carries the possibility of unwelcome resonance with the vegetation pattern or of a directional bias determined by the initial placing of the grid. The theoretical merits and demerits of these two basic models, together with the compromise model of restricted random sampling, have been discussed fully by Greig-Smith (1964), and we are currently attempting to examine their relative tolerances by a series of empirical tests. Figure 1 shows the partial results of a pilot study using random and gridded point samples on a vegetational map (P.F.M.Smartt, unpublished). The measured areas (each expressed as a percentage of the whole) of eight predetermined vegetation types were used as the control, and the two sampling models were compared over a range of sample numbers in terms of success of recovery of the areal proportions. The plots of the summed squares of deviations about the absolute values show that the systematic model stabilized quickly at about the 125-sample mark, but then showed little improvement with increasing sample number; in contrast, the random model fluctuated much more irregularly until the 300-sample mark was reached, but then began to steady to a better performance than the grid. After the 100-sample mark, both models in fact showed an acceptable level of accuracy (i.e. within two standard deviations of the actual value) for all the

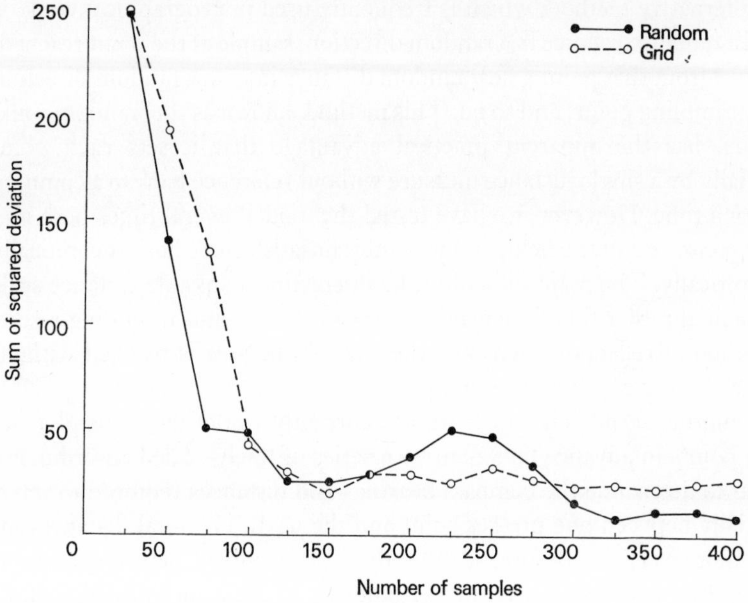

Figure 1. Relative error of random and gridded samples summed over 8 vegetation types. (Sum of squared deviations calculated as $\sum_{i=1}^{8} (x_i - p_i)$ where x_i=observed proportion and p_i=expected proportion.)

vegetation types considered separately. Our experience so far, therefore, is that there is not a great deal to choose between the two basic methods in terms of absolute performance, so that a choice between them must rest on other criteria such as ease of operation and general flexibility.

(i) Models for random sampling

The commonest model in use for random sampling in a spatial context is that which pre-sets each sampling point independently by the use of random pairs of map co-ordinates. This system has certain theoretical limitations. It is not completely devoid of bias because of the finite spacing of the co-ordinates, and with large interco-ordinate spacing the system can degenerate into a systematic random technique with the samples merely becoming a random sub-set of a very limited grid defined by the co-ordinate spacing. It also has practical disadvantages in that the larger the scale of the exercise, the more difficult it becomes to identify the position of the co-ordinate intersections in the field.

An alternative method, which is frequently used in geographical work, is to travel a random distance in a random direction, sample at the point reached, then travel from that point a new random distance in a new random direction to a new sampling point, and so on. This method, known as the 'random walk' technique, has the apparent practical advantage that it sets each point sequentially by a single distance measure without reference back to a common origin each time. However, we have tested the model several times both as a paper exercise and in the field, and we find it unsatisfactory both theoretically and empirically. The main difficulties lie in erecting a suitable distance scale for sequential rather than independent measurements, and in finding a suitable method of reflection each time the edge of the area is reached without bias towards oversampling the outer parts of the area.

A compromise model, which we are currently exploring, is to plot out random points in advance on a map by a series of finely-scaled co-ordinates, and then to determine the compass bearings and distances required to travel sequentially between one pre-set point and the next. The total distances involved, moreover, can be drastically reduced if a critical path technique is employed (i.e. if the model is further adjusted to allow an optimal path or paths to be traced between the sampling sites instead of visiting them in the precise order in which they were defined); although this carries the danger of a directional bias in recording error related to the direction of the optimal path, this disadvantage could well be outweighed in large-scale work by the great saving in travelling time.

(ii) Models for systematic sampling

With both random and systematic sampling, the total number of samples which can be taken is limited both by the time taken to visit and record each site and by the ceiling capacity of the computer programs needed to process the data afterwards. With random sampling, where the model is based on a system of equal chance, the sampling strategy and number of samples are independent of one another; any number of extra samples can thus be taken in the area without disturbing the basic sampling pattern, and the system is completely open-ended up to the ceiling imposed. In contrast, in systematic sampling by means of a grid, the samples are normally disposed in a closed system with regular spacing so that the number of samples is alterable only in an exponential manner.

Although a systematic model is therefore less flexible than random sampling in terms of straight sample numbers, its inherent regularity provides certain compensatory advantages. Whereas it is extremely difficult to adjust an irregular system by predefined means to take account of changes in the scale of the vegetational pattern from place to place, such adjustments can easily be made to a systematic model by using a simple inbuilt criterion to alter the spacing where required. For instance, provided the adjustment is made definitively and in full recognition of the consequences, there is much to be said for a model which deliberately deploys the available samples in such a way that areas showing the greatest vegetational heterogeneity are sampled more intensively than the rest.

The main problem in designing a flexible systematic model of this type is to define a suitable mechanism for adjustment. A possible solution is to use a proportion of the available samples to lay down a skeleton grid, and then to use the information in this grid to set up a series of secondary grids of different spacings within the primary pattern according to some predetermined heterogeneity measure. This measure involves calculating the heterogeneity independently between pairs of adjacent samples and relating this either to the internal heterogeneity of the primary grid samples as a whole or to some fixed external heterogeneity criterion. In the former case, all samples from the primary grid must be collected and analysed before the secondary grids can be laid down, while in the latter the secondary samples can be organized section by section during the actual course of the primary recording. Thus, although the former method is theoretically better in that the measure used is based on greater total information, it could be less efficient on practical grounds for large-scale work in that it is essentially a two-stage operation in the field with consequent duplication of travelling time to different parts of the region.

We have so far carried out a number of small pilot experiments based on the second model, using quite simple measures which can be evaluated from tables in the field. The measures have been variously based on physiognomy, floristic richness and diversity, and the results have been sufficiently promising to warrant further work along these lines.

B. Sample size

Once the general sampling pattern has been decided and the samples have been defined with reference to each other, each sampling site becomes an entity in its own right for recording purposes. A further model is then required to set the size of each sample as a basis for such recording. Here the multifarious methods which have been suggested in the past ultimately reduce to two basic systems: either a predefined area may be used, irrespective of the nature and composition of the vegetation at the sampling site, or a predefined number of plant entities (e.g. life-forms, species, individual plants) may be recorded, irrespective of the area needed to accommodate them. The first is essentially quadrat sampling in its various guises, while the second is a formalized version of the multiple-nearest-neighbour model recently advocated by Williams *et al.* (1969) and derived from various systems of so-called plotless sampling.

With either method, the initial definition may be made in absolute terms quite independent of the nature of the vegetation at each site, so that *either* the sample area, *or* the number of plant entities to be recorded, is fixed at a constant value before the survey is begun. Although this has the advantage of ensuring comparability between the samples in at least one property, a model with an inbuilt absolute restriction could prove unduly rigid over a range of diverse vegetation types. A more flexible model would be one which takes into account the properties of the vegetation actually encountered at each sample site, and in this sense a model based on a species-area curve (or its equivalent for other entities) has much to recommend it. If such a curve can be constructed independently for every sample site, then the size of the sample required in each case should be theoretically determinable from the curve by predefined criteria. However, to find an appropriate and universally applicable parameter may prove extremely difficult in the light of the conflicting theoretical and empirical evidence for such curves demonstrated by Hopkins (1955, 1957), and most of the species-area models used in traditional phytosociological work are insufficiently rigorous to be generally recommended.

Further critical work on the species-area curve is clearly desirable, and we have recently begun a theoretical and empirical re-examination of its properties. As an alternative to the use of this particular model, however, we have also made a pilot attempt to devise a flexible model for sample size related to certain physiognomic properties of the vegetation. Such a model could well prove useful eventually as part of a general survey system based on vegetational physiognomy rather than floristics, but at present it is not well enough defined or tested for any conclusion to be drawn.

C. The plant material

After the spatial properties of the samples have been determined, all subsequent recording of the vegetation at the sample sites is dependent on these properties. The plant material associated with a particular site can obviously be categorized in a number of different ways, depending on the primary purpose of the survey. The traditional use of species in 'general purpose' work stems from the fact that they are generally accepted entities with characteristic properties which automatically import a range of other information. Moreover, they form a finite set of variables whose definition is set externally by the current taxonomic position, so that the investigator is relieved from defining his own categories.

Against these advantages, the use of species has a number of disadvantages which become increasingly pronounced with large-scale exercises. To recognize species accurately, especially in unknown vegetation, requires previous taxonomic training and much time spent in sheer identification; it therefore precludes the use of unskilled assistants in data-collection and puts up the cost of the exercise. More fundamentally, from a theoretical angle, the various sample sites invariably show qualitative as well as quantitative differences in their species complement, so that absences as well as presences become important. With floristically heterogeneous areas, this leads to marked asymmetry in the final data matrix in that the presences are usually far exceeded by the absences over the data as a whole, and the matrix is thus difficult to handle by standard numerical techniques. Moreover, unless the data-matrix is constructed in relation to the full complement of species on the earth (thereby enormously extending the number of absence records), it is difficult to standardize the results of one regional survey against those from another area.

The problems involved can be tackled along two independent lines, each with its own particular set of possible models. On one front, the use of species recording can be accepted as a primary requirement, and work concentrated on the exploration of numerical models able to cope with the basic asymmetry of the species-based data matrix; several investigations have been made along these lines (cf. Field 1969), and further discussion of this complex and rather intractible numerical problem is not possible here. The alternative and more revolutionary approach would be to abandon species recording and to explore instead the erection of new categories of plant material which are numerically more amenable as well as possessing the general properties required for large-scale survey work; certain recent work on the use of life-forms and other physiognomic categories in a numerical context is a step in this direction (cf. Webb *et al.* 1970), but otherwise the possibilities remain largely uninvestigated.

A possible derivative from a model based on physiognomic attributes lies in the use of categories defined in relation to individual organs rather than whole plants. We have recently tried at Southampton (S.L.Attanapola, unpublished) to define a strictly limited set of 'eco-organ' categories based on easily recognizable vegetative features of organ status. So far, the eco-organ model has been examined purely in a qualitative context, where experiments run in parallel with species-based controls have given quite promising results for a number of test-communities: the eco-organs consistently provide a much more symmetrical qualitative data-matrix, and the ecological results from subsequent numerical analysis of the data have in most cases been as good as, if not better than, those from species-based records. Table 1 shows the percentage fill of the raw data matrix for both species and eco-organ records from four test-communities covering very diverse vegetation. At this stage of the work, a strictly circumscribed set of some 80–90 eco-organ categories was being used, and if this number can be reduced further to about 50 easily defined, universally applicable, and ecologically meaningful entities, a very useful model may emerge.

Table 1. Percentage fill of raw data matrix using (*a*) species and (*b*) eco-organs in four test communities. (After S.L.Attanapola, unpublished.)

Test-community	General features of vegetation	% Fill of data-matrix	
		Species	Eco-organs
1. Pigbush, New Forest	Oakwood, pinewood, heath	19·9	60·4
2. Matley, New Forest	Oakwood, wet and dry heath, valley bog	12·6	49·4
3. Wheatfen, Norfolk	Reedswamp, fen and carr	16·8	52·8
4. Hantana, Ceylon	Savannah grassland and tropical forest	19·3	47·2

In any data embodying qualitative differences, a further problem of data asymmetry arises if quantitative measures are required for the variables: entities which are present can take a positive value, but there is no compensatory negative value obtainable in the field for those that are absent. Whether these negative values can be calculated in retrospect from the information in the positive part of the matrix still remains to be seen, but we know of no appropriate theoretical or empirical model at the moment which could help to overcome the difficulty.

Even though this particular problem appears somewhat intractible, the question of the use of quantitative measures in survey work is so urgent that it seems appropriate here to examine other aspects of the situation. The main difficulties reside in the selection of appropriate measures for use in the field, and of suitable transformations of the quantities obtained. In the following discussion, it will be assumed for convenience that the plant material is to be recorded in the form of species, although most of models are equally applicable to physiognomic attributes.

Since different quantitative measures can yield very different sorts of information when the data is analysed, the choice of the most appropriate model for a 'general purpose' survey is no mean task. A vast range of possible models exists founded on such concepts as cover, frequency, density, biomass, and so on, to say nothing of the ill-defined scales of cover-abundance much used in phytosociological work. Apart from the fact that the usefulness of certain of these measures is by no means clear, many of them are so time-consuming to apply in the field that they may be regarded as completely impracticable for large-scale work.

A useful introduction to some of the mathematical properties and interrelationships of the commoner quantitative measures is given in Greig-Smith (1964). With recent rapid advances in numerical techniques, however, a number of other properties may need to be defined before any quantitative data can be analysed with confidence. For instance, it must be realized that if the sampling unit itself is defined by absolute area rather than by species number, all measures based on cover by orthogonal projection will have a ceiling value for the sample as a whole irrespective of the number of species involved; other measures, like certain forms of frequency record, may be bounded for the individual species but not for the whole sample, so that the ceiling value is then related to the species number and will vary in sympathy; yet others, like total biomass partitioned between the species, may be theoretically unbounded and liable to such massive numerical fluctuations that these tend to dominate the analysis. Unbounded and partially bounded data may thus require a degree of standardization to compensate for variation in range of values between the different samples; conversely, completely bounded data like cover or percentaged measures may require some other form of transformation designed to counteract the effects of the constraint.

We have investigated certain of these quantitative models empirically on a test-community specifically selected to show marked quantitative differences as well as a qualitative pattern. Using a standard-sized quadrat throughout, the species were recorded in terms of cover, frequency and biomass as well as straight presence-or-absence. In addition to calculating a correlation between various pairs of data types from their Euclidean distance matrices, the different hierarchies produced from the data sets by the same classificatory technique were compared by the cophenetic correlation method of Sokal and Rohlf (1962). Partial results from the complete exercise are given in Table 2,

Table 2. Correlations between various data types.
A=raw data correlation; B=hierarchy correlation.

		Data type					
Data type		Qualitative	Frequency	Biomass	% Biomass	% Cover	% Cover (arcsin)
Qualitative	A	1·00	0·83	−0·07	0·49	0·18	0·48
	B	1·00	0·69	−0·15	0·36	−0·04	−0·04
Frequency	A		1·00	0·02	0·65	0·33	0·61
	B		1·00	−0·07	0·38	0·04	0·28
Biomass	A			1·00	0·10	0·59	0·39
	B			1·00	−0·05	0·43	0·43
% Biomass	A				1·00	0·63	0·82
	B				1·00	0·23	0·81
% Cover	A					1·00	0·90
	B					1·00	0·38
% Cover (arcsin)	A						1·00
	B						1·00

in which it is seen that the highest correlations occur between percentage biomass and percentage cover with arcsin transformation, and between frequency and the qualitative records. This latter correlation is particularly interesting in view of the fact that, according to Greig-Smith (1964), ecologists tend to favour frequency determination as a useful and easily operated compromise model for quantitative recording. Since straight frequency measures and presence-and-absence data share the same property of dependence on species number, the high correlation is not surprising with a sampling model involving variable numbers of species; and if the empirical correlation can be sustained over more extensive tests, the situation could support the view frequently expressed that a qualitative model is adequate for most primary survey work.

The very low correlation values seen elsewhere in Table 2 suggests that the properties of the various quantitative measures are so diverse that extremely careful consideration must be given to choosing an appropriate model in situations where some form of quantitative measure is considered essential. Unless the primary function of the model is precisely defined in advance and matched to the appropriate measure, the model actually chosen may reflect personal discrimination rather than anything else. The same caveat applies to any transformation of the data deemed to be necessary before the main analysis, and it is therefore imperative that the effect of the transformation should similarly be clearly understood.

Some indication of the extent of personal bias which could be involved was shown by a further simple experiment with the different data sets from the test-community discussed above. Two of the workers responsible for the field collection of these data produced their own subjective classifications of the vegetation from field impressions alone, and then compared their results by cophenetic correlation with the different classifications produced from the different data sets by a given numerical method. The results are shown in Table 3, from which it is seen that the classification of one worker (P.S.)

Table 3. Comparison of two separate intuitive hierarchies (P.S. and J.B.) with computed hierarchies covering a range of different data types.

	Data types					
	Qualitative	Frequency	Biomass	% Biomass	% Cover	% Cover (arcsin)
P.S.	0·40	0·38	0·00	0·84	0·27	0·97
J.B.	0·75	0·83	−0·13	0·42	−0·06	0·31

shows greatest affinity with the control classifications based on percentage biomass and on percentage cover with the arcsin transformation; in contrast, the classification of the other (J.B.) is clearly most influenced by frequency and qualitative differences. Moreover, the two workers concerned showed a correlation with each other of only 0·43, and this in itself is enough to support the initial contention in this paper that precisely defined models are essential for data collection as well as for data analysis.

III. Data analysis

With the recent upsurge of interest in the use of computer-based techniques, the number of theoretical and empirical models available for data analysis seems to have greatly outstripped the number concerned with data collection. This is particularly true in the classificatory field where, according to the most recent review (Cormack 1971), the current swell of publications is estimated at over 1,000 a year. As Cormack indicates, this interest in classificatory models stems from a natural human desire to pigeon-hole information. This is not a bad thing in itself provided the information is subsequently used, but difficulties arise in primary survey work when the nature and extent of such usage can only be estimated rather imprecisely. In such circumstances, the only possible alternative to burying the undigested records in a data-bank is to try to condense the information into a set of simplified entities which are sufficiently well-defined to be useful afterwards in a variety of contexts. The very fact that the human race in general is more attuned to classification than to any other form of abstraction suggests that a classificatory model is likely to be the most generally acceptable simplification technique. This assumption, therefore, will form the basis for the subsequent discussion in this section.

A. Classificatory strategies

(1) General

The recent great proliferation of classificatory models has led to a certain amount of terminological confusion in the literature. Cormack (*loc. cit.*) makes a useful three-fold primary distinction between hierarchical classification, partitioning and clumping, in which the classes are mutually exclusive in the first two models but can overlap in the third. Since overlapping classifications can prove confusing for pigeon-holing purposes, one of the two models leading to disjoint groups is probably to be preferred for general-purpose work. In the remaining choice between hierarchical and partitioning techniques, the former is probably the more flexible as far as the user is concerned: the very fact that the available information is eventually displayed in the form of a tree, rather than as a reticulum of interrelationships at a pre-set level, provides for subsequent direct entry into the system at any level of simplification required without the necessity for re-analysis. A great deal of nonsense is often talked about distortion of the data by an hierarchical technique: *any* manipulative technique will by definition 'distort' the original data in one way or another, and the only choice which is open concerns the form, direction and extent of the distortion.

The chief types of hierarchical classification are too well known to require much further description here. Briefly, the main distinctions lie between agglomerative and divisive models on the one hand, and between monothetic and polythetic models on the other: the former distinction concerns the strategy of the actual construction of the hierarchical tree, while the latter concerns the number of attributes used directly in the definition of the classes. The more important properties of these various models have been discussed in a vegetational context in an earlier paper (Lambert & Dale 1964), and the immediate requirement here is to assess how far the models are appropriate for large-scale work.

Most of the existing models in the literature are of the polythetic-agglomerative type. These have the disadvantage that, since the manipulative process begins at the bottom of the hierarchy by grouping the individual elements, distortion of the original data increases progressively upwards; distortion is therefore greatest in the important upper classes. For large-scale work, where the size and complexity of the data are usually such that only the topmost classes are likely to be retained, the progressive upward distortion could well produce extremely suspect groupings at the level of critical use. In contrast, divisive methods form the uppermost classes first, and the greatest degree of distortion therefore occurs in the minor groups at the bottom.

Although a divisive model is thus to be preferred on theoretical grounds, its greater efficiency has to be paid for in terms of computing time. To operate the model at maximum precision requires a 'best split' to be identified from all the possible splits which could be made at every stage. The astronomical number of calculations needed thus puts the model completely out of reach for all practical purposes even with very small-scale work. Nevertheless the general properties of the divisive model are so attractive that even a relatively crude version could still have advantages over other models.

One of the ways in which a divisive model can be modified to cut down the computation time is to institute some form of 'directed search' in which only a small proportion of the possible splits are examined. For instance, one possible solution which has been tried with some success is to confine the search of those splits which can be made monothetically in relation to the properties of the separate attributes taken one at a time, as in association-analysis (Williams & Lambert 1959 1960) and related models. Although such models are liable to generate quite serious misclassifications in polythetic terms, and are moreover somewhat difficult to define for use with quantitative data, they can deal with reasonably large qualitative data-matrices and have been frequently used in vegetation survey work to date.

An alternative approach is to try to define an efficient form of directed search aimed specifically at division on a polythetic basis from a strictly limited subset of the possible polythetic splits. In spite of the attractiveness of the idea, surprisingly little critical work appears to have been carried out along these lines, and we have therefore made a major effort in this field at Southampton over the last two years. The results will ultimately be published in full elsewhere, but a brief interim account is necessary here in view of the general relevance of the work to the present theme.

(ii) Polythetic divisive models

According to Cormack (1971), hitherto the only feasible suggestion for a polythetic divisive model has been the dissimilarity analysis of Macnaughton-Smith *et al.* (1964). We have so far devised and tested four other strategies, known domestically as MIM, AXOR, STEP and PAIR. MIM and AXOR both rely on previous extraction of axes from the data matrix by either principal component or principal co-ordinate analysis; they differ in that MIM then becomes an iterative regression model operating on the axis scores, while AXOR uses the axes as constructs to define a unique order of search and returns to the raw data for discrimination. In contrast, STEP and PAIR both operate directly on the raw data from the beginning; STEP accumulates a splinter group by a form of stepwise regression and is similar to Macnaughton-Smith's original dissimilarity model, while PAIR is basically an iterative clustering model adapted to provide a hierarchy of dichotomising groups. It is difficult to compare the theoretical properties of these four models without providing more detail of their actual mechanics. All that can usefully be done here is to give the results of certain empirical tests on their respective performances with various sets of real and simulated data.

One of the first essentials in these tests was to assess the performance of the models against the 'perfect solution' provided by the 'best split' strategy. Since it is computationally possible to find the absolute best dichotomous split on a given criterion in very small data sets (i.e. of not more than about 12 individuals), a series of such sets can be used to set up a series of 'targets' in terms of achievement of the perfect answer. Table 4 gives the total scores, in

Table 4. Percentage scores for 'best splits' for different polythetic strategies.

Strategy	% 'best splits'
MIM	90·2
AXOR	98·7
PAIR	100·0
STEP	90·2
Agglomerative equivalent	69·5

terms of numbers of hits on the target, for the four divisive models over 82 such tests; the table also includes the score for the equivalent agglomerative model, which was tested in mirror-image alongside the others for comparative purposes by using the topmost fusion of its accumulated hierarchy.

From these scores, together with other subsidiary information from the tests, AXOR and PAIR were selected for much more intensive trials using a range of artificial and natural data-sets of different properties and dimensions. Since the size of most of the data sets involved made it completely impracticable to provide an absolute control, assessment of the performance of the models in these further tests could be made only in comparative terms. For instance, AXOR and PAIR were tested in parallel on the qualitative and quantitative test-community data mentioned earlier in this paper, and the relative efficiency of discrimination of the two models compared at places where their hierarchies diverged.

The overall results from the various tests to date have so far suggested that AXOR seems likely to prove the generally most efficient and reliable model. It is known to be culpable in one particular theoretical situation, but the type of failure is such that it is unlikely to have serious consequences in practice. Since AXOR is expensive in computer time and storage, it is only feasible at the moment to contemplate using it on fairly small matrices (i.e. up to a maximum of about 100 individuals). Although the programming possibilities are still being explored, the immediate intention is to use it mainly as a control against which other cruder models of greater capacity can be tested. With large-scale work in mind, we are currently examining one such model which is essentially a compromise between a monothetic and polythetic divisive strategy, and this particular method could probably eventually be programmed to accommodate data-matrices of very much larger dimensions.

B. Coefficients for discrimination

The usefulness of a particular model for a particular purpose depends not only upon the basic strategy but also upon the coefficients involved. The coefficients can be designed to accommodate almost any quirk of either the data type or of the investigator himself, and the resulting position is bewildering. Figure 2 shows the effect of using five different coefficients with the same agglomerative strategy on the same data set, and from this and numerous other similar examples in the literature (cf. Crawford *et al.* 1970), it is clear that different models can display entirely different properties. Without some guide-lines to direct the choice, the decision as to which is the best model to use in a particular situation could therefore become an extremely haphazard process.

There have been numerous attempts to compare the theoretical proper-

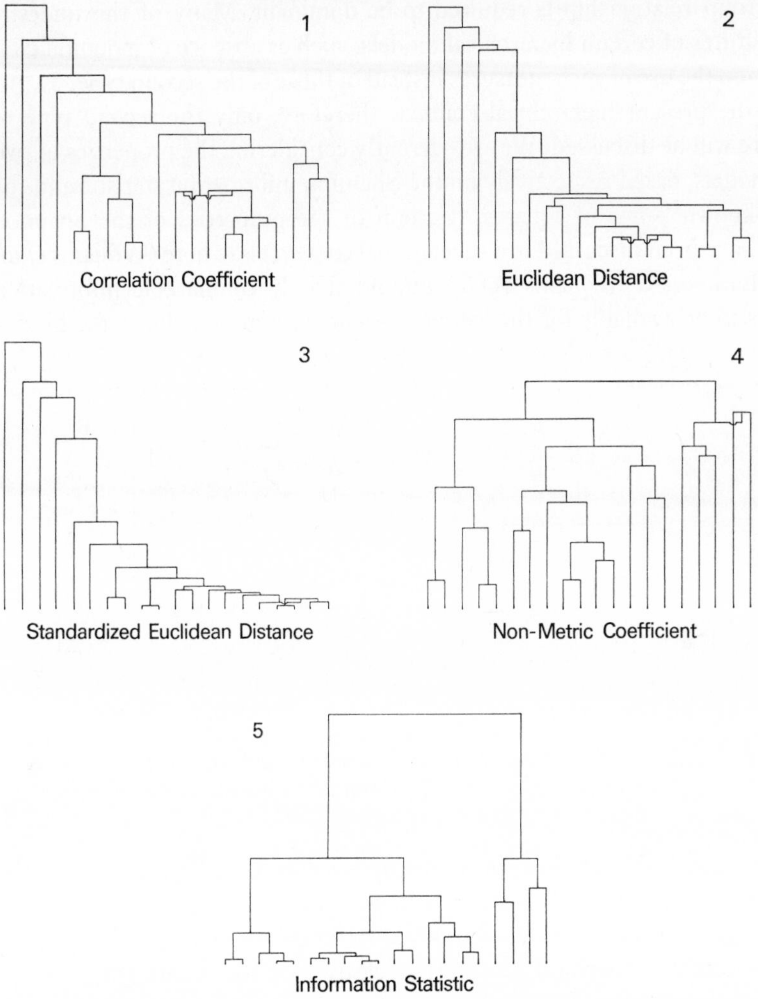

Figure 2. Different polythetic-agglomerative hierarchies (centroid sorting) produced by using five different coefficients (after Williams, Lambert & Lance 1966).

ties of different coefficients (cf., for instance, Williams & Dale 1965), and such comparisons are inevitably coloured by the background setting. For present purposes, the basic distinction appears to lie between measures which take into account only the matching properties of entities considered separately, and those which include an element related to the properties of any larger constellation of which they form a part. Most of the measures in current use are of the former type, and these are adequate for systems where the interest centres on reticulate relationships. It is frequently not realized, however, how far such models are inappropriate in hierarchical systems, where the sub-group/

supergroup relationship is required to be dominant. Many of the undesirable features of certain hierarchical models, such as absence of monotonicity, stem from the use of such measures instead of those of the second type.

In the present hierarchical context, therefore, only the second type of measure will be discussed. We are currently considering the properties of two such models, based respectively on the Shannon information statistic and the better-known 'sums of squares'. Certain of the properties of the Shannon model have been discussed in a vegetational context elsewhere (Williams *et al.* 1966, Lambert & Williams 1966), but no strictly comparable information appears to be available for the 'sums of squares'. Figure 3 shows the hierar-

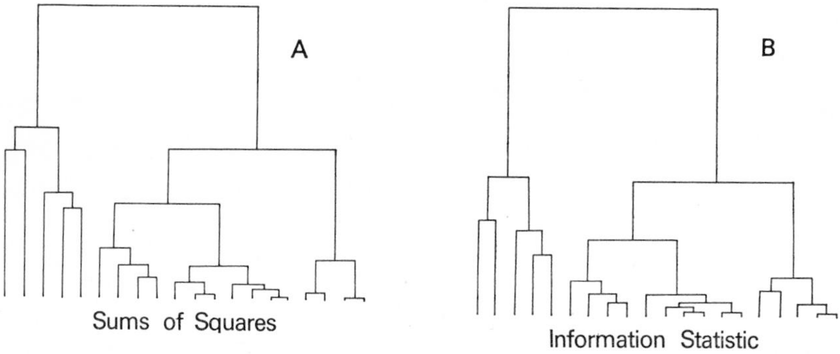

Figure 3. Comparison of polythetic-divisive hierarchies (AXOR strategy) produced by using (A) sums of squares and (B) the Shannon information statistic as the coefficient.

chies produced for the same set of qualitative data as in Fig. 2 by using the AXOR strategy with each coefficient in turn, and in each case the residual quantity (i.e. Δ I or Δ SS) has been used to define the course of the analysis and the actual hierarchical levels. It is clear from this figure that the two measures have similar, but not identical, properties, and we still have to isolate the critical differences between them.

The results of a further test designed to investigate empirically the degree of accord in the behaviour of the two measures with different types of data are given in Table 5. In this experiment, 100 different sets of data, each of sufficiently small dimensions for all possible splits to be explored, were constructed artificially from random number tables and each then modified to give four different data types: fully quantitative, partially quantitative, fully qualitative, and quantitative dichotomised to qualitative. The comparative behaviour of the two coefficients was assessed first in terms of percentage of identical 'best splits' obtained for each data type, and secondly in terms of where the best split for one coefficient occurred in the order for the other when all possible splits were arranged in optimal order for each.

Table 5. Comparative behaviour of the Shannon information statistic (I.S.) and 'sums of squares' (S.S.) over a range of simulated data types.

		Total % of different 'best splits'			
		Type 1	Type 2	Type 3	Type 4
Overall differences		12	29	16	10
Distribution of differences					
In terms of S.S. 'best splits'	1st S.S.=2nd I.S.	8	15	11	7
	1st S.S.=3rd I.S.	2	8	2	2
	1st S.S.=4th I.S.	2	4	3	1
	1st S.S.=5th I.S.	0	1	0	0
	1st S.S.=>5th I.S.	0	1	0	0
In terms of I.S. 'best splits'	1st I.S.=2nd S.S.	8	16	10	7
	1st I.S.=3rd S.S.	4	7	4	2
	1st I.S.=4th S.S.	0	2	1	1
	1st I.S.=5th S.S.	0	3	1	0
	1st I.S.=>5th S.S.	0	1	0	0

Type 1=Full quantitative data
Type 2=Mixed quantitative/qualitative data with zeros positioned randomly
Type 3=Full qualitative data converted directly from type 2
Type 4=Full qualitative data from type 1, dichotomised at mid-value.

One inherent disadvantage of the information statistic is that it is not easily convertible to cope with quantitative data, and in particular we know of no satisfactory device to enable it to deal directly with unbounded quantitative measures. In contrast, sums of squares can be directly related to Euclidean distance measures and correspondingly the model partakes of some of their properties. Although we have so far been unable to define the precise theoretical relationship between the information statistic and the sums of squares, our experience so far suggests that the latter is likely to prove the more flexible model for general use.

C. *Use of classificatory models with vegetational data*

In analysing vegetational data it is usual to regard the sampling sites as 'individuals' and the species (or other plant entities) associated with them as the attributes. A 'normal' classification will therefore group the sites in terms of the species they contain, rather than the species in terms of the sites in which they occur; the groupings can thus be transferred directly to a map and assessed in a regional context. Although an 'inverse' analysis can equally well be performed on the same data (Williams & Lambert 1961), the results will have no immediate regional relevance unless the two analyses are superimposed. Since information of value can reside on both site- and species-analysis, a model which combines the two should have wider applications than either model operated separately. Models for extracting vegetational 'noda', defined by both species and sites, are still in a very imperfect form, and a great deal of theoretical and empirical work is still required on this front. An early attempt by Lambert & Williams (1962) is crude and computationally clumsy, and a later effort by Tharu & Williams (1966) has never been fully explored.

One of the advantages of a nodal model is that it can be used to cut down noise generated by species/site interactions as well as that contributed by each of the elements independently. With the relatively coarse techniques which often have to be employed in large-scale work, the amount of noise involved could prove a serious encumbrance in attempts to interpret the final results. Any model designed to pin-point the central species/site relationships and eliminate the others is therefore likely to be valuable.

It is frequently forgotten that one of the most important functions of an analytical technique is to act as a magnifying glass to focus attention on those features where most interest resides. Complaints that a particular technique obscures 'real' vegetational relationships usually mean that an inappropriate model has been used for the particular purpose in hand. For instance, if a classification of the data has been specifically asked for, then a really efficient model is one which produces clear cut boundaries by an appropriate maximising function, rather than one which merely rearranges the data without emphasis. The five different vegetational hierarchies shown in Fig. 2 provide an excellent illustration of this principle. The first four examples all use coefficients unsuited to an hierarchical model and the resulting picture is obscure; in contrast, in the fifth, where the coefficient and strategy are matched and reinforce each other, a clear-cut picture emerges which is readily interpretable.

IV. Discussion

Perhaps one of the most important lessons to be learned in any attempt to construct a model system is the extent to which the success of the whole depends upon the smooth integration of the constituent parts, right down to the minutest detail. However, in primary survey work, as in any exploratory system, it is usually impossible in practice to foresee all the contingencies which may arise during the course of the investigation. An elaborate system which is completely rigid from its inception may therefore fail dismally through sheer lack of adaptability. Where two or more possible models are in mind for any part of the project, therefore, it is usually advisable to choose the most flexible, provided the flexibility is not accompanied by loss of definition.

An insistence on rigour in definition will always pay dividends in that the effect of any adjustment to the model can be anticipated and the necessary compensatory action taken. The degree of rigour required, however, is not necessarily the same in every part of the system. Any section of the work which is central to the whole should obviously be as rigorous as possible, since any looseness here could escalate; for less fundamental parts of the exercise, a cruder model can often suffice.

In general, the better the model, the more expensive it is to operate. With limited resources, therefore, one possible course of action would be to concentrate the major part of these resources on the most critical parts of the work, rather than aim for coverage of equal quality throughout. Thus a highly efficient but time-demanding sampling model and a highly effective but computationally expensive analytical technique might be employed to erect a very precise vegetational classification from the minimum number of samples required to attain stability, and then additional samples allocated to it secondarily by a more economical discriminant technique. We have just begun to construct a model along these lines, using random sampling for the critical basic exercise and a gridded system for the derivative operation after the vegetation units have been defined. The function of the random model is to extract vegetational information from the area as effectively as possible, while that of the subsequent grid is to reimport the derived information into the area and display it spatially to the best advantage. The use of random and gridded sampling sequentially in this manner, rather than as alternatives, in fact employs the practical properties of each type of sampling pattern to maximum advantage: the complete absence of any restriction on sample number in random sampling enables the stabilization process to be carried out efficiently, while the greater spatial control possible in the systematic model makes it the better equipped to provide the ultimate spatial coverage required. The success of an operation of this kind depends on finding a satisfactory method of stabilizing the primary vegetational classification, and models for this are currently being examined. Once the vegetation-units have been stabilized and economically defined, all subsequent samples can be allocated to the system one at a time by an appropriate form of simple discriminant analysis; there is thus no longer a need for computer programs of vast capacity to handle in a single operation all the samples needed for adequate spatial coverage, and the system becomes virtually open-ended as far as the actual computation is concerned.

It will be realized that any system involving an initial definition of vegetational entities, followed by allocation of additional samples, in fact approaches very closely in principle to the procedure of traditional phytosociologists. In fact, a parallel with various facets of general phytosiological thinking can be identified, in one guise or another, in practically every model for vegetation survey that has been discussed. The defects of current phytosociological systems lie not in the underlying concepts, but in the lack of rigorous definition in the models used. If the main sources of imprecision and inefficiency can be identified and removed, and some of the more restrictive assumptions discarded, then a model system could emerge with very valuable properties for future large-scale vegetation survey.

Acknowledgements

Much of the work on which this contribution is based was carried out under an S.R.C. grant for research into multivariate methods of analysis of ecological and other data. Many thanks are due to the members of the research team—Mr J.Barrs, Mrs S.E.Meacock, and Mrs P.F.M.Smartt—who were responsible for many of the examples quoted and who have also helped greatly in the preparation of this paper. I should also like to acknowledge the contributions made over the last few years by various postgraduate and undergraduate students of the Botany Department at Southampton; in particular, I should like to mention the undergraduate dissertations of R.L.Quick (1968), J.Beazley (1970), J.E.A.Grainger (1970), and M.T.Jackson (1970), where I have quoted freely from the results where appropriate.

References

CORMACK R. (1971) A review of classification. *J.R. Statist. Soc. A.* (in press).

CRAWFORD R.M.M., WISHART D. & CAMPBELL R.M. (1970) A numerical analysis of high altitude scrub vegetation in relation to soil erosion in the Eastern Crodillera of Peru. *J. Ecol.* 58, 173–91.

FIELD J.G. (1969) The use of the information statistic in the numerical classification of heterogeneous systems. *J. Ecol.* 57, 565–9.

GREIG-SMITH P. (1964) *Quantitative plant ecology*. London.

HOPKINS B. (1955) The species-area relations of plant communities. *J. Ecol.* 43, 409–26.

HOPKINS B. (1957) The concept of minimal area. *J. Ecol.* 45, 441–9.

LAMBERT J.M. & DALE M.B. (1964) The use of statistics in phytosociology. *Adv. ecol. Res.* 2, 59–99.

LAMBERT J.M. & WILLIAMS W.T. (1962) Multivariate methods in plant ecology. IV. Nodal analysis. *J. Ecol.* 50, 775–802.

LAMBERT J.M. & WILLIAMS W.T. (1966) Multivariate methods in plant ecology. VI. Comparison of information-analysis and association-analysis. *J. Ecol.* 54, 635–64.

MACNAUGHTON-SMITH P., WILLIAMS W.T., DALE M.B. & MOCKETT L.G. (1964) Dissimilarity analysis; a new technique of hierarchical sub-division. *Nature, Lond.* **202**, 1034–5.

MOORE J.J., FITZSIMONS S.J.P., LAMBE E. & WHITE J. (1970) A comparison and evaluation of some phytosociological techniques. *Vegetatio* **20**, 1–20.

Sokal R.R. & Rohlf F.J. (1962) The comparison of dendograms by objective methods. *Taxon*, **11**, 33–40.

Tharu J. & Williams W.T. (1966) Concentration of entries in binary arrays. *Nature*, Lond. **210**, 549.

Webb L.J., Tracey J.G., Williams W.T. & Lance G.N. (1970) Studies in the numerical analysis of complex rain-forest communities. V. A comparison of the properties of floristic and physiognomic-structural data. *J. Ecol.* **58**, 203–32.

Williams W.T. & Dale M.B. (1965) Fundamental problems in numerical taxonomy. *Adv. bot. Res.* **2**, 35–68.

Williams W.T. & Lambert J.M. (1959) Multivariate methods in plant ecology. I. Association-analysis in plant communities. *J. Ecol.* **47**, 83–101.

Williams W.T. & Lambert J.M. (1960) Multivariate methods in plant ecology. II. The use of an electronic digital computer for association-analysis. *J. Ecol.* **48**, 689–710.

Williams W.T. & Lambert J.M. (1961) Multivariate methods in plant ecology. III. Inverse association-analysis. *J. Ecol.* **49**, 717–29.

Williams W.T., Lambert J.M. & Lance G.N. (1966) Multivariate methods in plant ecology. V. Similarity analysis and information-analysis. *J. Ecol.* **54**, 427–45.

Williams W.T., Lance G.N., Webb L.J., Tracey J.G. & Connell J.H. (1969) Studies in the numerical analysis of complex rain-forest communities. IV. A method for the elucidation of small scale forest pattern. *J. Ecol.* **57**, 635–54.

Organization and management of an integrated ecological research program—with special emphasis on systems analysis, universities, and scientific cooperation

GEORGE M. VAN DYNE *Natural Resource Ecology Laboratory, Colorado State University, Fort Collins, Colorado, USA*

Summary

This paper reviews a large-scale, long-term, interdisciplinary, integrated ecological research program. The magnitude of these studies is indicated by the example that the grassland study now has more than 90 senior scientists involved, representing about 30 academic institutions and federal agency groups, and an annual budget of near $2.0 million is spent in field, laboratory, and armchair research. We are using results obtained from field and laboratory ecological experiments, analysis of the literature, and natural resource data as a major source of inputs to our programs. Our programs attempt to integrate these data and information sources through a systems analysis approach. The output will be comparative analyses, implemented through use of simulation, optimization, and structural models. We feel there will be a payoff from these studies toward the development of a new ecological theory and toward development of resource management principles.

Specific attention is directed in this paper not to the large volume of data generated in these studies, but instead to discussion and analysis of some of the developmental characteristics of the programs. I have also attempted to provide suggestions for this type of study, on a national and international basis, as they carry forward during and following the life of the I.B.P. Some approaches in mathematical modelling and analysis are discussed with special emphasis on simulation models. The use of the state variable approach involving systems of differential equations, is referred to and generalized modifications are included to allow the inclusion of the real-life complexities of ecological systems.

The paper flows in a generally chronlogical sequence: introduction and discussion of integrated research, the first 15-month research segment, the next 16-month grant, and general discussion of problems and needs.

Introduction

In June 1965 I lectured to a group of ecologists on the problem of 'systems ecology research'. In an essay based on this talk I considered the inadequacy of current ecological theory with respect to rapid solution of some of today's environmental problems and analyzed and critiqued the reliance of some scientists upon analogies from physics for the solution of ecological problems (Van Dyne 1966, Cox 1969). At that time there was a dearth of quantitative ecologists; most ecologists were isolated and not well supported, and there were hopes that we would be able to develop centers wherein 'critical masses' of systems ecologists would migrate and find suitable niches.

I hypothesized that the system ecologist would have to 'guess' at the fundamental cause-and-effect relationships in a system, and he may even have to guess about the basic variables. These guesses or hypotheses would be tested by formulating them into models, running the models, and then testing the output of the models against the quantitative and qualitative behavior of a real world. These tests of hypotheses would require some varied and powerful tools, such as computers, for their implementation. It was suggested that theory and experiment could be combined, sensu Holling (1963), into a 'systems analysis' in order to reduce the great number of hypotheses that might be tested about the system. Even though powerful tools would be used, generally it would be only pencils, paper, and discriminating thought rather than the sophisticated tools to choose among alternative hypotheses. I suggested that the systems ecologist would have to develop the knack of feeding on negative information; he may have to perturb the system (followed by measurements) to get new insight for a model.

Systems Ecology Pre-International Biological Program (IBP)

It seemed that the major hurdle to start systems ecology research in 1965 was the lack of precedent in funding detailed and integrated research on total ecological systems. Probably no single organization was working then in depth on complex, total-systems ecology problems. Such effort seemed to be beyond the role, objectives, or structure of existing organizations. Universities, national laboratories, and state and federal experiment stations each had some unique resources and capabilities for studying such problems and, based on my experience in working in such organizations, I attempted to summarize their advantages and disadvantages in the above essay.

Most ecological research conducted at universities generally has involved one investigator, or a few at most, on a part-time basis on problems of limited extent. Extensive and intensive interdepartmental cooperation is exceptional. Various fund sources were available, but usually amounts were insufficient to attract permanent personnel and to support long-term ecosystem research. Graduate students were used efficiently, but there was a lack of continuity in long-term environmental work. Although students were making considerable contribution, by the time they gained confidence and became capable of independent efforts in a project, they would graduate and be lost.

I felt that applied ecologists, such as in federal or state experiment stations, often had controllable research areas on which long-term work was in progress. But still, there was a relatively high rate of turnover of personnel. My interpretation of their situation was that the work had become restricted to only one, or at most a few, phases of the total ecosystem problem, with funds becoming even more limited in recent years.

At that time, it was noticeable that much of the ecological research was being done in several national laboratories where scientists had available many services, tools, and consultants often lacking by the university or experiment stations. Yet, much of the work was highly directed and oriented to specific needs of the funding agency. I felt that a total-system, inter-disciplinary approach usually was not highly developed in their research. Costs were high because of the nature of their work. Because of the limited number of national laboratories, some important biomes were not included.

In 1965 I was working in a national laboratory within the Deciduous Forest biome. Therefore, I felt safe in making suggestions about work in another biome under the then proposed International Biological Program. I suggested that the IBP called for 'systems ecology research, both experimental and theoretical'. I felt the IBP could provide an incentive and a means for ecologists from universities, experiment stations, and national laboratories to work cooperatively and share funds, research areas, and talent. As an example (and in one paragraph!), I suggested how ecosystem research could be done on the semi-natural grasslands of the North American Great Plains. This was interesting because about one year later I moved to Colorado State University and still later became involved in International Biological Program research on grasslands. The present paper gives me an opportunity to test what has happened since this project started against the ideas I presented five years ago about ecological research in the grasslands.

A somewhat *chronological development of this integrated research program will be presented here*. However, before initiating this discussion, I should note that the US IBP Grassland Biome study is one of six planned biome programs covering the major biome types of North America. These include studies on grassland, desert, deciduous forest, coniferous forest, tropical forest, and tundra. At this writing, intensive field investigations have been initiated in the USA in five of these biomes. The programs differ in many respects, but much of what I say about the Grassland Biome study will have its counterpart situations within the other biome programs. I believe there is some generality in these observations and conclusions.

Objectives

This paper reviews the evolution of management and organization of an integrated research program—the Grassland Biome study of the US contribution to the International Biological Program, and frequent reference will also be made to the other biome studies in the US IBP Analysis of Ecosystems program. Throughout this discussion, reference will be made to the implications of this type of research program on the training of new scientists, the organization and operation of existing scientists, and the advantages and disadvantages of various institutional structures for such integrated research. I will criticize our own program, the agencies which support the research, and the institutions in which the research is housed. These *criticisms are made in a constructive sense* and should be viewed in that light.

The objective herein is not to whitewash our failures, nor to chastize our critics, nor to platitudinize to our supporters—instead, it is to report it how it is. It is important to review the program in detail at this time to help determine which, if any, of the original opinions were correct: (i) those who said it couldn't be done; (ii) those who said it had already been done or they were already doing it; (iii) those who said it could be done, quickly and easily; and (iv) those who said it could, should, and would be done, but that we would never live to see it!

Please see the ACKNOWLEDGEMENTS at the end of the paper. Programs of this type require the enthusiasm, ideas, and hard work of many individuals, and some of the names are provided there. There must be frequent and free interchange in this kind of program and I have clearly drawn on these people. Many of the ideas presented herein have been extracted from various internal working papers and reports, the origin of which is due to the total group effort. Finally, because I have been heavily involved in the origination, development, and conduct of this program, much of what I write could represent a *biased personal interpretation.*

Integrated ecological research

Some definitions

The definition and structure of an integrated research program will be clarified if I describe my use of the terms 'systems', 'systems approaches', and 'ecosystems'.

The term 'system' implies an interacting, interdependent complex. The 'systems approach' to problem solving is not new, but unfortunately, it is utilized all too infrequently in ecological analysis (Van Dyne 1969*a*). A systems approach is a systematic consideration of problems wherein analysis is stressed as well as intuition. As an example, consider a systems analysis of grasslands, but first define grasslands. The term 'grasslands' is interpreted broadly to include planted pastures as well as native rangelands, and the latter to include a wide variety of conditions from semiarid to subhumid.

In a system analysis of grasslands we seek an optimal course of action, e.g. in management, by systematically examining the objectives, costs, effectiveness, and risks of alternative management strategies. We may design additional management strategies if those examined are found to be insufficient. A resource manager may treat his resources as a system, especially in multiple use management of natural resources. But evaluations developed by a single individual are largely intuitive and are usually inadequate because of the complexity of the issues and the numbers of pertinent factors to be considered. In utilizing a systems approach to grassland problems, we must (i) combine, condense, and synthesize a great amount of information concerning the components of the system; (ii) examine in detail the structure of this system; (iii) translate this knowledge of systems components, functions, and structures into models of the system; and (iv) use the models to derive new insights about the operation, management, and utilization of the system.

I shall use Tansley's definition (1935) of the term ecosystem, 'a system resulting from the integration of all living and non-living factors of the environment'. An ecosystem, as a unit of landscape or of seascape, applies for a defined segment of space and time. It is a complex of organisms and environment forming a functional whole.

Integration of research

The International Biological Program allows a new and different type of ecological research within the United States. A substantial part of the US contribution to the IBP involves team studies of ecosystems. This is best characterized by the term 'integrated ecological research' which implies some unique characteristics.

The basic concept here is that a few, carefully selected large-scale systems are being studied. These systems are far larger than have been comprehensively studied previously. These studies are 'integrated' because many scientists work together cooperatively in carefully defined team structures attacking problems pertinent to the 'man-environment-ecology' interrelationship. Obviously, these activities must be coordinated with similar programs as part of the IBP around the world. In the United States these integrated programs are referred to as 'IRPs', i.e., integrated research programs, and they are the core of the US IBP efforts. These IRPs do not represent the total US IBP effort, but they are handled under special funding arrangements and do not compete primarily with regular Federal support programs. Most of the 'IRPs' are defined for the life of the IBP, but it is expected that many of these kinds of activities will continue beyond IBP. This might be interpreted from former President Johnson's statement proposing that this type of effort (i.e. IBP) be of permanent concern to the nation.

Because of the complexity of ecosystems, and because of the skills and understanding required, probably no one scientist can encompass all the required specialities and knowledge to undertake a thorough study of a complete grassland system. This necessitates having an interdisciplinary team; it requires development of systematic procedures for studying grasslands (Coupland, Zacharak & Paul 1969). Mathematical models provide a framework within which to coordinate the efforts of the interdisciplinary team members and to help make meaningful the individual contributions in a total-systems sense. With many individuals involved in the study of a single system, there must be a great deal of collecting and sharing of basic data. Scientists must participate as a team and must include services to others as a part of their activity. There must be a careful blend of field and laboratory studies in measuring systems components and processes. The interaction of a large number of specialists and the smooth running of a project requires guidance by a scientific committee group toward project rationale and management.

The importance of integrated research programs can be clarified with an introduction to some of the concepts in the original plans of the Analysis of Ecosystems program, a group of programs in the US IBP. These studies are to be conducted in different biomes (grassland, desert, tundra, coniferous forest, deciduous forest, and tropical forest) and at least one site in each biome will be the location of an intensive study. At this location, the study will be as complete as modern technology and resources permit. At such an intensive site different sub-areas receiving different treatments will be investigated. Thus, the study is not simply an intensive survey but, in effect, an experiment on whole systems. In each biome there will also be supporting studies on other sites, variously called comprehensive network sites, validation sites, etc. The work within each biome will be integrated not only on the intensive site but also in the comprehensive sites, and data and information will be synthesized over both of these and finally across different biomes. These efforts, of course, imply a greatly different approach to integrating research than has been evident in the United States in the past! The integrated research programs must be *collective* efforts not *collected* efforts.

Ecological objectives

The overall purpose of the IBP is to examine 'the biological basis of productivity in human welfare'. Internationally, a major objective is to study organic production on a world-wide basis. But we are concerned both with productivity and its utilization. The global objectives will not only include the acquisition of information necessary for the development and testing of ecological theory, but also that such theory will eventually have usefulness to man in his understanding of ecosystems and thereby his intelligent management and use of them. Immediate goals include the analysis of such aspects of ecosystems as energy flow, nutrient cycling, trophic structure, spatial patterns, interspecies relations, and species diversity. Further goals include the understanding of ecosystems processes by which observed characteristics of flow, cycling, density, diversity, etc., are achieved and maintained. Early in our research planning we identified a series of about 30 broad ecological questions for which we would seek answers through field, laboratory, and armchair research. Typical questions that might be asked of the system include the following.

- What is the effect of the relative abundance of consumers upon rates of nutrient cycling and of energy flow?
- How is the consumption of plant material apportioned among the various herbivores? How do these proportions change seasonally? From year to year?
- How do the primary producers vary in numbers and productivity in relation to variations in seasons and weather?
- Are the rates of nutrient cycling or energy flow dependent upon species diversity or life-form diversity at any given trophic level? In what way is 'information' processed and stored in an ecosystem? Do plant patterns reflect in a quantitative fashion the stresses imposed upon the system?
- Is there a relationship between pool size of a nutrient within the system and the rate of cycling?
- How do relative proportions of nutrients and energy within the total system vary among the different trophic levels with different methods of manipulating the system?

It should be clear, by review of the above kind of questions, that securing *adequate answers will require information about the total system.* Furthermore, seeking answers to these questions requires inputs from a number of scientists because of the sheer amount of work involved, and from different kinds of scientists because of the specialties involved. Thus, the focal point in the research is to improve our understanding of entire systems. No matter how narrow or detailed a given project may be, the relationship to the whole system will be the dominant theme. Also, *there must be a mechanism of synthesizing this information into a whole* as the program progresses. We find models are useful tools here, as indicated above and as discussed in later sections.

One might question 'how do the above stated ecological objectives of an integrated research program differ from the objectives of "mission agencies" of our government?' Most programs in the mission agencies supporting biological research deal with more specific and smaller projects, not the large-scale integrated studies of the IBP type. Perhaps the difference is more in approach. Mission agency personnel are involved in the IBP integrated research projects, but usually in a limited part rather than in the whole project. A group including members from all the major federal mission agencies called the 'Interagency Co-ordinating Committee' was established to review, evaluate, and integrate with the IBP integrated research programs. It has been possible in a few instances to reprogram some mission agency activities so that they are more directly related to the IBP-type research which is primarily university-based and organized.

Planning and initial research in the grassland biome study

Early organization

The original plans for the IBP Grassland Biome study were implicitly outlined, to some degree, at a meeting in Massachusetts in October 1966 at the time the overall Analysis of Ecosystems program was first framed. At this time, a new approach and new level of organization of American ecology was envisaged. Out of this meeting arose suggestions for a new scheme of co-ordination, stimulation, assistance, and direction of ecological research.

Further meetings were held by the Analysis of Ecosystem program work groups in February and May 1967. At the May meeting, systems analysis was stressed and a prototype proposal was presented for grassland research. This proposal [developed by Van Dyne, Hansen, and Doxtader (Colorado State University), Moir (then University of Colorado), and Bement (Agricultural Research Service, USDA)] called for coordinated research at several sites on the Great Plains. Sites for grassland research from Canada to Mexico were examined in August 1967. The final site for the intensive study was selected by a task force at a working meeting in Colorado in October 1967. The 7-man task force was assisted by a group of consultants including 9 individuals not having potential involvement in the grasslands study, and 19 scientists and administrators from universities and organizations having potential direct involvement. This group developed a 43-page report which was later presented to the National Academy of Science Committee and accepted as a prototype plan.

Similar organizational meetings were held in 1968 in each of the other biomes (e.g. Deciduous Forest at Atlanta, Georgia in January; Coniferous Forest at Seattle, Washington in February; Desert at Logan, Utah in February

1968; Tundra, two locations in Alaska in October; and Tropical Forest at Ann Arbor, Michigan in October; ref. Smith & Kadlec 1969). Most of these meetings have been organized and funded through the central Analysis of Ecosystems grant which was first directed by Dr F. E. Smith, then of the University of Michigan. The intensive working meeting on grasslands concerned the entire biome program in general and the intensive site study in particular. At the October 1967 meeting I was asked to be Biome Director. The plan called for detailed studies at one location wherein as many components of the system would be examined as possible. This 'intensive site' research was to be supported by studies at several other sites in a 'comprehensive network' program.

The general plans for the grassland research project were presented to the National Committee of the IBP in Washington, D.C., in late October 1967. The plan was endorsed by the group, and suggestions were made on how to proceed. A meeting was held in November 1967, and subprojects were discussed by 61 potential participants. Subproject proposals for the grassland studies were solicited from interested investigators for the intensive site and for the comprehensive network. Working meetings were held by small groups to discuss proposed research, and some 10 pounds of individual and group proposals were submitted to me! These proposals were evaluated by a committee of experts from around the nation in a meeting in late November 1967. Surviving proposals, or parts thereof, were rewritten, and I condensed and combined them into a single, cohesive 371-page proposal. This proposal was submitted to the National Science Foundation requesting $1·9 million for a 12-month period beginning April 1968. The program was discussed widely and heatedly throughout the granting agencies, and especially in the Interagency Coordinating Committee of the IBP. In Spring 1968 I was asked to redesign studies and rebudget for $700,000 for a 15-month period, instead of the 12-month period. Some funds were received beginning in June 1968, and eventually over several grant segments, they totalled $851,000 for the first 15 months.

Our original plans, based on answering eventually the broad ecological questions outlined in the preceding section, called for detailed total-system research. Even as naive as we were at the outset, however, it was clear that we would not be able to study *all* things about *all* grasslands. So major boundaries were established for our research.

We decided to focus our major effort on studying intraseasonal rather than interseasonal dynamics of grasslands. We put initial relative priorities on the variables we would trace through the system. Thus, we put decreasing priorities on studies of biomass, energy, nitrogen, phosphorus, sulfur, . . ., and eventually other elements.

We recognized it would not be possible to find enough scientists in a given area, of diverse disciplines, and having time and interest in cooperative ecosystem research to do such studies at very many places, even if the money were available. Thus, we selected the *design concept of one intensive site and a network of comprehensive sites.*

Our long-term plan calls for a minimum of two sites in each of the following kinds of operationally defined grasslands: shortgrass prairie, mixed prairie, tallgrass prairie, desert or semidesert grassland, Northwest bunchgrass, annual grassland, and mountain grassland. In such a design we recognize the long-term potential of comparisons with the mixed prairie grassland research on the Canadian IBP site at Matador, the developing Mexican IBP site at Campana, the US IBP desert shrub-grassland site at Santa Rita, potential US IBP coniferous forest sites, especially in the ponderosa pine-bunchgrass zone, and potential US IBP alpine-subalpine mountain herbland sites.

We recognized at the outset that on any site if one simply studied the system in great detail one would simply be conducting an intensive survey and would only gain a limited amount of information about relations of structure and function. Therefore, we decided that in each major field study we would impose a minimum set of treatments or stresses on the system and then measure abiotic, producer, consumer, and decomposer factors within each treatment. So we selected what we felt would be the most important stresses in grasslands, which were also measurable within our financial framework, for inclusion in our design. On each major field study we investigate replicated areas where there has been very limited or no grazing by large herbivores over many years or ever vs. intensive or heavy grazing by large herbivores over many years. Thus, the level of herbivory, although not identical on all sites, is the principal stress. In most instances the large herbivores are domestic animals, but on one of our sites the native bison was the meaningful herbivore.

Level of herbivory is the minimum stress on all sites, but especially on the Pawnee Site other treatments are included. There, for example, the present treatments include replicated plots or pastures wherein the following treatments are imposed on the system: zero, light, moderate, and heavy herbivory by cattle for more than 30 years; added moisture, either seasonlong or in specific periods; added nutrients as a 'shock' to the system, i.e. 100 kg/ha of nitrogen applied annually; season-long moisture and nutrients in combination; and organism removal from the system, i.e. areas where selected producers or consumers are removed by hand or chemically without destruction of the remainder of the system.

The amount of funds received in our first grant was far less than the amount requested; considering time and dollars, less than 30 per cent of the requested budget was obtained in the first grant segment. This brought the first face-to-face confrontation with real-life, suboptimal budgets! The following guidelines for suggestions were considered in developing the revised budgets and in reallocating funds to individual subprojects and many subprojects were delayed and many were consolidated with others.

- Studies in the comprehensive network program were delayed until the intensive site project was brought up to near full levels of funding.
- Studies on aquatic phases were delayed until studies on the major terrestrial systems or components were initiated.
- A significant fraction of the required field facilities and instrumentation was to be developed so that work would progress rapidly in a subsequent year.
- Some effort was to be spent on preliminary modelling, on synthesis of information of grassland ecosystems, and on using these models and syntheses as a basis for further allocation of efforts in the following year.
- A critical core of some projects were funded at a reduced level, primarily to allow only instrumentation and preliminary studies.
- A minimum of new staff or graduate students were committed to the research program initially.

The subprojects selected in the major research proposal to receive funding in the first year were grouped into 11 key areas. These project areas were selected not only because of their criticality to the total ecosystem study, but also because of the criticality of maintaining high interest of key scientists. Some 60 senior scientists were in the research during the first 15-month segment. In addition to the Ph.D.-level senior scientists, there were approximately 5 junior scientists, 30 graduate research assistants working toward either a master's or doctoral degree, and 5 field and laboratory technicians.

Synthesis as a research activity

Synthesis was one of our goals within the program's first phase. I feel it is historic fact in environmental biology that, in most instances, a comprehensive and exhaustive survey of the literature on a given topic is not made prior to initiation of field or laboratory research in that area (Van Dyne 1969*b*). The trend in many recent technical journals also seems to be to decrease in scope or eliminate the detailed literature review. In the past when a review has been made, it usually was done at the time of summarizing results and writing research papers. The problem is further complicated by the fact that the amount of factual information in many areas of ecology has grown exponentially within the last two decades. Furthermore, this factual information is distributed widely in many journals, bulletins, books, and internal reports of various organizations.

Perhaps it is only natural that individual investigators seldom undertake exhaustive reviews because of the magnitude of the task. But in the 'age of the information explosion', reviews and more important syntheses of information must be undertaken to organize the large and diverse body of literature that characterizes ecology. I felt that even though gaps exist in our knowledge, it would be desirable to bring together the information we have about specific kinds of ecosystems. Such activities must be planned as part of the research effort. I felt this was important even though, conventionally, organizations and agencies supporting field research, or even theoretical or analytical research, seldom have stimulated or even condoned detailed syntheses prior to the conduct of field, bench, or laboratory investigations.

For the above reasons, a synthesis project was organized in the initial efforts of the US IBP Grassland Biome study. This information synthesis project had as its major goal the assembly and synthesis of data on concepts relative to grassland ecosystems. Important secondary goals were the familiarization of the grassland ecologists with the rationale and procedures of systems ecology. This series of synthesis workshops also served as a creation of the common ground for association between various grasslands specialists. Many investigators expressed desire to participate in initial field studies but could not because of funding limitations. Many of them participated in the synthesis study and maintained a strong interest in the ecosystem program through these workshops.

We attempted to synthesize information on grassland ecosystems in general, particularly those in North America, and especially interpret this information relevant to our proposed field, laboratory, and analytical research projects. Through this review, we wanted to really find out what we knew about these ecosystems, what remained in doubt, and we also used the review to give us some guidelines on priorities for future researches.

A 'synthesis' is a combination of separate elements or parts into a whole. Our synthesis was undertaken in a total ecosystem framework. To facilitate this review, a general first-approximation 'picture model' was defined as a framework of reference (Fig. 1). This picture model, essentially a 13-compartment model, was segmented and some 32 separate review papers were developed, presented, discussed, and eventually published (see Dix & Beidleman 1969, and supplement 1970). Firstly, emphasis in these reviews was to examine the literature for estimates and variances of intraseasonal dynamics of biomass, energy, and mineral elements in ecosystem components or compartments. Secondly, the external or driving forces acting upon grassland ecosystems were examined and characterized. Thirdly, some of the major structural properties were examined. Little emphasis was placed, however, on the mechanisms of transfer or matter from compartment to compartment.

Models are a form of synthesis and word, picture, and mathematical models are used freely in our studies. There is a strong mathematical thread running through the fabric of the Analysis of Ecosystems integrated research

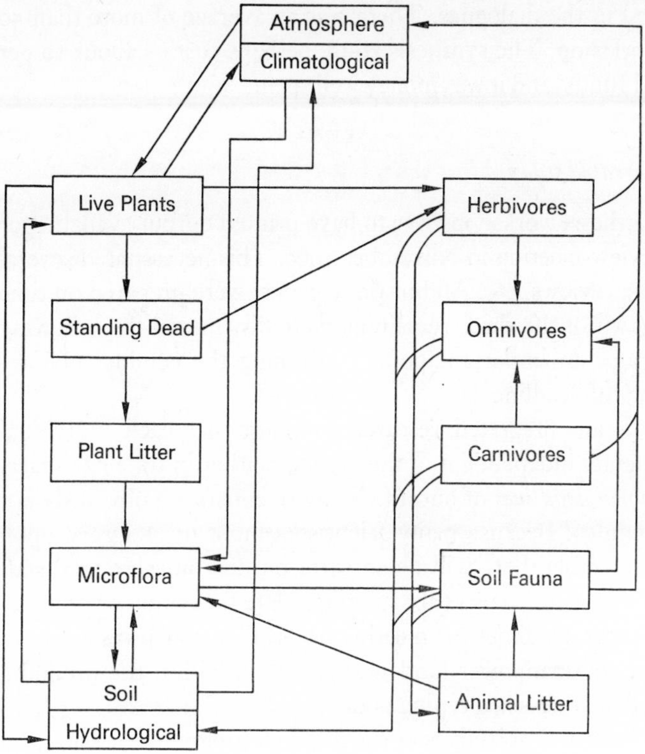

Figure 1. The first approximation 'picture model' used in developing budgets in Spring 1968 and in developing topics for the synthesis workshops beginning in Summer 1968 in the US IBP Grassland Biome study. The boxes represent compartments in the system or driving or exogenous variables. The arrows represent pathways of flow of matter and energy.

program. This was true too in our synthesis project. In the first workshop a paper was presented to review examples of modelling grasslands and to discuss cautions, considerations, advantages, and needs of modelling. The final paper in the last workshop also focused on mathematical models, in part incorporating into a quantitative framework many segments of information discussed in earlier papers.

An important methodological aspect of organizing these workshops was to present the review papers in five two-day sessions in isolated locations. Under these informal conditions, there was a great deal of interaction among participants. The presentation of each paper was accompanied by a consideration of the topic by a discussion group. In all of our workshops, papers were presented on all trophic levels within the ecosystem. In all workshops, and in all discussion groups, scientists of many disciplines were present. In total, 106 individuals from 35 institutions and agencies representing 20 states and 4 nations participated in the dialogues. There was an average of more than 50 persons at each workshop. The synthesis study perhaps cost us about 10 per cent of our original funds, and it was money well spent.

Deadlines in an integrated study

One goal of the synthesis workshops was to have printed output available for our first annual review meeting in November 1969. This necessitated several deadlines for drafts, reviews, etc. All but three papers were prepared on time for a final report (two of the three remaining were distributed the following year). This illustrates an important point concerning the conduct of integrated research, that of deadlines!

It is obvious in an integrated research program that each individual scientist cannot operate independent of the others, neither in time nor space. All must *meet common deadlines* of output. Many scientists are obviously not disciplined to schedules! Because many scientists cannot meet agreed upon deadlines, it is predictable there will be an extra burden on secretarial staff to help regain lost time when papers are submitted late for typing or preparation and yet must meet deadlines for meeting dates. This, too, gives a hint to the organization of the manpower (and womanpower) within the program, i.e. an energetic and flexible-sized typing pool.

The evolving needs and objectives of the research program necessitated a new framework or organization of the research studies, as compared to that outlined in October 1967. This was reflected in a new organization chart developed in the first 15-month research period. First, a numerical scheme was provided for the coding of individual efforts consistent with a PPB-approach to budgeting and analysis. The studies were organized to enable separate totals for the grassland program, program phases, areas within phases, and projects within areas. The numerical scheme provided in the organizational chart was used in a computer program in budgeting, reporting, and analysis. (Originally, in the Analysis of Ecosystems program plans were made to allow cost analyses in all the biome studies. The 1000 series was used for budget categories in the central program, and the 2000 series was assigned to the grassland study.)

Within the first six months of our research, it became obvious that we must look more critically and systematically at long-term planning and setting of deadlines. A major effort in management of research in the Grassland Biome study is the accounting of all the individual projects, the scheduling of events, the assessment of results, and so forth. In order to assure coordination of events, a preliminary and generalized activity schedule was developed. The dependence of each event upon another and preliminary estimates of starting time, optimal ending time, pessimistic ending time, etc., on each job were made for a five-year program. The major groups of events included for such categorization were modelling efforts, organizational efforts, informational synthesis efforts, Pawnee Site development efforts, laboratory facility efforts, and conceptually (at that time) but not yet in an analysis, the comprehensive program effort. This network of events was analyzed by critical path methods (CPM), specifically by a program review and evaluation technique (PERT).

PERT is a particular CPM for planning and analyzing projects which have some initial uncertainty about scheduling, certainly a case in our biome study! Events are entered as nodes connected by lines for branches such as is used in a network diagram. The path joining nodes is an ordered set of branches. A project is represented by a time-oriented graph, letting the nodes represent events in time and the branches represent activities requiring time to complete. A critical path of the network is that which will provide the total time required for the completion of the project, i.e. the longest sequence of events in the project. The duration of these activities determines the duration of the project.

The predicted critical path in the program was in the modelling efforts, but was restrained by the systems analysis block in organization, specifically by compartmental measurement phases. At the outset, we contemplated the completion of a final report in 1,250 working days for a five-year first phase of our program. Yet the calculated duration of the project in the PERT analysis was 1,740 days! Obviously, there can be some trade-off between time and dollars. Yet, using our estimates of funding and completion times for activities, it was clear from this analysis this program could not be completed in time. What was feared by many, i.e. that the late start and slow funding in the US IBP, was that IBP would not be able to meet its commitments within the original planned time span through 1972. This preliminary analysis suggested that the final report for the initial phase of our grassland research program could not be completed until between December 1973 and June 1974. Interestingly, and coincidentally, the IBP has recently been lengthened in many countries through June of 1974! We still have a chance!

The initial run of this PERT program detected the critical path in the project so that we could direct our attention to the activities along that path. This was an *extremely useful management exercise* even though the initial 225 events we selected represented only a small part of the events which must be scheduled in a 5-year project of the complexity of this one.

The management of integrated research programs in the IBP encounter a number of difficulties because of the environment in which they operate. For example, university scientists are not accustomed to having their efforts put into an operations research analytical framework such as the PERT analysis indicates above. However, the aversion of the academic scientist to having their research 'programmed' can be overcome once the importance of the exercise is explained to them and that there is an absolute requirement of integrating newly discovered knowledge on a constant basis. Still, the dilemma of time has not been fully faced in the organization and management of these programs, as will be referred to in a later section.

Initial systems analysis considerations

Some aspects of systems analysis have been discussed above. We define for purposes of our project organization the term 'systems analysis' to include the orderly and logical organization of data and information on grassland ecosystems into models. As noted above, a model may be a word model, a picture model, or a numerical model, and all three kinds are necessary in our study. In our systems analysis efforts, we are determining what are the components or properties of grassland ecosystems, what are their functional interrelationships, and how are these interrelationships affected by external or driving forces. Initially, it is necessary to take a simplified view of the system and to assume that many of the interractions or interrelationships may be disregarded. This simplification seems to come with difficulty to most biologists. Since systems analysis is a new tool to many biologists, a few comments are in order.

Systems analysis is not a cure for poor data. It is very useful as a means of not losing information in complex problems, but it cannot improve upon the quality of the original data. On the contrary, it may indicate when data are unnecessary, or when they are insufficient. Although models may appear extremely complex, they are usually built up from simple mathematical statements and statistical distributions which represent the functions, interrelationships, and values attributed to the real world ecosystem. These statements are translated into computer language to permit computer simulation of the system. Numerical simulation is the only known technique which is capable of representing the complexities of the real system. Some of the many approaches to modelling and analysis of ecological systems have been discussed elsewhere (Van Dyne, Innis & Swartzman 1971), but it is important here to consider model impact on project organization. Primary emphasis will be given only to dynamic models here.

One useful systems analysis procedure is sensitivity analysis of models. Sensitivity analysis is a method of determining the relative sensitivity of a model (i.e. system of equations) to changes in its constant coefficients, input functions, or the form of the equations of which it is composed. Sensitivity analysis may be performed by a systematic variation of a particular part of the model, while other parts retain their normal values. Contemplating making sensitivity analyses of complex ecological models leads to a consideration of the hardware, personnel, and organizational needs for such research. To make sensitivity analyses of complex models requires decisions about what variables to vary, in which combinations, and to what degree. Obviously, a high-speed computer is necessary. A great deal of time is required in running back and forth between the 'drawing board' and the computer. This predicates the need of a remote computer terminal rather than a centrally located, batch-processing computer system.

As examples of the quantitative relations are formed (i.e. model hypotheses) and modified, it is necessary for analysts and biologists to interact and check out these relations. There will be constants or parameters to be determined at this stage requiring only biological intuition, or perhaps a little library research, on the part of analytical specialists or biological specialists, or both. Whatever hardware is used, it must be possible to provide rapid graphic response to changes of parameters and equations so as not to lose the continuity of thought involved in an interactive working session between a biologist and a modelling specialist. This dictates another type of hardware especially necessary for this type of research, i.e., some rapid even though crude plotting system.

The corollary of the above considerations is that much of the modelling work must be done on a day-to-day basis in direct concert with the experimentalists conducting field and laboratory projects. Thus, only parts of the modelling effort can be let to contractors, consultants, etc. Most of the modelling must take place where there is a critical mass of both modellers and experimentalists to be efficient and rapid. Yet for political and other reasons, there is some desirability to disperse modelling efforts, and this is done in some of the biome programs (see Van Dyne, Innis, & Swartzman 1971).

In our preliminary work we recognized that ecological models have several features: ecosystem components, driving forces, parameters, processes, and controls. Our original conventions are listed in Table 1. Subsequently, as will be noted below, we have changed some of our formulations and concepts, but the items in Table 1 provided an initial focus for field and laboratory activities. Details of some of our initial modelling considerations are given by Van Dyne (1969*a*); an example is shown there of a total-system energy flow model in a difference equation format along with a computer program listing.

There has been considerable discussion in the ecological literature about ignoring simplicity of mathematical form and development of models for ecological phenomena. Over-concern for simplicity of form has led to a sacrifice of reality for the sake of generality and precision of expression (Levins 1966). Nevertheless, if approached in the proper manner, a generalized and concise mathematical notation has much to offer in terms of understandability, organization, and ease of manipulation and solution. Somewhat apologetically, we put forth in late 1968 an initial mathematical notation for describing and analyzing ecosystems. We realized it lacked formality, and in part, it may be incorrect mathematically, but it was a place to start! We have

Table 1. Conventions used in model formulation in early US IBP Grassland Biome studies on dynamic or simulation models. These conventions, developed late in 1968, were used in organizing ideas for differential equation system models.

A Vector of Ecosystem Components or Compartments (variables)

$v =$

v_1	live plants, above ground
v_2	live plants, below ground
v_3	standing dead vegetation
v_4	soil flora and fauna
v_5	humus
v_6	soil nitrogen compounds
v_7	litter, animal and plant
v_8	litter, flora and fauna
v_9	consumers
v_{10}	surface water
v_{11}	soil moisture
v_{12}	available soil minerals
v_{13}	unavailable soil minerals
v_{14}	soil hydrogen ion
v_{15}	litter moisture

A Vector of Driving Forces or Input Functions

$e =$

e	=
e_1	radiant energy
e_2	wind
e_3	atmospheric temperature
e_4	atmospheric humidity
e_5	atmospheric CO_2
e_6	external surface water
e_7	atmospheric O_2
e_8	atmospheric N_2

A Vector of Parameters or Properties

$p =$

p	=
p_1	soil temperature
p_2	litter temperature

A Set of Control Functions

$S =$

S	=
s_1	plant community effects
s_2	animal community effects
s_3	microbial community effects
s_4	human manipulation effects

(s_i represent subsets)

A Set of Processes

$Q =$

Q	=
q_1	photosynthesis
q_2	ingestion
q_3	assimilation
q_4	growth
q_5	reproduction–death
q_6	decomposition
q_7	emigration–immigration
q_8	physical processes
q_9	mobilization–immobilization

(q_i represent subsets)

continued work along this line, as discussed in later sections. Note here, however, that an operationally useful notation and methodology must cope with the use of maximum and minimum functions, step functions of independent and dependent variables, piece-wise linear equations, switching between formulae on the basis of variable values, and various combinations of these.

Some initial 'hang-ups'

There was a great initial lack of flexibility in the granting organizations toward the funding of our program. For example, no contingency funds were nor have yet been allowed. Yet, in any large effort of this type and of this time span, probably in the order of 10 per cent of the total funds should be available to be allocated as specific needs arise. In a program of this size, it is extremely difficult to predict in advance, and to the nearest dollar, the funds for each individual scientist or institution. This can be done, of course, but to a great degree it is only a paper exercise and an inefficient use of time.

The granting agencies did allow me, as the principal investigator, to subcontract to other individuals for work. This was not without mixed blessings, however. It required considerable effort to write subcontracts in specific terms to insure the work to be done, but not so 'iron-clad' so as to eliminate participants! In retrospect, I feel some of our contracts were not specific enough.

One difficulty encountered in the financial relationships in the first 15-month period was that grant funds became available actually in different segments. Therefore, it was impossible to plan with certainty at one time for most efficient use of the total funds finally obtained. In fact, a separate supplementary proposal was required three months after initiation of the study to get all of this money. Such difficulties and administrative activities were not foreseen by the early organizers and discussers of integrated ecological research programs. The management needs in such programs have been greatly underestimated, as far as I can tell, in all of the biome programs. Yet, 'management overhead' in this integrated research is much smaller than for an equivalent dollar volume of individual project grants to isolated scientists. There is a considerable efficiency in grouping projects from a budgetary standpoint, but even so, inexperience leads to underestimating the management costs.

One of the initial hang-ups not predicted by those enthusiastically involved in developing these large-scale biome programs was the 'personal factor'. In each biome there seems to have been realized, in initial efforts, an unforeseen magnitude of psychological hang-ups of participants toward working together. This is slowly being overcome in each instance through various management activities and plans. For example, in the US IBP Grassland Biome study, in the first 15 months, the series of five 2-day workshops held to review and synthesize information about grassland systems was useful in this respect. These workshops, held in isolated locations, each included participation of about 50 scientists. In this environment, there was much personal interchange, discussion, and 'getting to know each other'. A relatively good *esprit de corps* has been established, largely through these initial workshops, and it is a challenge to devise procedures to maintain this working attitude.

Personnel management problems in interdisciplinary, large-team, varying-organization research programs are not well understood. There are not many case examples to examine, but it is noteworthy that universities can boast of only few significant multidisciplinary accomplishments (Handler 1970, 1971). Multidisciplinary research can be done in the university environment, but it may require special structures and can provide certain advantages in goal-oriented programs. It is a well known fact that the mission agencies often have turned to the universities for certain special services, i.e., solution to segmentable problems. It is equally important to note that such service usually is rendered at less than full cost (DuBridge 1967). Thus, there can be considerable economy of effort if the universities can participate in the research program.

Let us analyze the people problem for large-scale, interdisciplinary ecological research because it is people that produce. Furthermore, we must seek creative and original scientists to participate in the team research program. Many creative scientists have characteristics which must be considered in the team-environment research. Creativity is a complex product of compulsion, audacious initiative, and sheer industriousness which will thrive only in certain climates (Kerr 1963). The creative scientist operates best in a psychological climate which encourages him to question, criticize, disagree, argue, and debate—but there is a limit before these activities become destructive. One of the management tasks is discerning this limit before it is passed. Passing the limit may have almost irreversible effects. Creativity also thrives when the rules of the game are known to all the participants—thus the age-old problem of adequate communication. But rules can't be written too tightly. Free choice rather than force typifies productive situations, and the way to get the best out of any scientist is to allow him to work on what interests him (Weaver 1969). A too tightly designed frontal attack is seldom as effective in basic research as a more broadly based approach capitalizing on individual interests, curiosities, and enthusiasms on issues basic to the problem at hand. So another important management task is to recognize and support a proper balance between research segments of low and high probability of immediate success and application to the overall theme. Creativity is fostered if individualism is encouraged yet we have recognized that the solution to the 'ecosystem equation' will require many scientists working together. Although highest productivity often comes from a single individual working alone, a single scientist cannot complete the task, and there are examples in science where intellectual interaction stimulates creativity (Kerr 1963). Thus, one management problem is to design a program to have opportunities for many scientists to work alone and for others to work in small groups. Research has shown that small groups have higher morale than large groups.

One general characteristic of creative individuals is their irregular or eccentric work schedules. This leads to problems when schedules must be met, certain facilities must be shared, and communication must be insured. For example the problem of meeting deadlines was discussed above. The communication problem is of special importance in a team research program where the scientists may come from more than two dozen geographic locations. In the American university environment the research organization is usually horizontal (Jensen 1969). There may be some vertical structure within a group organized around a senior professor with a large number of graduate students and possibly postdoctoral fellows, but there is seldom more than two levels in the hierarchy. In many research institutes and industrial laboratories there often is considerable vertical structure in the organization. There are certain advantages to either organizational framework, but under either structure research will be facilitated if there is ample lateral communication, especially and preferably among investigators of different disciplines. This, too, is a key management problem in these interdisciplinary, university-based, ecological research programs.

Even before we were first funded it became clear that the communication task was no small one. I wrote literally several hundred pages of letters, memoranda, prospecti, and proposals before we started the study. Soon after we received funds we initiated a Newsletter to assist in communication. This US IBP Grassland Biome study Newsletter is issued periodically, it contains reports of various meetings and items of wide interest, it lists various publications produced in our work, and it is now distributed to over 1300 scientists interested in the program.

A second structuring of research

In a program of this type there should be continual self-examination and change within the limits of budgetary constraints and procedural constraints imposed by the granting agency. Several of the shortcomings detected within the first 15-month segment of research were wholly or partially corrected within that time segment. However, most changes (and especially those requiring more funds or major reallocation of funds) come at times of initiating a new grant period. In this section, I will discuss our second grant phase, a 16-month period, but some consideration of the lead time is important here.

The grant cycle time problem

It takes from several weeks to months of evaluating subproject results before one can confidently suggest major changes in organization. These must be incorporated into a preliminary draft of a progress report-continuation proposal prior to within-biome analysis. Then revisions must be made prior to between-biome evaluation, which is accomplished largely by review of leaders, or designates, of integrated research programs (primarily biome studies) within the US IBP, i.e. a PROCOM or program committee review (Smith & Kadlec 1969).

I believe the most effective form of the IBP PROCOM review has been assembling the revised proposal after intrabiome review, circulating it to the between-biome group by mail, and then bringing the review group together for at least one full day to discuss the report-proposal. The purpose of this review is to make constructive criticism of the program, primarily concerning its technical and organizational content, but also to some extent its budgetary credibility. In the future, I hope and expect there may be more emphasis on the interbiome compatability and strengthening the character of these documents. The now weighty document must be revised again prior to submitting it to the granting agency. Note, by now many parts of the document have been revised at least three times. Considering that biome-type report-proposals have ranged from about 100,000 to 200,000 words, you can see the purely mechanical part of the job is no small task! We have used successfully magnetic tape typewriter systems to make the task bearable. And we have been most fortunate to have an understanding, enthusiastic, and efficient secretarial group. (One encouraging trend is that the granting agencies have accepted progressively shorter proposals, perhaps in self defense because the size of the progress reports in toto has grown at least exponentially!)

After our report-proposal reaches the National Science Foundation, we lose track of it! I know they sometimes receive all or several of the following inspections: internal review, mailed to panel reviewers, mailed to ad hoc reviewers (sometimes foreign), sent to other federal agencies for review, etc. Recently, we requested site reviews by their task force so we could get constructive as well as destructive comments from the granting agency. However, we are still learning, and there are more things we can do to improve the site review procedure for all parties concerned. After a proposal of this dollar magnitude clears all external and internal reviews and after highly probable revisions to a lesser budget, then it is submitted to the National Science Board for approval. The National Science Board is a 24-member advisory group reviewing the general policy of the National Science Foundation. Grants of $500,000 or larger must be approved by this Board. These meetings occur only a few times each year, and there is a required lead time for items to be included on their agenda.

Consider the implications of the typing, reviewing, and evaluation times required for a single year's grant. In January 1969 we submitted our report-proposal for a grant period beginning in September 1969 (April and the following January were the respective dates for the next grant). Thus, we have been requested to submit a report-proposal some eight to nine months prior to funding the next segment. Remember, too, probably a minimum of three revisions of the document are required intra-program before it is sent to the granting agency. When we are given a grant for only a 12-month period, we have much less than three to four months to accomplish work in the current grant before requesting the next one! Thus, although we are attempting within-the-year turnaround on research information within the biome (as discussed later), we can produce a progress report-continuation proposal containing essentially only one-year old information. A minimum of a two-year grant period would be much more realistic and efficient for this type of program once it has passed its initial growth phase.

Some changes in the second grant period

A detailed progress report-continuation proposal was submitted to the National Science Foundation in January 1969. This 700-page report and proposal called for numerous changes in the program and reported on some of the findings of the early efforts, from both a research standpoint and from an organizational and management standpoint. Note, however, that this report was submitted after only 7 months of the initial grant had transpired. This proposal presented a preliminary model (to be discussed below) as a basis for outlining the expanded program. Consideration of this model was helpful in justifying the intensive site study as well as the necessity of a comprehensive network program. Plans were included to expand the program to some 80 scientists and mechanisms to provide a cohesiveness among them. This proposal requested \$2·2 million for a 16-month period for 80 scientists, and emphasized the 'whole ecosystem' approach. The procedure for arriving at such a large budget request is of considerable interest.

Original requests from scientists for this grant period were in the order of \$3 million. Internal evaluation of the individual budget requests resulted largely in reductions, but with a few additions, and by eliminating considerable duplication and frills, the resultant total was about \$2·55 million. The surviving projects were reviewed in an early December meeting by the US IBP Grassland Program Scientific Coordinators, Analysis of Ecosystems project personnel, and two specialists on systems analysis. They trimmed it to \$2·2 million. The proposal was then reviewed before a mid-December meeting of Biome Directors, Analysis of Ecosystems project personnel, and US IBP International Coordination Committee personnel. In late December a 5-man review team was brought in for the US IBP Program Coordination Committee, and the resultant final request budget was arrived at, \$2·225 million for 16 months. This sequence is illustrative of the general procedure and tight time schedules used in evaluating budgets in these large, interdisciplinary ecological research programs.

We obtained our second grant for the period of September 1969 through December 1970 for a total of \$1·8 million, about 80 per cent of our request.

This increase in funding brought additional scientific personnel to the program, allowed initiation of a network study, and completed most of the construction and developmental activity on the field sites. This increased complexity required us to reorganize our organizational chart, to devise and revise a computer program to give us the budget information necessary for a proposal and for accounting, and attempt to streamline our contractual procedures.

During this grant segment, we completed our central processing and analysis laboratory. This facility is staffed with program technicians and programmers, and the facilities are also used by various investigators and their graduate students, both on-campus and off-campus personnel from throughout the program. We have found considerable program-wise efficiency in the purchase and use of equipment and in the hiring of personnel to accomplish such program efforts.

The central program staff was increased. The original plan, based on the October 1967 meeting of experts, was to have a Biome Director, Pawnee Site Project Director, and Comprehensive Network Project Director. The third of these positions was filled in October 1969 to complete the then-expected management team. Also, early during this grant segment, we were able to add a systems engineer and a part-time biometrician to complement our part-time computer-ecologist specialist. This largely completed the early planned mathematical group. Yet, as discussed below, our plans along this line also were inadequate. Data and reports began flowing rapidly during the second grant segment. We increased our secretarial support force and increased use of magnetic tape typewriter systems to facilitate the development and completion of the many reports and manuscripts on research efforts.

The addition of the network program brought with it attendant difficulties and costs in holding meetings, increases in travel, and the need to look at new means of communications. The difficulty of this problem is illustrated in Fig. 2 showing locations of many of the IBP grassland research sites in North America. Fortunately, Denver, Colorado is within ready driving distance of the US IBP Grassland Biome headquarters, and this is a major transportation center in Western North America. We evolved a system in which various meetings are held in motels at the Denver Airport. This allows many of the Fort Collins scientists and network scientists to travel in, attend a full-day meeting, and to travel home on the same night.

It had become clear that a large number of meetings were required to keep all participants (Biome, Pawnee, and Network) focusing on the central theme. It is important that each scientist knows in effect 'what the man on his left and the man on his right are doing and how it relates to him.' This means that considerable time must be spent in explaining results to one another, in questioning each other's methods, results and interpretations, and finally in

Figure 2. Locations of some of the major IBP grassland research field sites in North America. These sites include tall, mixed, and shortgrass prairies, desert and semidesert grasslands, mountain grasslands, Northwest prairie grasslands, and annual grasslands. The 'order' of the site refers to the kind and amount of experimentation conducted there. At first and second order sites studies are conducted at all trophic levels.

each scientist aiding in a synthesis toward a whole. In order to keep scientists thinking about the total program in this type of study, it would appear that something in the order of a dozen meetings of various types are needed per year. For a given scientist, these may include the grassland ecology seminar series we initiated for local Colorado-Wyoming and regional participants, review meetings, working meetings, and so forth. Additionally, many discussions are held in subject matter areas or between small groups of physical and biological specialists and mathematical analysts.

Recently, in planning some of the activities for 1971 studies in the US IBP Grassland Biome study, it became clear that for several scientists some

25 per cent of their funded time would be involved in 'personal interactions' (I prefer this term over 'meetings'!) Perhaps, the relative sequence of time spent on such activities for different groups of individuals is:

interbiome + international management > intrabiome management > integrators > coordinators > project scientists > programmers, technicians, graduate students

Unfortunately, time in the first two categories includes a high proportion in airplanes and in airports! For the third and fourth categories there should be a high percentage of the interaction time in small-group sessions. For the last two categories much of the interaction time is in large-group sessions in information meetings. Not too long ago, I had the opportunity to be involved 'hands on' in field, laboratory, and armchair efforts, but now I am restricted primarily to the first two categories above. Although these are necessary and I hope useful activities, I feel frustrated because I have not lost the urge for hands-on involvement.

A technique used to facilitate information exchange and 'togetherness' was the development of a series of grassland ecology seminars. These seminars rotate among academic institutions in the Colorado-Wyoming area from Southern Colorado State College at Pueblo on the southern end to the University of Wyoming at Laramie on the northern end, a distance apart of about 240 miles. Seven of the seminars were held in the program's second phase, i.e. the 16-month period. The seminars provide an opportunity for the participants to visit during a social hour and the evening meal and then to hear two presentations on grassland research. Attendance has averaged more than 50 scientists and graduate students.

Objective analysis of models and experiments

In a program of this magnitude and diversity, we are continually faced with making management decisions based on much-less-than certain information (!) on priorities or relative usefulness of different components of the program. We did some of this in our initial studies in a PERT analysis, as described above. This decision-making process is extremely difficult because of the many subjective elements and because of the number of interactions among the different phases of the program. With respect to the systems analysis phase of our work, one might ask 'given the experimental results, which of the modelling efforts would be most beneficial?' Alternatively, one can ask the *dual* of this question, 'given the kinds of models projected, which of the experimental projects are most useful?' Furthermore, especially in many of the experimental project areas, there is a critical amount of financial support which is needed before the project can even be attempted. Even though the decisions are difficult, and even though the answers will be partially subjective, it is important to attempt at least a semi-quantitative evaluation of the usefulness of individual research projects, e.g. on the Pawnee Site, in the Comprehensive Network, and in individual modelling efforts. We devised a scheme to guide us early in 1970 for interrelating these elements of the management decision. This was discussed along with evaluation of modelling ideas (Swartzman 1970) and is described below.

Evaluating individual models

Any resource allocation must be based on the ultimate value of the project goal and its subdivisions (separate models). Since it is difficult to objectively measure the value of the individual models, we must settle for a subjective measure based on certain guidelines, as follows: (i) the value of a model is relative only to another model and has no interpretation in absolute units such as dollars or Ph.D.-man-hours; (ii) a model for a whole ecosystem or a multitrophic level or multi-component model is more valuable than a model for a process or small part of an ecosystem; and (iii) a model (at any level) having high resolution is more valuable than a model with low resolution.

We shall let W_j be the relative value or importance of the jth model. Given that the necessary experimental data are available, a particular model will have a certain probability of completion or success. We define this probability as M_j for the jth model and emphasize that M_j is conditional upon completion of the required experimental work. M_j is not a certainty (i.e., $M_j \neq 1$.) since there are other random events, e.g., availability of analysts and programmers, or probability that the experimental data can be successfully interpreted in terms of the model, upon which success of a model is dependent. Notice that the set of probabilities M_j is based upon events which are not independent, i.e., the successful conclusion of the kth model would require the success of models i and j if model k is a superset containing i and j.

Evaluating experimental projects

Suppose that the ith subcontracted experimental project has probability E_i of successful completion. Let S_{ij} be defined as follows:

$$S_{ij} = \begin{cases} 1 & \text{if the } i\text{th project will contribute data necessary for the } j\text{th model} \\ 0 & \text{otherwise} \end{cases}$$

We shall assume that the probabilities E_i are based upon independent events.

Then the absolute probability, P_j, of successful completion of the jth model is

$$P_j = M_j \cdot \prod_{\substack{i=1 \\ S_{ij} \neq 0}}^{n} E_i$$

where n is the number of projects. A measure of the value of the ith project, V_i, is

$$V_i = \sum_{j=1}^{m} S_{ij} P_j W_j$$

where m is the total number of models. Based on this measure, we can provide a provisional estimate of the dollar value of each experimental project, $\$_i$, under the restriction that the total amount of money available to the experimental aspects of the program is $\sum_{i=1}^{n} \$_i$.

$$\$_i = V_i \frac{\sum_{i=1}^{n} \$_i}{\sum_{i=1}^{n} V_i}$$

Numerical example

Long-term plans called for three models of minimal, intermediate, and maximal resolution for each of four trophic levels. We also planned three Pawnee Site whole ecosystem models with varying degrees of resolution, and two generalized grassland ecosystem models for purposes of comparing sites nationally and internationally, thus $m = 17$. For experimental projects we grouped studies into the following experiments.

Abiotic	
Gross meteorology	1
Soil physical	2
Soil chemistry	3
Producer	
Standing crop, aboveground	4
Standing crop, belowground	5
Plant physiology and phenology	6

Consumer	
Wild	
Animal biomass and numbers	7
Animal diets and consumption	8
Domestic	
Consumption per animal unit	9
Biomass per animal unit	10
Decomposer	
Activity of soil microorganisms	11
Activity of soil fauna	12
Dead organic soil constituents	13
Above ground dead material	14

Using the guidelines listed above and realizing that the research manager assigning probabilities is cognizant of such factors as an experimental scientist's past performance, difficulties in making measurements of certain types, difficulties in interpreting data, etc., we constructed the following hypothetical matrix (Table 2).

The model weights were assigned on an intuitive basis taking into consideration such things as possible uses for the model in solving economic and environmental problems, etc., the resulting relative values (V_i) are reduced, in the $V_i/\Sigma V_i$ column, to a basis of 1·0 total value for the whole experimental project.

Since there are 14 experimental subprojects in this calculation, comparison of $V_i/\Sigma V_i$ with $1/14 = \cdot 0714$ will indicate whether the project is more or less valuable under the above scheme than under an assumption of equal value for all subprojects. Using the non-parametric Chi-square test on the V_i's, we get a statistic of $48\cdot 9 = \sum_{i=1}^{14} (V_i - 542/14)^2/(542/14)$ which exceeds the 99th percentile critical value (27·69) with 43 degrees of freedom. This indicates that the model gives results significantly different from the assumption of equal project values.

The ranking of the 14 experiments (according to value 1 ≡ most valuable) gave results not out of line with what one would expect, i.e. aboveground herbage measurement was the most important project and soil fauna the least important. The weights and probabilities were assigned only once before calculations were made and were not readjusted to give 'more reasonable' results for purposes of this presentation. This information was used subjectively in planning budgets.

Table 2. An example of a procedure for quantitatively interrelating experimental projects and mathematical modelling activities in a cost-effectiveness analysis.

	Model Names																	
		Pawnee			General Grass-land		Abiotic			Producer			Consumer			Decom-poser		
	Model Resolu-tion	1	2	3	1	2	1	2	3	1	2	3	1	2	3	1	2	3
Experi-ments	E_i	1	2	3	4	5	6	7	8	9	10	11	12	13	14	15	16	17
1	·999	1	1	1	1	1	1	1	1	0	0	0	0	0	0	0	0	0
2	·99	0	1	1	0	1	0	1	1	0	0	0	0	0	0	0	0	0
3	·98	0	1	1	0	0	0	0	1	0	0	0	0	0	0	0	0	0
4	·95	1	1	1	1	1	0	0	1	1	1	1	0	0	0	0	0	0
5	·90	0	1	1	0	0	0	0	1	0	1	1	0	0	0	0	0	0
6	·98	0	1	1	0	1	0	0	1	1	1	1	0	0	0	0	0	0
7	·80	0	1	1	0	1	0	0	0	0	0	0	1	1	1	0	0	0
i 8	·90	0	1	1	0	1	0	0	0	0	0	0	0	1	1	0	0	0
9	·98	1	1	1	1	1	0	0	0	0	0	0	1	1	1	0	0	0
10	·99	1	1	1	1	1	0	0	0	0	0	0	1	1	1	0	0	0
11	·80	0	1	1	0	0	0	0	0	0	0	0	0	0	0	1	1	1
12	·85	0	0	1	0	0	0	0	0	0	0	0	0	0	0	0	1	1
13	·93	0	1	1	0	1	0	1	1	0	0	0	0	0	0	1	1	1
14	·95	1	1	1	1	1	0	0	1	0	0	0	0	0	0	1	1	1
	M_j	·90	·85	·75	·90	·85	·99	·95	·90	·99	·90	·85	·99	·90	·80	·90	·80	·70
	W_j	5	20	50	10	20	1	2	3	2	5	10	1	4	8	4	6	8
	P_j	·79	·46	·28	·79	·48	·99	·87	·65	·93	·75	·71	·77	·63	·56	·64	·48	·42

	V_i	$V_i/\Sigma V_i$	Rank	Experiment
1	49·33	·0909	5	Abiotic—Gross meteorology
2	36·49	·0673	10	Abiotic—Soil physical
3	25·15	0464	13	Abiotic—Soil chemistry
4	59·31	·1093	1	Producer—Standing crop, aboveground
5	36·00	·0664	11	Producer—Standing crop, belowground
6	47·46	·0875	6	Producer—Plant physiology and phenology
7	40·57	·0748	8	Consumer—Wild animal biomass and numbers
8	39·80	·0734	9	Consumer—Wild animal diets and consumption
9	52·42	·0966	3	Consumer—Domestic consumption per animal unit
10	52·42	·0966	4	Consumer—Domestic biomass per animal unit
11	32·00	·0590	12	Decomposer—Activity of soil microorganisms
12	20·24	·0373	14	Decomposer—Activity of soil fauna
13	45·29	·0835	7	Decomposer—Dead organic soil constituents
14	55·40	·1021	2	Decomposer—Aboveground dead material

$\Sigma V_i = 542{\cdot}55$

$$X^2 = 48{\cdot}876 = \sum_{i=1}^{14} \frac{\left(V_i = \frac{542{\cdot}55}{14}\right)^2}{542{\cdot}55114}$$

Deficiencies of the method

The method includes several assumptions and perhaps modifications can be devised to circumvent some of these deficiencies: (i) Models are assumed to be 'discrete objects' when in fact a high resolution model might develop in small steps from a lower resolution form; (ii) A model is assumed to be possible only when all of the necessary data are available. In fact, failure of a specific project would mean substitution of sections of the model receiving data from that project so that the whole model still would be possible although at a slightly lower degree of resolution; (iii) It is unlikely that an entire experimental project would fail. More likely is the case that certain types of data would become unavailable out of the several collected by any one experiment; (iv) The model probabilities do not take into account the interdependence of different models. Thus, Pawnee model 2 might be constructed by interfacing the four trophic level models at the 2 stage. A modification of the formula for P_j might be able to take this into account, but more detailed model interrelations would have to be specified.

A comprehensive diagrammatic framework for dynamic models

A major achievement during this grant segment in the systems analysis area was the development in Spring 1970 of a preliminary diagrammatic framework for depicting the main functional and structural aspects of grassland ecosystems. In contrast to previously used 'box-and-arrow diagrams', we converted information about ecosystems into a hierarchial representation allowing clear presentation of a great amount of detail. Such diagrams are used to structure the various processes that influence energy flow between compartments in the system (Swartzman *et al.* 1970).

The organization of a hierarchical diagram is from the left to the right along each figure segment. On the extreme left are the outputs of the particular compartment while to the right of these elements are those intermediate variables (processes), parameters (descriptors), or forcing functions which influence those elements to the left of them. For example the producer biomass is directly influenced, among other factors, by the physiological functions of reproduction, mortality, and photosynthesis.

Elements in a diagram on the right side of each hierarchical chain generally fall into one of three categories. First is that of forcing functions which are inputs from other compartments which may, for purposes of a hierarchical diagram at hand, be assumed to influence the compartment relatively independent of what occurs within the compartment. Second, parameters and descriptors are generally determined from the field data. Third is the feedback element which occurs if an element which appears toward the left-hand side of the diagram also happens to exert a direct influence on some element to the right of it. Indirectly, then, it is exerting an effect on itself, and thus, may be used as a feedback element.

Our hierarchical diagrams are *not unique*; other organizational schemes may also be developed which are equally or more valid. The major purpose of these diagrams is to provide a framework within which a set of dynamic models can be developed. Such a set could completely represent the ecosystem and would focus on the interactions between various compartments. Once a set of models is developed using these diagrams, the models can be interrelated using a matrix, and the matrix of interrelations can be used to place the models in the relative perspective of importance (see, for example, the matrix interrelating models and experiments in Table 2).

Varying and evolving roles of participants

By this stage of development in our program (Autumn 1969), we had recognized separate roles of work in subproject research (Pawnee or Network studies), scientific coordination (by a 9-man group of experimentalists), systems analysis, data and sample processing, biome management, and interbiome and international coordination. Some individual scientists participate in more than one of these roles. Yet a new understanding of the role of participants was developing.

A few external consultants were brought in at our November 1969 annual review meeting. They provided us with important perspective on the organization and structure of our program. Their ideas, which were developed independent of, but reinforced, our own, were then discussed with university administrators and with granting agency personnel. One factor emerged clearly. Answers to questions about ecosystem structure and function require detailed integration of experimental studies, of literature, and synthesis of parts into the whole (note the difference between collective vs. collected studies). This is not accomplished readily by graduate students, by technicians, by programmers, or by support staff. It requires the dedication of knowledgeable and imaginative senior scientists functioning almost entirely in an '*integrating capacity*' rather than in a '*data-generating capacity*' or a '*coordinating capacity*'. We were completely missing such scientists! Furthermore, all but three of our scientists at that time were dedicated to subprojects or phases rather than being dedicated essentially to the total program.

Our experience and analysis, which was confirmed externally, showed between 10 and 20 per cent of the scientists in a program of this type should be dedicated to the program per se rather than to individual subprojects. This would permit 80 to 90 per cent of the scientists to focus primarily on individual subprojects. Although there are exceptions, probably about 75 per cent of the time of the program-oriented scientists should be devoted to the biome study, whereas the project-oriented scientists should dedicate perhaps 25 per cent of their time to a subproject in the biome study. Both groups are necessary for the program to function effectively. There are several major administrative, public relations, coordination, and developmental activities, as well as the details of scientific integration and synthesis, that require the attention of a critical mass of scientists dedicated to the program for the majority of their time.

The structure that results from the above plan is somewhat intermediate between that of a national laboratory or experiment station arrangement, where essentially each scientist works full time in a given organization, and that of the university situation, where research is primarily only a part-time effort. In contrast, the ability to utilize the scientific expertise of a variety of fields necessary for an ecosystem study capitalizes upon availability for summer efforts by many university investigators. The university environment for housing a program such as the US IBP Grassland Biome study maximizes the use of graduate students, thus providing training in a broad sense, as well as high quality research assistance. Since the program is housed in the university, perhaps no scientist should dedicate 100 per cent of his time to the program. Scientists are attracted to the university environment because of the opportunity to participate in some teaching and graduate student direction. Even if the program were not housed in the university environment, probably 100 per cent commitment to a single activity would lead to tedium, inefficiency, and stagnation.

With these considerations in mind, a new staffing plan and arrangement of existing duties was developed. Basically, we now plan for at least a dozen senior scientists in our program to dedicate about 75 per cent of their time to the biome study. These programmatic efforts include program management, scientific coordination, direction of service activities, and detailed integration and synthesis across subprojects by and between trophic levels.

Critical needs of university-based integrated research programs

Further analysis of our study suggested there were three components to consider in providing an organization that would insure success of the program. We have recognized above that a certain *critical mass* of scientific investigators dedicated to the program is needed, but they must be housed in a *critical space*, and they must operate under a *critical management* or administrative structure.

There are advantages and disadvantages of different types of institutions relative to the conduct of integrated ecological research on total systems. Lack of rapidity of change is one of the disadvantages of university systems. However, universities do change, albeit slowly and painfully, and there are recent indications and pressures that changes will become more rapid. Consider the 'typical' land grant institution where a wide variety of scientists interested in ecological research are available in the natural science, agricultural, engineering, and natural resource disciplines. These scientists, of course, have access to a wide variety of facilities, skills, and knowledge in a large number of separate departments. As noted above, however, we feel some 10 to 20 per cent of the scientists in an integrated research program should be more or less 'dedicated' to the program rather than to individual subprojects. Now, consider this in relationship to the past National Science Foundation granting procedure. Most of our scientists in a university are on a 9-month academic appointment in most of the colleges listed above except agriculture. On a 9-month appointment, the investigator may obtain 2-month summer salary segment through an NSF grant. In the past, if he were to pay part of his academic year's salary, that was to be cost-shared on a 50:50 basis. I believe that in most land grant institutions, personnel on academic appointments are more or less fully saturated with formal undergraduate and graduate teaching, advising, and other university activities. This leaves little time for research and much of their research is 'bootlegged'. In some special cases, of course, teaching loads are light, and facilities and support are provided for research, but this seems to be a less and less common case. Therefore, if interested scientists are to dedicate the necessary time, perhaps 75 per cent of their year, to the total program activities, they are in conflict with university procedures. This soon hits them where it hurts the most, i.e. in the paycheck, for they are judged by a different set of standards. Let me elaborate.

In many of the land grant universities, the departmental chairman is a key figure in the evaluation of promotion and salary of the individual scientist. It is my impression that the departmental chairman will evaluate a scientist first on his undergraduate teaching load and effectiveness, second on his formal graduate teaching work, third on departmentally-oriented research, and lastly on interdisciplinary research. I admit this is a simplified view, but the priorities fall along those lines. Consider a scientist who is spending perhaps nine months a year on an integrated research program. Even though he may be doing a good job in that activity, he is, in effect, working toward a goal that is broader than that of the department, broader than that of the college, and probably broader than that of a single university. Thus, the conflict arises.

Many universities have organized various institutes or centers which are interdisciplinary in nature. Certainly not all of these centers have been effective, and there are certain key factors affecting their success. In order for multidisciplinary training programs to succeed, apparently, some special factors must be considered. The requirements seem similar for research programs. The following statements are presented here, with self-evident meaning, from a review in *Science* of an Office of Science and Technology report (Steinhart & Cherniack 1969) concerned with training programs related to environmental quality.

> 'Curriculum, faculty rewards, and most of the research have been controlled within the departments representing the narrow academic disciplines. These departments grow narrower and more numerous year by year as the advance of modern science results in increasing specialization.'
> '. . . those few centers or other units found to have genuinely effective multidisciplinary programs all had two things in common —they had substantial influence or complete control over faculty hiring, promotions, and other rewards, and they enjoyed flexibility in introducing new programs. Also, in most cases the successful programs were found to have the direct support of one of the university's more senior administrators who could help provide resources and protect the programs from "traditionally minded faculty members".'

Logically, if scientists who are working on interdisciplinary programs can be associated with such centers, they can be evaluated more effectively according to the total program rather than to a particular department's goals.

We have been fortunate to have good working relationships with several departments and colleges at Colorado State University, yet we have felt it desirable to transfer the scientists working in this central part of our program into an interdisciplinary center. Therefore, our university Administration, through approval of the State Board, formally established the Natural Resource Ecology Laboratory within an interdisciplinary center, the Environmental Resources Center. Each of the senior scientists on regular faculty appointments who are in our laboratory also have a joint appointment with some academic department. Graduate research assistants are funded by our program but obtain their degree through an academic department. Support staff of various types including post-doctorals, programmers, technicians, secretarial-clerical, etc., may be fully funded within the Natural Resource Ecology Laboratory.

NSF's review panel (mentioned above) and their National Science Board and other outside consultants recognized the need of having a critical mass of staff dedicated to the program. The NSF approved the addition of several new scientists to our program in integrative capacities. We are now negotiating positions for each of these individuals jointly with some departments. Approximately three months of the academic year should be in an established academic department and the remainder of their time in the Natural Resource

Ecology Laboratory. This still presents some difficulty in hiring dedicated senior people.

The new faculty member is, of course, most interested in the nature and challenge of his work and the facilities and opportunities for conducting the work. However, there are some important 'secondary' considerations including the four major items of salary, rank, tenure, and sabbatical leave! These latter factors are interrelated in the following ways. Certainly, there is a clear, generally direct relationship between salary and rank. There are also important interrelations between salary and tenure and between salary and sabbatical. In many institutions there are several classifications of academic appointments. The regular appointments are ones in which the scientist is hired in a full time, academic teaching role. This automatically includes rather clear-cut responsibilities, the opportunity to earn tenure, and the opportunity to accumulate time for sabbatical leave. Likewise there are temporary appointments which bring no sabbatical or tenure and for which there are clear-cut responsibilities. Intermediate or special appointments are found in many universities in which the individual is largely working on research but may or may not have a tenurable position or qualify for sabbatical leave. It is in this latter case that most of the integrative scientists might find themselves in many universities. In effect, they must answer to two masters. University scientists, in many fields, feel tenure is quite important and are willing to accept lower salary rates for tenurable positions than otherwise. Sabbatical leaves, providing at least half-time pay, are often earned every seventh year in many universities. If this sabbatical privilege is not available, the scientist should receive greater financial compensation.

Now consider the problem of the duration of a long-term integrated research program and its effect on personnel in interdisciplinary centers in universities. The analysis of the needs of an integrated research program call for senior investigators. Yet when approaching the usual academic departments, that is not the typical man who is wanted. Most departmental members, and some chairmen even, seem to want to hire a 'fresh' Ph.D., put him at the bottom of their 'departmental peck order', shove off unwanted jobs and courses on him, and, in this way, elevate their own relative positions! Relatively few faculty seem to want to bring in senior and highly competitive men. Perhaps they are a threat to the security and status quo. Consider also the funding of the research programs as described above. Generally, we get our funding only one year at a time; there is no guarantee of long-term commitment by the granting agency. Therefore, the individual scientist is justified in seeking a higher than average salary. Yet, if his goal is to eventually merge into a full-time departmental appointment, this higher salary is a detriment.

The above problems are difficult but resolvable. I believe the success or failure of these long-term integrated research programs rest on the type of integrative scientists I have described above and the activities of integration and synthesis. There seems to be no shortage of scientists who are willing to go to the field or lab and generate more data. There are relatively few who are willing to condense and integrate these data. Can such interdisciplinary/multidisciplinary scientists be found? Can other than relatively senior people do this job? It is possible that post-doctoral students can function effectively in these integrative roles, but it is rarely possible for graduate research assistants to accomplish the task because they lack time and experience. In any event, continuity is important and therefore, a senior scientist expected to be with the program for a number of years should be in charge of each of these areas. This, then, rules out visiting scientists, post-doctorals, and graduate research assistants as the key individuals. Perhaps we are seeking 'super ecologists' for these integrative roles?

A large-scale program of this type must make provisions for lateral as well as vertical communication, evaluation, and feedback. Lateral communication is facilitated by the many working, reporting, and planning meetings held each year by the activities of the scientific coordinators, and by the specific role of the integrators.

Scientific integrity and coordination should carry across all field and laboratory phases of the research program as well as in the systems analysis efforts. As noted on our current, abbreviated 1971 organizational chart (Fig. 3), there are *needs for scientific integration, integrity, and balance across each of the abiotic, producer, consumer, and decomposer areas.* Naturally, there are interactions and overlaps between these trophic levels. More than one scientist may work in each of these levels in integration and coordination and especially in synthesis. Some of these activities can be accomplished by contributions from short-term people, such as visiting scientists or consultants. Whoever the scientific integrators are, they should be particularly competent in multiple subject matter areas but also with major strength in one of the four trophic levels.

These integrator scientists must also assist in the scientific public relations of the program and evaluation of studies. They can make their input in at least three major ways; each integrator could and should talk to individual scientists in experimental projects, to Directors of each of the other four program areas, and report to the Integration Director (Fig. 3). They can also participate as members of the mathematical modelling teams. They should be particularly helpful in bridging the communication gap between the program analysts and experimenters.

In addition to the personnel, space, and organizational requirements for these large-scale programs to be housed in a university system, the financial and accounting system merits consideration. In these biome studies the entire grant goes to a single institution, and the Biome Director is responsible for its

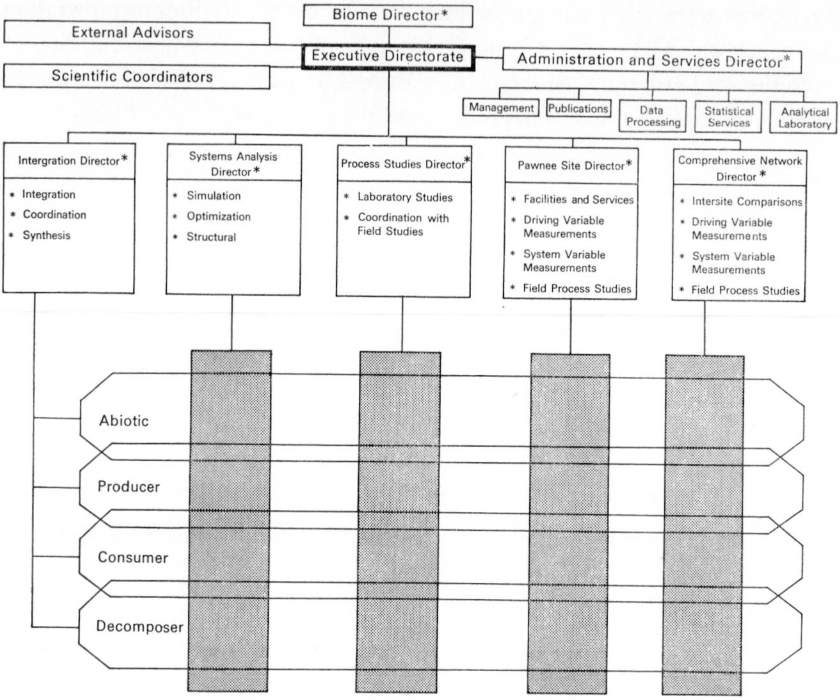

Figure 3. The 1971 US IBP Grassland Biome organizational chart illustrating the need for a broad management and administrative responsibility (executive directorate), scientific integrity across all program phases (integration activity), support activities (administration and services), and well grouped experimental activities (Systems Analysis, Process studies, Pawnee Site studies, and Comprehensive Network studies).

allocation and use. He, in turn, may subcontract a large percentage of funds for specific researches to individual scientists in various institutions and agencies. It is necessary to follow this scheme in order to have a cohesive study. But, we have found many of the bookkeeping and accounting systems of many universities, especially including my own although it is improving, are not sufficient to the task. Externally to some, the amount of funds in these programs may seem large, but we feel the number of valid requests and needs greatly exceed the available funds. Therefore, we must allocate and use our funds judiciously and we must have rapid and accurate accounts of expenditures and incumberances, expecially because our grant now is on a 12-month basis. The universities or organizations involved must provide adequate accounting systems, which are paid for in part of the indirect cost charges, or else take the chance of losing the program from the institution.

Originality v. repetitive mediocrity

Early in the IBP studies there was fear among some that there would be inadequate opportunity for scientists to express 'individuality' because they must participate in an at least partially goal-oriented program. Projects must mesh in time and space, results must be produced, reduced, and integrated at regular intervals, and much of the work must have a very high probability of success. Therefore, there was fear that many scientists must work as 'super-technicians' subjugating their individual interests to that of the common cause. I believe that these fears have largely been dispelled. A central core of highly dependable work must be carried on, but as we dig deeper into ecosystems, we find more and more important interrelationships must be examined for the first time anywhere. There is much opportunity for originality and innovation.

In one year a study may be exploratory on one site, but in the subsequent years it may become a standardized investigation across several sites. A great deal of thought and planning must be put into both the 'test phase' and the 'operational phase' with respect to a given type of measurement. Techniques must be derived which are applicable across a wide variety of situations within the biome to make it useful in comparative studies. This alone is a task sufficiently great to challenge the imagination and inventiveness of many scientists. Even relatively well established techniques and procedures must be interpreted in the local situation in a way to provide the greatest overall programmatic benefits. This generally requires the insight and experience of a senior scientist and cannot be accomplished only with technicians or graduate research assistants.

Can individual scientists be challenged by such activities? Many can, but other scientists feel they must have a little 'scientific overhead' in their subcontract to justify their participation. Of course, the independence of the scientist is conditioned by the general availability of research funds throughout the nation! I estimate that approximately 25 per cent of our program's scientists would not be able to secure funds independently from agencies such as the National Science Foundation even in years of good 'research fund climate'. Perhaps 40 per cent could secure individual grant support in years of good funding, but not in bad. The remaining 35 per cent probably could secure individual research funds even in bad years. But this is realism, for not everyone is an Einstein of ecology! A program, of course, is only as good as its participating scientists. But *it is important that it be judged as a total program, not as an isolated individual project.*

It is clear that the interactions promoted by these interdisciplinary programs are highly beneficial in 'retreading', or at least modifying, the ideas of many individual scientists. There are dual benefits because many highly innovative and productive scientists can profit also. In the usual case, a scientist is limited to his own 'two hands', or some small multiple thereof, in carrying out his ideas. If the scientist is secure emotionally, if he is imaginative, and if he is skillful in communicating his ideas, he in effect has gained a large group of hands to help carry out his ideas! Thus, both the 'idea-generator type' and the 'idea-needer type' are benefited by cooperation in an interdisciplinary, large-scale program.

Division of labor and the reward system

As we continue now to plan coordinated attacks in research and management on natural resource ecology problems, we find that integrated, interdisciplinary teams are essential. We also must note that to accomplish the goals of these programs there must be a division of labor (Van Dyne 1971*b*): (i) some team members must focus on *collection of samples and data*, perhaps without participating in detailed analyses; (ii) some must focus on *sample and data analysis and modelling*, perhaps foregoing field and laboratory studies; (iii) other members must concentrate on *synthesis of results*; and (iv) yet others must work in *program management*, *development*, *and coordination.* This division of labor and specialization of effort is well established in many industrial areas, in physical science, and in engineering efforts. Perhaps this approach is used sometimes, but is not always successful, in research and management programs of some of the state and federal resource management agencies. But seldom is this cooperation or philosophy evident in university efforts.

The lack of quick acceptance of this division of labor by biologists and natural resource scientists stems largely from the history of their fields. Most of the major biological and resource accomplishments in the past have been made by a scientist working alone, or at most, a few scientists working together. But they did not often tackle problems of total-ecosystem complexity. Furthermore, most of today's university professors in biological and resource areas have not participated in team efforts. So they often hold sacrosanct the concept of highly individual contributions to science. Perhaps even more influential is the fact that the backgrounds and attitudes of the university administrators often are the same. And they control many of the rewards in the system. Thus, most students who mature in the university environment, either as an undergraduate or as a graduate, come to the real world somewhat unprepared to contribute fully to team research or management. They do not realize that often they can make greater contributions through services to the group than through entirely individualistic efforts. For example, sample and data processing and analysis efforts may often be the limiting factor in ecosystem research. Team members devoting their time and energies to such efforts can make significant contributions for the overall effort. Eventually, our reward system in universities will better recognize program efforts, and more students will learn this and will more effectively participate during their career in large, integrated research and management programs.

Types and roles of modelling

Ideally, our work starts primarily with the development of experimental studies of ecological, physiological, and physical processes. Additionally, detailed examination is made of the literature and other sources of information. Along with the output of the process studies, this information is used in developing three major types of mathematical models—*simulation*, *optimization*, and *structural*. Greatest emphasis to date in most of the biome studies has been placed on simulation models. For example, in the US IBP Grassland Biome study we are developing a series of simulation models of various degrees of resolution and precision. But we are also initiating work on optimization models and work on structural models (Fig. 4).

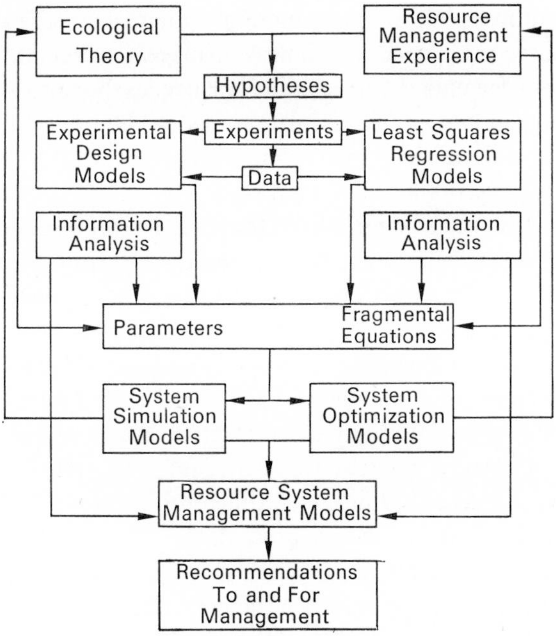

Figure 4. Many of the long-term benefits to man of the large, interdisciplinary biome-type studies will come through improved natural resource management. Mathematical models of various types will be needed to help bridge the gap between ecological theory and management experience, detailed ecological research studies, and management tactics and strategy. Some roles of three types of models are shown here—statistical (i.e. part of 'structural'), optimization, and simulation.

Many of our experiments, based on the hypotheses derived from ecological theory and resource management experience, result in data which are analyzed by either experimental design models (i.e. conventional analyses of variance models) or least squares prediction models. These statistical models represent only one type of 'structural' models. One of their primary purposes, in a total resource system analysis, is to derive parameters and fragmental equations used to structure either simulation models or optimization models (Fig. 4). Much of what has been said in this paper has been concerned with the organization aspects of a research program to collect data for development of simulation models and data to validate these models. Yet, optimization models must also be included to systematically and efficiently consider man's impact on ecological systems, and there are indications in most of the biome studies that these models will be developed. In the US IBP Grassland Biome study we are accelerating our efforts in 1971 in this phase, both independent of an integrated with simulation modelling. Eventually, we will need to combine systems simulation models with system optimization models into what are called in Fig. 4, 'resource system management models'. As noted early in this paper, I feel these large, interdisciplinary research programs will result in both development of new, improved, or more quantitative ecological theory and to better resource management. Resource management requires not only an understanding of the underlying ecological principles, but also an ability to predict system dynamics (e.g. via simulation models) and to select from a large number of alternatives those management strategies which will maximize benefits to man (e.g. via optimization models).

Simulation models are constructed largely from the results of process and literature studies, and from accumulated experience about ecosystems. The process studies may be in the field, growth chambers, greenhouse, or metabolism apparatus. The studies are designed to give the form of, and parameter values in, equations giving the rate process as a function of other variables. The models are largely systems of simultaneous differential and algebraic equations. These equations, when solved, yield predicted output or performance of the system. Simulation models, especially, must be tested or validated against results from *independent* field experiments. Therefore, field experiments are essential to the overall modelling process.

Field validation studies for simulation modelling may be of two types. First, and generally more frequently, we measure the components, i.e., the compartments, in the system during the year. Here we seek a time sequence of measurements on particular compartments in the field to test against results from the model. We measure driving variables in the field and use these data to drive the models and see if we can produce with the models the same results that have been produced in the field. Second, field validation studies may be of individual processes themselves. However, the investigators and the program can be accused of '*circularity*' if improper design is followed. If one uses the field measurements of driving variables along with field measurements of processes to design and run a simulation model, and then tests the output against field measurements of the components, from the same system in which the processes were measured, then circularity occurs. Therefore, in any given year, measurements on the field sites provide driving variables and validation of components to be tested against the model derived from the literature, general knowledge, and perhaps independent process measurements from previous years. This procedure, leaves at least some gap in the circularity loop.

The above discussion shows another implicit component in our modelling philosophy; there must be feedback between field studies, laboratory studies, and armchair studies. For any given year, the models are built, in a large part, from independent data and information, driven with field-measured driving variables, simulated in the computer, and the results tested against data from separate field studies on the components. This leads us each year to re-designing the field studies and redesigning the model. But it appears we may be limited to one turnaround a year with such feedback. Through several such turnarounds we hope to develop, test, and modify models until we are satisfied with their general performance characteristics. Then it is possible to use probabilitistic generators derived from analysis of long-term records for driving variables and to conduct a great variety of experiments on the models themselves. This will be useful in testing basic ecological theory and in deriving resource management implications (Fig. 4).

A special problem with modelling is that the literature is beginning to proliferate with 'quick-and-dirty' mathematical models on small-scale ecological problems or topics (e.g. Donnelly & Armstrong 1968, Goodall 1969; Van Dyne 1969*b*, Patten 1971, Timmen 1970). But, because of the model format or the narrow approach, the computer programs of these models have not produced output on large, real-life problems. Most early attempts have focused on deterministic, linear, constant-coefficient systems of equations and have utilized either incremental or difference equation approaches. Frequently, such models attempt to fit biology into a standard mathematical form as opposed to letting the biology dictate the mathematics! A large proportion of the existing models are little more than computer programming and analysis exercises which fail to relate to realistic biological mechanisms. The main factor of importance here, however, is that programs for such models run! It is also clear that as one goes to more realistic models, i.e. *as one makes them more mechanistic and data-based, the amount of effort required goes up by several orders of magnitude*. Thus, it is possible for one individual in a short period of time to sit down and develop, say, perhaps a system of ten first-order, constant-coefficient linear differential equations for a model, and use a constant input as the driving variable. But a model with perhaps 50 differential equations, and many of them nonlinear, with many mechanisms involved, and with complex driving inputs, may be several man-years of work (Fig. 5)!

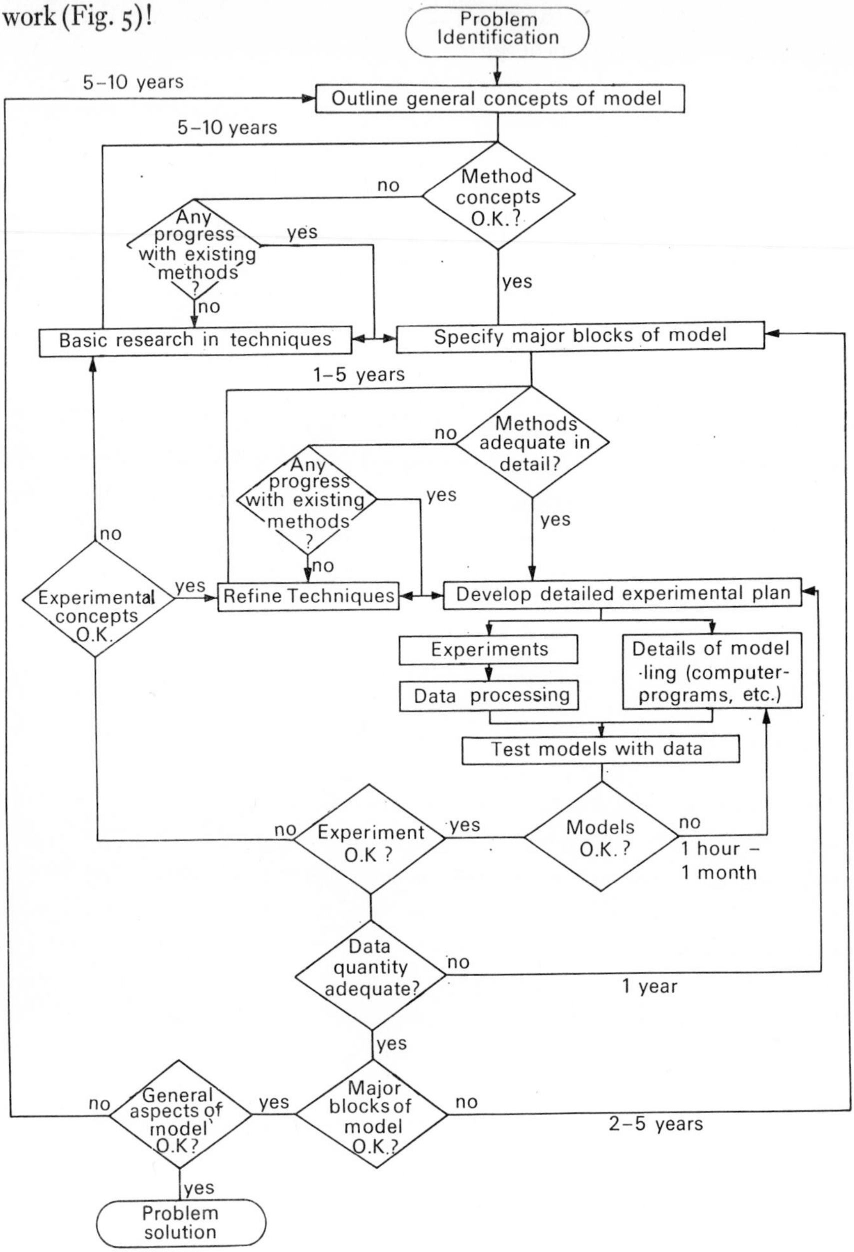

Figure 5. An example of some of the steps, feedback, and time considerations in the model–experiment–application area. At best, time estimates are nonprecise and are based on estimated present state-of-the-art capabilities in many of the biome studies. Time is often represented here in man-years.

The discussion above makes one pause to reflect upon the overall magnitude of the experimental and analytical effort. The question might be asked, 'how many years and man-years might be required before "precise" models might be developed for natural resource ecosystems?' Perhaps a pessimistic view is represented in Fig. 5. Development of the model, once the problem has been identified, involves many steps including outline of concepts, examination of available methods for implementation, design of major blocks of the model, relating models to experiments, construction and implementation of the model, and testing the model against experimental data. Many of these efforts can go on concurrently, and there are many 'turnarounds' available in the overall effort. After an initial round of model design and experimentation, the model should be tested against the experimental data and then redesigned to guide the second round of experimentation. The necessity of rapid turnaround is apparent here. Theoretically, the modelling effort should be far ahead of the experimental effort. The time requirements shown in Fig. 5 represent early 1970 estimates based on state-of-the-art capabilities, but hopefully, these times will be greatly reduced as we gain more experience, manpower, and organization in national and international efforts of this type. I feel that no biome program reached in 1970 a critical mass of combined systems analysis capability, experience, and operational format to enable the interchange needed to reduce these time estimates significantly.

Examination of the literature and recent unpublished work shows that most groups working in ecological modelling are beginning to accept the state variable approach generally utilized in the engineering sciences. The tendency appears to be toward the inclusion of nonlinearities, stochastic elements, and discontinuous variable differential equation systems. Nevertheless, neither the biologists, nor the engineers, nor the analysts operating alone in these programs seem to have a clear idea of how to handle ecological systems in a meaningful way! The overall progress in ecological modelling seems limited by at least three factors: (i) In most modelling groups there has been *insufficient time and effort* as well as subcritical mass of dedicated, capable personnel; (ii) In most instances the models (although pedagogical) have been *too simplistic* and *too data-poor* to produce meaningful results. Where data are available from concurrent field and laboratory experiments, the ecological and analytical manpower is badly fragmented by a variety of tasks; (iii) There has been an almost complete *lack of working interaction*, between many biological modelling groups. Also, a serious shortcoming is that there is even less interaction with the large pool of physical science analytical manpower which has recently become available throughout our nation because of cut-backs in many governmental contracts.

Benefits of integrated research to teaching

We have already observed that these integrated research programs have had unique contributions to teaching in at least three different ways. First, participation by scientists in these integrated research programs provides them with fresh and unique classroom examples. Particularly, he can show the relevance of his particular subject matter to the total system by his knowledge of the total program, his activities in it, etc. Second, we have found that faculty members involved in these integrated research programs have been stimulated to offer seminars and special problem courses on topics related to their program research. For example, an interdepartmental seminar sequence concerning dietary competition among herbivores was organized for graduate students (Haug, Van Dyne & Hansen 1969). Data and insight gained from the research program help provide a focus and realism through first-hand examples to the seminar sequence. Third, we have found entirely new courses have been developed in universities by participating scientists; such courses would not have been possible only a few years ago (e.g. see Van Dyne 1969*c*). For example, a course sequence regarding ecosystem analysis procedures was recently initiated where the entire function of the sequence is on tracing energy and nutrients as they flow through the system. Thus, course work is becoming more integrated and interrelated rather than packaged into discrete taxonomic units, a widespread problem even in ecology!

Unsolved problems and future needs

We need to continue our introspective reviews and not become lax about accomplishments. This is particularly important in trying to increase the year-to-year turnaround so that results of one season can be compiled, subjected to at least preliminary analysis, and used in the planning of the next field season's activities. The role of graduate student participants v. technician participants is important here.

Graduate research assistants v. technicians

One difficulty with graduate students in this short turnaround schedule required in our work is that much of their work is really never examined or analyzed in detail until completion of a thesis or dissertation, which may be two or three years in preparation. Therefore, although much can be done by graduate research assistants in integrated research programs, their efforts and time phasing should be considered with respect to general program needs. On the other hand, graduate students often are highly motivated, original, and imaginative and especially provide a tie to the latest literature which a technician does not and which many scientists also fail to do.

These considerations make it particularly critical that the staffing plan for workers in these biome studies be carefully examined. Thus, in the US IBP Grassland Biome study, now we are attempting to make certain exactly who are the graduate students involved in the research, when is their expected completion, and if their thesis or dissertation will delay other activities. In working with the principal investigators, i.e. the contractual scientists, we try to stress the point that there are many projects which cannot be considered good thesis or dissertation research projects. In some instances, a new area of work is investigated as such a study, but then when the techniques and procedures become well established it perhaps would no longer qualify as a thesis, or especially a dissertation, and the work should be conducted independent of such efforts. This does not mean the work is any less useful or less essential to the program, but it is just not compatible with the graduate student's needs. On the contrary, much work with a high probability of successful completion is essential for the overall study.

We have made some mistakes in assigning certain critical project areas to graduate students for thesis or dissertation topics. The major characteristic of such an area is that it is one that will involve the direct day-to-day input from several scientists, thereby it violates one implicit characteristic of a dissertation, i.e. independent and original contribution by the student. Yet, we must carefully examine what we in the universities have been using as criteria for thesis and dissertation research. If the student has the capability, and indeed he will be a rare one, to carefully work with a group of senior scientists and integrate and evaluate their input into a single project it could certainly still count as an original effort, and perhaps also his part of it could be considered an independent input.

Information storage and retrieval

One of the disappointing features in the biome studies thus far is the lack of development and implementation of an adequate information storage and retrieval system. Data and information are beginning to accumulate at an astounding rate. An example of the rate of data and information accumulation can be shown by examining the US IBP Grassland Biome Technical Report series. Some 132 Technical Reports were produced from November 1969 to October 1971, and we are still in the latter part of a 'tooling-up' phase! These documents contain information of a preliminary nature and are prepared primarily for internal use in the US IBP Grassland Biome study. They contain information which generally is not for use in the open literature at the present stage; it is not for use in the open literature prior to publication by the investigators named unless permission is obtained in writing from the Director of the US IBP Grassland Biome study. These reports include individual data, sometimes with analyses, as well as detailed descriptions of field and laboratory and analytical methods. Some reports also include description of experimental sites and vitae of scientists. We feel these reports are a necessary prerequisite to the successful information and storage retrieval system. The major value of these reports within the program is that detailed description of methods and the tabulation and presentation of data will precede by months, or even years, the eventual final publication of a segment of research in the open literature.

The volume of the work is somewhat frightening. Through Technical Report No. 132 they have averaged 43 pages each. Yet, they do not contain full data sets. If the data to be summarized require more than about 40 pages of computer output listing, then only examples are extracted from the magnetic tape records and included in the Technical Reports. The reports are produced in a mimeograph form in limited numbers and circulated among scientists within the program, to cooperating groups, as well as circulated to the director of each of the other biome programs.

Many of our individual scientists are starting to consider synthesis efforts and, thus, request usage of the technical reports. Although some individual scientists are attempting to piece together work in their own area, few are attempting to tackle the complexity of whole trophic levels or whole ecosystems. Obviously, the integrator scientists mentioned above have a key role here, but they too will be swamped with information unless more adequate information storage and retrieval systems are soon developed.

One of the key ideas early in the development of the US IBP Analysis of Ecosystems program was to plan for and include interbiome modelling and syntheses. This also requires adequate information and storage retrieval systems. The tendency in individual biomes, however, is to tackle first the accumulation of data and information (in as well-designed a framework as possible). They have generally allocated low priority to the long-term storage and retrieval problems.

Development of adequate information storage and retrieval systems is no small task. Evidently, adequate systems are yet to be developed for the type of information being derived in ecosystem studies. Preliminary estimates suggest the cost for development of such a highly sophisticated system may be almost equivalent to one or two years operation in a given biome!

The information storage and retrieval problem is particularly critical when one considers the importance of the area and previous efforts in this field. Overall, to my knowledge, most proponents of information storage and retrieval systems at the ecological and environmental level have 'oversold' their case. None seem to have come up with a system that is nearly as workable and useful as they originally proposed. Yet large amounts of data are being collected in each of the integrated research programs and NSF has concerns about the distribution and utilization of these data. Adequate information transfer procedures between these NSF-supported programs and other governmental agencies, whether they be Federal or state or local, have not been implemented. Likewise, to this date and in my opinion, adequate procedures have not been established for sharing and transferring data on an international basis. Although the researchers and the NSF realize the international importance of IBP, it appears that initial funds have gone largely into initiating national field and laboratory experimental efforts in the integrated research programs and the long-term needs of the studies have been subjugated to day-to-day expediency. The full potential of many of the integrated research programs, such as the biome studies, will not be fully realized until such international efforts are implemented. This will be discussed in more detail below but one encounters a special difficulty here because of the limited sources of funding for truly international cooperative efforts. The IBP Terrestrial Productivity Grasslands Working Group, through its international coordinating committee, has been investigating the problems of developing an International Synthesis Center to carry out many of the above kinds of activities. Even within one biome, it is evident that data from many kinds of environments worldwide are needed to develop, test, and extend the applicability of models of the system, as a form of synthesis. It is projected that these efforts would have value for not only development of ecological principles, but also application to resource management (Fig. 6). Preliminary discussion is underway with the FAO and with UNESCO because of an interest of the international agencies in the results of information syntheses about grasslands. A cooperative IBP-UN effort offers a unique opportunity to demonstrate a new concept in international cooperative research.

The information storage and retrieval problem in the Analysis of Ecosystems program perhaps is only a specific case of a more general problem. That is the problem of lack of foresight, of lack of adequate future strategy. Perhaps because we have been so busy in initiating activity in each biome, we have procrastinated when it comes to doing something effective about many interbiome efforts. Perhaps it is because of lack of continued, strong, central leadership—but that may not have been possible in any event because each biome is funded separately. Perhaps it is because there is a great deal of uncertainty in the challenging concept of making the Analysis of Ecosystems program idea work and eventually pay large dividends to mankind. Maybe we have hidden selfish interests about our own personal short-term or long-term future and intentionally or unintentionally play the 'delaying game'. Only time will tell.

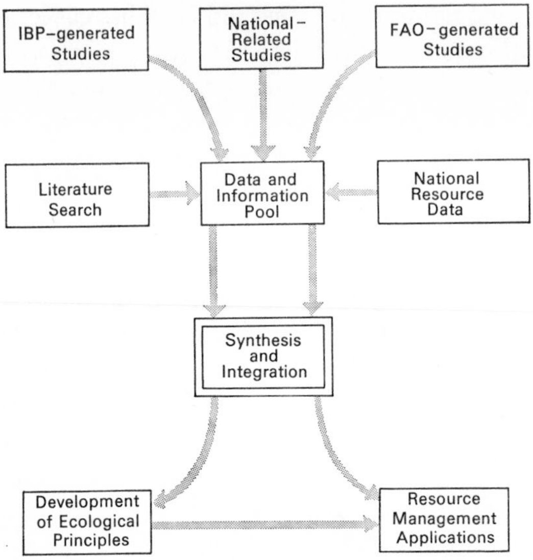

Figure 6. An example of the interrelationship between IBP studies, national related non-IBP studies, and externally-generated studies to the overall international problem of developing resource management plans. Mathematical models will be an important form of synthesis here (as is shown in Fig. 5). This figure is based on developing plans by the IBP PT Grasslands Working Group.

Increasing cooperation of state and federal groups

Another point needing strengthening in many of the biomes at our national level is increasing the degree of interaction and ties to federal- and state-supported research programs related to ecology. Although many of the US IBP Analysis of Ecosystems research sites are on federal or state lands, and although, in a few instances, there is great participation of federal and state scientists from their own support, in most instances the integrated research program has probably been considered 'non-central' to many of these groups. However, recent legislation has paved the way for many federal agencies to provide direct funds for participation in the International Biological Program. I think we will see much greater participation of the Interagency Coordinating Committee, at least through key agencies, in orienting some of their agencies' work to IBP objectives and to complete integration with IBP supported programs of some phases of their research planning and activities. Although there is great need for further strengthening the degree of interaction of various state, federal, and university groups, the progress to date is beginning to be encouraging. To me, at least, this indicates that one of the major implicit goals of the IBP is being accomplished. That is, in a sense, the IBP is testing whether or not workable procedures can be developed for management of large but dispersed scientific undertaking in the environmental area.

I think there is a good climate for increasing cooperation between IBP Biome groups and federal agencies particularly in the synthesis and integration activities, e.g., see the center and lower part of Fig. 6; specifically, ''we' need them and 'they' need us!' Suggestions for optimal solutions to problems at this scale, i.e. for national or international systems-complexes scale, will require all the ideas, experience, and energy we can gather for the task. I hope the federal agencies will see the way clear to intensify cooperative activities on field sites and to help plan and share in the examination of already-available data and information. To do this, I feel these agencies may need to assign some of their personnel to some of the biome centers to work with the university and state experiment station personnel on a day-to-day basis.

One early approach in the US IBP studies that was considered was for each designer (individual or group) of an integrated research program proposal to direct part of the proposals to appropriate agencies for funding and indicate priorities. The International Coordination Committee was to function as an intermediary between the program directors of integrated research programs and the agencies. At one stage in the development, this plan was to be supplemented by NSF having a separate fund to support and fill serious gaps left by the other agencies. Early experience with this approach proved it entirely unworkable at the time and place it was attempted. IBP projects were exposed to haggling between the agencies and early in 1968 this almost provided the cause for complete 'disenchantment' with the IBP by some of the program directors.

Improving services to participants

I feel an important aspect of our current US IBP Grassland Biome study structure is to provide increased central services to the participants in the project (Fig. 3) and thereby providing several side benefits. Central service activities help us in coordinating the work, insure high-quality analyses of data and of samples, and insure *more efficient use of equipment and support personnel.* By bringing data and samples into the central laboratory, we get another 'fix' on the rate of generation of information in the program. By bringing reports and papers into a central office for final processing, we are helping to prevent problems of data loss (unintentional or otherwise). More important, in this way we can *provide a more rapid communication of ideas and results* throughout the program. Providing these central services is not without some management cost. It is an important part of the project, but it takes time to direct, to make certain it runs smoothly, etc. Superficially, it may cause the central biome management to be criticized of 'empire-building', but I believe most of the participants of the program are increasingly taking advantage of these services and now are 'demanding' them! Another service to the participants is that the central program management provide the critical mass for and take the responsibility of developing progress reports, continuation proposals, carry the scientific public relations of the program, etc. In effect, this frees individual contractual participants from many responsibilities they would have otherwise. In doing this, however, the central program management is faced with a problem of lack of anonymity.

In the usual Federal agency–investigator relationship, proposals for research are reviewed in an anonymous manner. Most Federal agencies, including the NSF, are very careful in their replies to the potential investigator to paraphrase or rephrase comments from their review panel and other reviewers. I think this results in the investigator often being frustrated, sometimes angry, but not knowing who to strike out against! In the biome-type program, however, it is fairly clear who reviews the proposals. Program management personnel know contractual investigators on a first-name basis, and this makes deleting an investigator from the program extremely painful to both parties involved, whether it is because of inadequate financial resources or because of a change in emphasis in the needs of the program or a lack of production by the investigator. Thus, some of the services provided to the participants could, if not evaluated and handled carefully, lead to misunderstandings and antagonisms among scientists in the same general area of work and region.

Inadequate attention to international developments

International as well as national syntheses are necessary for IBP to be fully successful. The complexity of each plant-animal-decomposer-soil-atmosphere system is such that adequate interpretation probably is not possible by the usual means of analysis. Data from a large number of sub-projects must be synthesized by techniques of mathematical modelling. These techniques are only now evolving among too few mathematically-oriented biologists to staff all existing national studies around the world. Meanwhile, such talents are not being applied to interproject syntheses, on which many of the achievements of the IBP eventually must be evaluated, and no effort is underway to incorporate applicable data from IBP-related research of various countries. Once again, the priority has been to start studies in individual nations rather than develop intensive cooperation and planning for international synthesis centers. Fortunately, preliminary efforts are now underway in at least four international working groups towards the development of individual or joint centers for data storage and synthesis.

The size of the data and information pool required for international syntheses is quite large when one considers both the basic and applied data needed to test and develop ecological principles and to utilize these in resource management applications (see Fig. 4 and 6). Cooperative efforts between IBP scientists and international organizations utilizing such ecological information offers a unique opportunity to demonstrate a new concept in international cooperative research. Also, this is an opportunity to provide solutions to problems which are on a world scale and require the input of many scientists working in a diverse set of environmental conditions. A systems approach and models make it possible to apply information obtained from different parts of the world to the problems of resource management and sustained production at other locations. Particularly, these locations include developing countries where such information is scarce and difficult to obtain. Thus, information from the technically more developed areas can be applied to those lacking information needed for technical development. It permits the fullest exploitation of data gathered in countries where research facilities are limited.

Although we have progressed in the mathematical modelling within individual ecosystem studies and we have produced designs for both intensive sophisticated studies and extensive non-sophisticated studies, much remains to be done. Studies of both types are already underway which will provide parameters necessary for the mathematical models mentioned above. Data from several environments are already being collected, but data from many more environments are needed to extend the applicability of the models. Collection of most of these data makes relatively simple demands on instrumentation and techniques at various field locations but make an unprecedented demand on means of developing internationally acceptable approaches for coding, organizing, analyzing, synthesizing, and integrating the data. However, full discussion of the development and planning of international synthesis centers is beyond the scope of this paper; it certainly is a 'state of the art' problem.

Grant-time v. field-time turnaround: the one-year grant dilemma

I have illustrated and discussed the need for explicit schedules, but we are still learning by experience in this area. For example, even though a grant is obtained on a year-long basis, perhaps the contractual period should be reduced to 11 months. This provides time for a central biome office group to analyze technical reports from individual studies and to use this information in developing the coming year's contracts. An adequate data storage and retrieval system is needed to permit this year-to-year turnaround of data, analysis, and information in designing the next year's field studies. The contractual scheme requires that the granting agencies be more flexible in their future procedures. Typically, in these integrated ecosystem studies a progress report and continuation proposal may be submitted eight months before the initiation of the next grant year. In the past, there have been rather rigid requirements by the granting agencies that individual contracts be elaborated explicitly at the time the proposal is submitted.

Consider the case where the contract and grant period is the calendar year. In the northern hemisphere the major activity in the field in these ecological programs is probably during April through October. If adequate data analysis schemes, samples processing schemes, and storage and retrieval schemes are developed, then much of the information collected during the growing season can be analyzed by the end of the year and used in developing the next year's contract. Results of such analyses, of course, are not available perhaps four months into the year which has already been completely budgeted, and it is soon time when the next proposals must be submitted to the granting agency! Yet the granting agencies have been reluctant to allow complete freedom of choice in allocation of research funds at the project level. Still, the amount granted in most instances is considerably less than the amount requested. It may take two or three budgetary turnarounds before a final budget is reached. This final budgetary agreement may be reached informally only a few weeks ahead of the contractual period for the next year.

I do not have the absolute solution to the above dilemma. But the solution must include *greater local autonomy*, *opportunity for including contingency funds*, and *flexible management procedures* that have been developed in many research organizations. It is impossible to run what are at least semi-programmatic studies in a completely non-programmatic structure. Yet much would be lost if National Science Foundation support would be channelled into entirely action-oriented programs.

Need for continual introspection

We learn more-and-more clearly each year in our US IBP Grassland Biome study it is necessary to continually keep people talking and listening to each other, to spot interesting but not defendable studies as divergences from the main theme, and to *maintain a spirit of open and constructive criticism.* I am continually amazed at how much time is needed to give more than lip service to the idea of integrated research, to keep an interdisciplinary group focused on a main set of objectives, and to keep the program within the qualifications of an integrated research program.

Evolving modelling needs

Although we feel we made some progress in systems analysis during the second grant segment, we recognized several needs. It was clear that we continually needed time periodically isolated, for intense, day-long 'skull sessions' at which time we review and take stock of individual modelling efforts, relate them to experimental work, and so forth. During this grant segment we also felt we needed to map out an overall set of symbols and a master flow chart of systems analysis area efforts.

During our second segment of research we attempted to enumerate many of the submodels necessary for a total-system simulation model and to define and derive a common structure for describing these models. In this approach we identified the 'dependent variables' in the model (i.e. biomass, numbers, calories, etc.), the equation types (difference, differential, algebraic, or some combination thereof), expressed the model in general form using functional notation, and tabulated for both the input and output variables the notation, definition of variable, the units, and the source and value of coefficients or parameters or of the variable itself.

After evaluating a variety of models and modelling efforts, e.g. see Swartzman (1970), we decided we did not have sufficient manpower and resources to explore and develop simultaneously all the total-system, varying resolution dynamic models, and the trophic-level dynamic models described in Table 2 as well as continue initial efforts in optimization and structural modelling. Thus, most of the simulation modelling effort in the last two-thirds of 1970 was restricted to a total-system model of relatively high level of resolution, i.e. somewhere between Pawnee 2 and Pawnee 3 levels of Table 2. This model is now in an operational stage, although it requires a great deal of improvement before the output will be extremely useful in a quantitative sense to help guide field and laboratory experimentation (Bledsoe, Francis, Swartzman & Gustafson 1971). This model, PWNEE, is a time-dependent biomass model; it is a point model as no spatial aspects are considered. A set of 40 first-order differentials were developed to describe the major abiotic, producer, consumer, and decomposer compartments of a shortgrass prairie ecosystem. The six general types of components of the model are characterized by relatively standard and consistent notation, a first attempt to reduce and simplify the complexity of notation, designed to improve readability and enhance information transmission and retention. The components include: (i) driving variables, the forcing functions of engineering parlance; (ii) principal system variables, the primary dependent variables of the model; (iii) intermediate system variables, i.e. mathematical functions necessary to calculate the principal system variables; (iv) independent variable, i.e. time in various units; (v) dummy function arguments, special mathematical generalized functions useful as a special type of intermediate system variables; and (vi) the parameters, i.e. the coefficients which remain constant during any given running of the program. A summary of the numbers of different kinds of components in this first version of PWNEE are given in Table 3.

Table 3. A summary of the number of different components in PWNEE, a medium-high resolution level shortgrass prairie ecosystem model.

Component	Section of the program/model					
	Driving variables	Abiotic	Producer	Consumer	De-composer	Total
Driving variables	4	—	—	—	—	4
Principal system variables	—	3	7	17	13	40
Intermediate system variables	1	—	51	249	21	322
Parameters	5	2	116	217	59	399

The large number of components of the shortgrass prairie ecosystem simulation model PWNEE (see Table 3) illustrate the difficulty of the modelling task. One of our goals is to convert PWNEE (or develop a new and better structure) into a model of a high resolution level. Resolution will be increased by increasing the number of principal system variables and by increasing the degree of mechanism in the equations composing the model. Wherever possible we will validate model output through field measurements and verify the form of and parameters in the mechanisms in the model, i.e. the physiological, ecological, and physical processes operative in the grassland ecosystem.

Examination of the trends in simulation modelling alone illustrates the importance of the modelling team approach in these ecological researches as we recognized early in the US IBP Grassland Biome study. We have attempted to structure such a team from (i) *analytically oriented scientists* (i.e. those originally receiving their graduate training in mathematics, biometrics, and systems engineering), (ii) *experimentalists* (i.e. those scientists receiving their graduate training primarily in biological and physical sciences), and (iii) *systems ecologists* (i.e. hybrid scientists whose training and practice includes major amounts of both experimental and analytical components). The roles of these types of scientists and of these teams are discussed more fully elsewhere (Van Dyne 1969*a*, Hamilton *et al.* 1969), but some of the special problems of management of small-group, inter-disciplinary modelling teams are of importance here.

Hamilton *et al.* (1969) have commented on the common problem of the implicit assumption that assembling a group of individually competent personnel will automatically lead to an efficient and competent research team.

Such is not the case, however, for many competent individuals cannot be integrated into effective research teams. Also, most individually competent scientists have neither the interest nor the ability to coordinate the work of others. If there are constraints of time and budgets imposed on top of those involving the research problem itself even fewer scientists are able to do the coordination. As I have noted above, the management needs in these integrated ecological programs have been underestimated and especially in the area of systems analysis. Total-systems models cause a particular problem here. Let me explain.

At the present state of the ecological modelling art we have difficulty segmenting the problem in such a way that the pieces may be reassembled readily into a meaningful whole. Yet we must be able to isolate parts of a complex problem for separate study. Some modellers feel compelled to be abreast of detailed developments of all phases of the systems analysis task (and sometimes the experimental tasks), often to the point to cause their subtask work to suffer. I think this problem is more common to modellers than to experimentalists. Alternately, some modellers are more than satisfied to isolate a narrow segment of the problem for their attention without regard to the overall study. And again the program may suffer. Obviously, one of the major management problems is to strike a happy balance between the above two extremes. A major contribution to the solution of this modelling management problem would be to have a mathematical framework sufficiently complete to encompass the whole ecosystem, sufficiently flexible to allow change in structure and complexity, and yet sufficiently simple in structure so that it might be readily understood both by analysts and experimentalists. This would allow a broad overview of the problem to be obtained easily yet individual segments could be expanded or contracted as needs demand and information allows.

Present knowledge suggests that a compendium of mathematical forms of varying levels of precision for examples of these and other ecological processes would be extremely useful in accelerating progress toward improved resource and environmental management models. Experience to date also suggests that the resulting generalized models will take the form of differential equations of varying complexity to provide a maximum of generality within a well organized and well accepted mathematical framework. There will be a high degree of coupling among these equations.

Present experience also suggests that the general state variable approach used in engineering can be followed as a basic design:

$$\dot{x}(t) = A(t)x(t) + B(t)z(t)$$

$$y(t) = C(t)x(t) + D(t)z(t)$$

where

x is the state vector of the system

z is a vector of inputs or forcing functions

y is a vector of outputs

$A(t)$, $B(t)$, $C(t)$, and $D(t)$ are, in general, matrices of time-varying coefficients

In the usual engineering sense (e.g. Derusso, Roy & Close 1965), the matrices A, B, C, and D are composed of constant coefficients or time-varying coefficients. In the usual engineering sense, these coefficients are taken as deterministic as are the driving variables, z (e.g. Breipohl 1970). Under these conditions the solution is rather straightforward (e.g. see Kuo 1966):

$$x(t) = e^{At}\, x(0) + \int_0^t e^{A(t-\gamma)}\, Bz(\gamma)\, d\gamma$$

$$y(t) = C\, e^{At}\, x(0) + \int_0^t C\, e^{A(t-\gamma)}\, Bz(\gamma)\, d\gamma + dz(t)$$

where γ is the beginning of time interval of concern and e^{At} is the 'state transition matrix' in engineering terms (e.g. see Pottle 1966) and can be given by

$$e^{A(t-\gamma)} \simeq \sum_{k=0}^{n} \frac{Ak(t-\gamma)k}{k!}$$

But ecological experience suggests, however, that there may be probabilistic components both in the driving variables and within the system variables. Instead of constant-coefficient or time-varying coefficient matrices, in the ecological case these may be represented by matrices of generalized functions. These functions also include a great deal of nonlinearity, lag effects, and discontinuities (see discussion, for example, by Watt 1968, Pielou 1969, or Van Dyne 1971*b*).

Considering these ecological needs, we feel the general formulation might be written as follows:

$$\dot{x}(t) = A(x, z, t)x(t) + B(z, t)z + e(x, t)$$
$$y(t) = C\, x(t) + D\, z(t) + f(x, z, t)$$

where

A, B, e, and f are matrix or vector functions of the state variables, driving variables, and time

C and D are matrices of constants

x, y, and z are defined as above

In general, each component of the matrix function A will be a term which is representative of the various processes affecting the rate at which the corresponding component of x will vary. Each term will have the units of reciprocal time. The terms will generally have the form of a state or driving variable times a function of other state variables times a constant (for example, see the formulations used by Parker, 1968, or equations 18 and 21 of Bledsoe and Jameson, 1969):

$$\dot{x}_i(t) = \sum_{j=1}^{n} k_{ij}\, h_{ij}(x, z, t)\, x_j + g_i(x, z, t)$$

The typical form shown above indicates that the mass flows of an ecosystem depend on the size of certain state variables as regulated by generally non-linear functions of other state variables [note the equivalence, $a_{ij} = k_{ij}\, h_{ij}(x, z, t)$]. The functions constitute the information flows of the ecosystem. The model parameters, k_{ij}, and other coefficients involved in the functions h_{ij} and g_i will, in general, have not only central values but variances and more-or-less well-defined distributions making this a stochastic rather than a deterministic formulation. Alternatively, the parameters may be taken to be characteristic of points in space but constant with respect to time. They may be statistical or deterministic functions of spatial variables, making the resulting formulation capable of describing spatial as well as temporal variation of the state variables of the ecosystem.

Notice that the functions h_{ij} and g_i are shown as depending on the entire function which represents the state of the system rather than merely its value at a point in time. The proposed formulation is capable of utilizing generalized time lags, and the solution to the differential equations may have characteristics more complex than is the case for classically defined first-order ordinary systems (as defined by, for example, Ince, 1956). The equation system is properly a set of integro-differential equations since the effect of time lags are explicitly allowed.

The relation of output variables to the state and driving variables of the system allows for partitioning in space of the system to achieve higher resolution. For example, soil moisture or temperature might be characterized by differential equations for different soil depths with a weighted sum over depth as the corresponding output variable.

Ideally, mathematical forms developed in the project should be constrained only to the extent of mutual compatibility so that results could be used to assemble a total ecosystem model. For example, rather than aim at a 'general predation equation', perhaps we should develop mathematics describing separately the predation process such as in the coyote, eagle, and weasel with separate mechanisms where required by the biology of the species. Generality of results would only come at a later stage by observing commonalities in subsets of processes. This is analogous to the manner in which Holling (1963) developed a predation model first by observing the process in particular situations (i.e. mantis, stickleback, and cat) and later synthesizing the separate mechanisms into a general predation theory. Pragmatically, much more rigidity of mathematical forms might be valuable at least initially, i.e. until the total system model is running well.

Because mathematical functions useful in ecology are not simple (they include a great deal of nonlinearity, lag effects, and discontinuities) nationally, and internationally, we immediately need increased experience of a large number of ecological analysts and experimentalists in developing these complex differential equation systems. The payoff is large for combining such differential equation simulation models with optimization models providing output useful in natural resource management decision making on a worldwide basis (see Fig. 4).

Acknowledgements

The US IBP Grassland Biome study upon which I am primarily reporting involves a large number of scientists; perhaps 200 scientists thus far have directly participated in some way. Rather than enumerate all the individual names, I acknowledge the entire scientific group (see Technical Report No. 24, Wright 1970). Although not received in time to allow a major reorganization of the manuscript, the paper and the author benefitted from constructive criticisms by S. I. Auerbach, J. Brown, G. S. Innis, P. L. Johnson, and S. B. McKee. Our research support has come primarily from the National Science Foundation through Grants GB–7824 and GB–13096. Other direct support has also been obtained from the Atomic Energy Commission. Indirect support has been obtained through cooperating agencies including the Agricultural Research Service, the Forest Service, the Soil Conservation Service of the US Department of Agriculture, and the Bureau of Sport Fisheries and Wildlife of the US Department of Interior. Additionally, in several instances, especially in our network program there has been indirect support through various state Agricultural Experiment Stations. Without this total indirect support, the direct support would not have carried the program; both are essential and are acknowledged.

References

BLEDSOE L.J., FRANCIS R.C., SWARTZMAN G.L. & GUSTAFSON J.D. (1971) PWNEE: a grassland ecosystem model. *US IBP Grassland Biome Tech. Rep. No. 64*. 179 pp.

BLEDSOE L.J. & JAMESON D.A. (1969) Model structure for a grassland ecosystem, pp. 410–37. *In* Dix, R.L. and Beidleman, R.G. (ed.) The grassland ecosystem: A preliminary synthesis. *Range Sci. Dep. Sci. Ser. No. 2*, Colorado State Univ., Fort Collins. 437 pp.

BREIPOHL A.M. (1970) *Probabilistic systems analysis*. John Wiley & Sons, Inc., New York. 352 pp.

COUPLAND R.T., ZACHARUK R.Y. & PAUL E.A. (1969) Procedures for study of grassland ecosystems, pp. 25–47. In Van Dyne, G.M. (ed.) *The ecosystem concept in natural resource management*. Academic Press, New York. 383 pp.

COX G.W. (ed.) (1969) *Readings in conservation ecology*. Appleton-Century-Crofts, New York. 595 pp. (about two-thirds of Van Dyne, 1966, is reprinted herein, see pp. 21–47).

DERUSSO P.M., ROY R.J. & CLOSE C.M. (1965) *State variables for engineers*. John Wiley & Sons, Inc., New York. 608 pp.

DIX R.L. & BEIDLEMAN R.G. (ed.) (1969) The grassland ecosystem: A preliminary synthesis. *Range Sci. Dep. Sci. Ser. No. 2*, Colorado State Univ., Fort Collins. 437 pp. (Supplement, 1970. 110 pp.)

DONNELLY J.R. & ARMSTRONG J.S. (1968) Summer grazing, pp. 329–32. In *Second Conference on Applications of Simulation, Proc.* Publ. by SHARE, ACM, IEEE, and SCI.

DUBRIDGE L.A. (1967) University basic research. *Science* **157**, 648–50.

GOODALL D.W. (1969) Simulating the grazing situation, pp. 211–36. In Heinmets, F. (ed.) *Concepts and models of biomathematics—Simulation techniques and methods*. Vol. I. Marcel Dekker, Inc., New York. 287 pp.

HAMILTON H.R., GOLDSTONE S.E., MILLIMON J.W., PUGH A.L. III, ROBERTS E.B. & ZELLNER A. (1969) *Systems simulation for regional analysis—an application to river-basin planning*. Mass. Inst. Technol. Press, Cambridge, Mass. 407 pp.

HANDLER P. (1970) Toward a national science policy. *BioScience* **20**, 971–7.

HANDLER P. (1971) The federal government and the scientific community. *Science* **171**, 144–52.

HAUG P.T., VAN DYNE G.M. & HANSEN R.M. (ed.) (1969) *Dietary competition among herbivores* (papers from a joint seminar, Fall Quarter, 1968). Colorado State Univ., Fort Collins. 121 pp.

HOLLING C.S. (1963) An experimental component analysis of population processes. *Mem. Entomol. Soc. Can.* **32**, 22–32.

INCE E.L. (1956) *Ordinary differential equations*. Dover Pub., Inc., New York. 558 pp.

JENSEN E.V. (1969) The science of science. *Perspectives Biol. Med.* **12**, 274–89.

KERR W.A. (1963) Creativity in engineering and industry. *J. Amer. Soc. Agr. Eng.* **44**, 663, 679–80.

KUO F.F. (1966) Network analysis by digital computer, pp. 1–33. In Kuo, F.F. and Kaiser, J.F. (ed.) *Systems analysis by digital computer*. John Wiley & Sons, Inc., New York. 438 pp.

LEVINS R. (1966) The strategy of model building in population biology. *Amer. Sci.* **54**, 42–3.

PARKER R.A. (1968) Simulation of an aquatic ecosystem. *Biometrics* **24**, 803–21.

PATTEN B.C. (ed.) (1971) *Systems analysis and simulation in ecology*. Academic Press, Inc., New York. Vol. 1. 607 pp.

PIELOU E.C. (1969) *An introduction to mathematical ecology*. Wiley–Interscience, New York. 286 pp.

POTTLE C. (1966) State-space techniques for general active network analysis, pp. 59–98. In Kuo, F.F. and Kaiser, J.F. (ed.) *Systems analysis by digital computer*. John Wiley & Sons, Inc., New York. 438 pp.

SMITH F.E. & KADLEC J.A. (1969) *Progress report—analysis of ecosystems*. Processed report from University of Michigan to the National Science Foundation. 26 pp.

STEINHART J.S. & CHERNIACK S. (1969) *The universities and environmental quality—commitment to problem focused education*. (A report to the President's Environmental Quality Council) Office of Science and Technology, Executive Office of the President. US Government Printing Office, Washington, D.C. 49+22 pp.

SWARTZMAN G.L. (Coordinator) (1970) Some concepts of modelling. *US IBP Grassland Biome Tech. Rep. No. 32*. 142 pp.

TANSLEY A.G. (1935) The use and abuse of vegetational concepts and terms. *Ecology* **16**, 284–307.

TIMMEN M. (1970) Computer experiments with a generalized ecosystem model. (Paper given at Symp. on Analysis and Synthesis of Ecological Systems, Argonne Nat. Lab., 26–28 October 1970).

VAN DYNE G.M. (1966) Ecosystems, systems ecology, and systems ecologists. *Oak Ridge Nat. Lab. Rep. ORNL* 3957. 31 pp. (see also Cox, 1969).

VAN DYNE G.M. (1969*a*) Grasslands management, research, and training viewed in a systems context. *Range Sci. Dep. Sci. Ser. No. 3*, Colorado State Univ., Fort Collins. 50 pp. (a 4000-word condensed version appears as 'A systems approach to grasslands'. *XI International Grassland Congress, Proc.* **11**, A131–A143).

VAN DYNE G.M. (ed.) (1969*b*) *The ecosystem concept in natural resource management*. Academic Press, Inc., New York. 383 pp.

VAN DYNE G.M. (1969*c*) Implementing the ecosystem concept in training in the natural resource sciences, Chap. X. In Van Dyne, G.M. (ed.) *The ecosystem concept in natural resource management*. Academic Press, New York. 383 pp.

VAN DYNE G.M. (1971*a*) Some problems and possibilities in Australian arid systems research. *3rd Australian Arid Zone Res. Conf., Proc.* pp. 46–69.

VAN DYNE G.M. (1971*b*) Aspects of quantitative training in the natural resource sciences. In Patil, G.P., Pileou, E.C., and Waters, W.E. (ed.) pp. 440–454. *Statistical ecology: Vol. 3: Many species populations, ecosystems, and systems analysis*. Pennsylvania State Univ. Press, College Park, Pa. 462 pp.

VAN DYNE G.M., INNIS G.S. & SWARTZMAN G.L. (1971) Some analytical and operational approaches to developing dynamic models of ecological systems. pp. 19–26. In M. Lillywhite and C. Martin (ed.) *Environmental awareness. Institute for Environmental Sciences, Proc.* 60 pp.

WATT K.E.F. (1968) *Ecology and resource management: A quantitative approach*. McGraw-Hill Book Co., New York. 450 pp.

WEAVER W. (1969) Basic research and the common good. *Saturday Rev.* 52, 17–18, 54.

WRIGHT R.G. (Compiler) (1970) Scientific personnel participating in the grassland biome study, June 1968 through January 1970. *US IBP Grassland Biome Tech. Rep. No. 24*. 278 pp.

Building and testing ecosystem models*

DAVID W. GOODALL *Ecology Center, Utah State University, Logan, Utah*

Summary

The building of dynamic models of ecosystems for predictive purposes is discussed in terms of the identification of state variables, the processes involved in their changes, and the factors influencing their rates of change. Model building is often facilitated by division of a system into subsystems. Only in simple models can parameters be estimated from direct observations on the ecosystem as a whole. Usually this can better be done by *ad hoc* studies of simplified systems.

The testing of an ecosystem model is not comparable with the testing of a hypothesis. The hypothesis that a model is a perfect representation of the ecosystem would never be seriously entertained, so its disproof is without interest. The question is rather what is the distribution of values predicted from the model around values actually observed in the field. Sometimes this can be determined on the same sets of data used in constructing the model. It is, however, usually safer to collect a separate set of data for model testing.

The precision will differ in different parts of the universe of reference, and from one variable to another. Choice between alternative models will accordingly differ according to which variables and what parts of the universe are considered. The optimization of a general-purpose model accordingly requires a weighting function which can be applied to these numerous and diverse measures of precision.

1. Introduction—predictive models

In the process of analysing and trying to understand the dynamics of eco-

* This paper was prepared by the author in his capacity as Director of the US/IBP Desert Biome programme, and its preparation was supported by NSF Grant No. GB 15886.

systems, with a view to predicting the changes they may undergo, increasing use is being made of models.

A model of a system is a representation of it in other terms, often symbolic, and almost always simplified. Modelling is a mapping process (as a geographical map is a model of the terrain it represents); the model and the real system can be mapped on to one another, and the model is a homomorph of the system modelled. Some models, like the map, are static, and represent the structure of the system modelled rather than the processes going on in it. For predictive purposes, however, it is clearly a dynamic model which is required—one in which the processes are represented as well as the static structure.

Not all dynamic models of ecosystems have predictive value. The familiar box-and-arrow diagram is a dynamic model, but it has a purely topological relation with an ecosystem, and, though it may be able to predict inequalities, it cannot represent quantitative relationships. This type of model may have considerable value as a pedagogic tool or as an aid to communication and discussion, but its lack of predictive power means that it will usually be incapable either of verification or disproof. In considering models of ecosystems, we will confine our attention to quantitative models which can be used for predictive purposes—and in general to such as lend themselves to computer simulation.

The first requirement in building such a model is to define the universe of discourse—the range of ecosystems for which the predictions are to be made. This will include not only the composition of the ecosystems and the geographical area in which they occur, but also the range of values for each of the variables in the system. Naturally, it will involve specifications of any modifying factors, including human manipulations, to which the systems may be subject, and the effects of which may need prediction.

To focus ideas, let us consider an ecosystem dominated by sagebrush (*Artemisia tridentata*) in the Intermountain region of North America. We are interested in the changes to which it is subject under differing seasonal rainfall and temperature, and as a result of shrub-clearing treatments and the varying grazing pressures to which it may be subjected—specifically, let us say, changes in vegetational productivity, ground-water recharge, and deer population. That is, we wish to make predictions of the changes in these quantities under specified sets of conditions.

Almost any procedure for making such predictions must be based on an underlying model, tacit or explicit. Conversely, the purpose of an ecosystem model is usually the prediction of future changes—very often as a guide to decision. Even where the stated goal is an understanding of the ecosystem, rather than a practical purpose, it implies ability to predict. The touchstone of understanding is always prediction—a model that makes incorrect predictions must be based on an inadequate or faulty understanding. The present paper will concern itself with methods of building such models and of testing their performance, in the sense of assessing the measure of confidence which can be attached to predictions based on them.

Except in the rare case where a model is isomorphic—where, for every distinguishable component, relation and process in the real system, there is a corresponding element in the model—a model is by definition an imperfect representation of reality. It may approximate to it, but never attains it. For ecosystems, where every element has its own individuality, isomorphism is inconceivable and one must be content with the imperfection of homomorphism.

2. Model building

(*a*) *The variables*

In building a model to represent quantitatively the dynamics of an ecosystem, one must first decide on a set of variables within the system whose changes one is concerned to follow—the desiderata. Some of these may be state variables—values characterizing the system at a point in time, and changing as a result of the processes going on within it. In our sagebrush ecosystem, the deer population is of this type. Other desiderata may be rates of change at a moment of time, or changes integrated over time (as ground water recharge or vegetational productivity). These values, and other similar variables calculated within the model, may be described as *endogenous*.

Also involved in the model will be a number of variables whose values are not affected by the operation of the system, but have an influence on it—rainfall, for instance, or air temperature. These variables controlling the system from outside, as it were, are often called forcing or driving functions; they may also appropriately be termed *exogenous* variables.

(*b*) *Processes and factors*

The changes going on within an ecosystem may be thought of in terms of processes, which may be subdivisible. In the sagebrush ecosystem, change in the deer population is a process, which may in turn be subdivided into the processes of mortality, reproduction, and net migration. Processes can generally be considered as changes in one or more state variables, or as contributions to such changes. Herbivory, for instance, is a process contributing to changes in biomass of herbivore and vegetation, and to changes in their mean composition. Photosynthesis, respiration and mineral uptake are other processes contributing to the changes in biomass and composition of the vegetation. Each process is affected by a number of influencing factors, which may include exogenous variables, state variables, rates of change in state variables, and rates of other processes. The rate of the process of mortality in deer may be affected by the factors temperature, predator population, and rate of food consumption, among others. In the model of the sagebrush ecosystem, consequently, the change in the variable representing the deer population will be expressed in part as a function of the variables representing these three quantities in the model.

(*c*) *Functional relations*

One of the simplest forms of model consists of a set of equations relating the values of each of the state variables at a particular time, or their changes over a period of measurement, to the mean values during that period of the exogeneous variables, together with the initial values of the endogenous (state) variables; such a set of equations might be represented:

$$y_{it} = f(\mathbf{x}, \mathbf{y_0}) \tag{1}$$

where over the period, the vector $\mathbf{x}$ contains the mean value of the exogenous variables over the period, and the vector $\mathbf{y_0}$ the initial values of the endogenous variables, while the value of the ith endogenous variable at time t is y_{it}.

In the particular case of the sagebrush ecosystem, one might select as a model:

$$y_{1t} = (a_1 + b_1x_1 + c_1x_2 + d_1x_1x_2 - e_1x_3 - f_1x_3^2 - g_1y_{30})y_{10} \tag{2}$$

$$y_{2t} = a_2 + b_2x_1 + c_2x_1^2 \tag{3}$$

$$y_{3t} = (a_3 + b_3x_1 - c_3x_3 + d_3x_1x_3 + e_3y_{10} - f_3x_3y_{10})y_{30} \tag{4}$$

where y_{1t} is the vegetational biomass at time t, y_{2t} the groundwater recharge between time o and time t, and y_{3t} the deer population, while x_1 is the total precipitation during this period, x_2 the mean temperature, and x_3 the mean cattle population; the other symbols are constants. Instead of the system above, one might select a model based on non-linear functions, say:

$$y_{1t} = [a_1 + b_1x_2\{1 - \exp(c_1x_1)\} - d_1x_3]y_{10} \tag{5}$$

$$y_{2t} = a_3 + \exp(b_2x_1) \tag{6}$$

$$y_{3t} = a_3 + b_3x_1 + c_3y_{10}\exp\{-(d_3x_3 + e_3y_3)/x_1\} \tag{7}$$

If one were dissatisfied with totalling precipitation and averaging rainfall over an indefinite period, since this might obscure the important effects of temporal variations, one might instead use as independent variables the values of these quantities over successive sub-periods, x_{1k}, x_{2k}, and x_{3k} where k took as many values as there were sub-periods. This might, for instance, be in the form:

$$y_{1t} = a_1 + b_1 \prod_k x_{2k} \{1 - \exp(c_{1k}x_{1k})\} - \sum_k d_{1k}x_{3k}y_{10} \tag{8}$$

$$y_{2t} = a_2 + \sum_k \exp(b_{2k}x_{1k}) \tag{9}$$

$$y_{3t} = a_3 + b_3 \sum_k x_{1k} + c_3 y_{10} \prod_k \exp\{-(d_{3k}x_{3k} + e_3 y_{30})/x_{1k}\} \tag{10}$$

Continuous variation with time in the values of the independent variables, however, cannot be taken into account in this form of model.

Dependent variables may appear on the right-hand side of the equations so that they interlock – y_{1t}, the vegetational biomass, might for instance be a factor affecting the deer population, and vice versa:

$$y_{1t} = y_{10}\{a_1 + b_1x_1 + c_1x_2 - d_1x_3 + e_1(y_{3t} + y_{30})\} \tag{11}$$

$$y_{3t} = a_3 + (y_{1t} + y_{10})(b_3 - c_3x_3) \tag{12}$$

In this case, it may be necessary to re-cast the equations before they can be used in a predictive model, so that they become explicit expressions for each state variable separately.

Models of the types described above are almost inevitably empirical. A model purporting to represent the mechanisms of system function is likely to be cast in terms of rates of change, rather than the values of state variables at points in time. Most such models consist of a set of differential equations, expressing the rate of change in a number of state variables as functions of one another, and of other characteristics of the system. In the case of the sage-brush ecosystem, for instance, one might express the dynamics as:

$$\frac{dy_1}{dt} = y_1(a_1 + b_1x_1 - c_1x_3 - d_1y_3) \tag{13}$$

$$\frac{dy_2}{dt} = a_2 + b_2x_1 \tag{14}$$

$$\frac{dy_3}{dt} = (a_3 + b_3y_1)y_3 \tag{15}$$

x_1 being defined as the rainfall over the previous month. Such expressions are often simpler in form than those expressing change over a fixed period, and often need include fewer variables on the right-hand side. They can moreover take into account continuous variation in time in the values of the independent variables.

Where a set of differential equations like (13)–(15) can be solved analytically, it can be converted into a form such as (5)–(7) where the value of each state variable at a fixed time is expressed in explicit form—provided the independent variables are constant over the period in question, or their rates of change can themselves be expressed as simple functions of time. It is, however, exceptional for a set of differential equations representing ecosystem dynamics to be integrable, so that direct conversion into a form such as (5)–(7) is often not practicable and one must resort to numerical methods.

(*d*) *Model expansion*

Almost irrespective of the intentions of the model builder, ecosystem models have an intrinsic tendency to expand, in respect of the number of state variables, processes and functions included. If the rate of change of each of the desiderata is expressed as a function, the evaluation of this function will require values of a number of other variables. Some of these may themselves be desiderata, others may be frankly exogenous variables, but there will usually be other variables which are neither. In (13) above, for instance, the variable x_3 (mean cattle population) might indeed be an exogenous variable if the cattle were introduced by a grazier, the ecosystem modelled was delimited by a fence, and the cattle suffered no mortality or reproduction during the period in question. But otherwise x_3 would be determined in part by processes going on within the ecosystem—by birth and death processes, and by cattle activity which affected the proportion of time spent within and outside the ecosystem. Thus, determination of a value for x_3 at a particular moment in time might involve a new function, say:

$$\frac{dx_3}{dt} = a_4 x_3 \left(1 - e^{b_4 x_4}\right) - c_4 \left(x_5 x_3 - \frac{x_6}{x_5}\right) \tag{16}$$

where x_4 is the food consumption of the cattle per individual, x_5 is the relative palatability of forage outside and inside the ecosystem, and x_6 is the population density in the whole area (assumed large) of which the ecosystem forms a part, and over which the cattle are free to move. x_6 will be another exogenous variable, but x_4 and x_5 will clearly change as a result of the functioning of the ecosystem, and hence could be treated as new variables in the model, with new functions to express their rates of change.

This process of expansion of the model need not continue indefinitely. One might, for instance, proceed to express these two new state variables as:

$$x_4 = a_5(1 - e^{b_5 y_1}) \tag{17}$$

$$x_5 = a_6 + \frac{b_6}{y_1}\frac{dy_1}{dt} \tag{18}$$

Thus, the variables in these expressions would already be included elsewhere in the model; the model would have come full circle and would consist of five endogenous variables (y_1, y_2, y_3, x_4, x_5), and four exogenous ones (x_1, x_2, x_3, x_6).

It would be possible, however, at the stage of model building represented by (16), to decide arbitrarily that variables x_4 and x_5, though recognized as endogenous within the ecosystem, will be treated as exogenous to the model. In other words, the values of these variables at any moment in time will be regarded as known by direct measurement, and they will not be derived within the model by functional calculations. This was, in fact, done tacitly in the model of (13)–(15) when x_3 (the mean cattle population) was treated as an exogenous variable.

A model is complete if all the variables appearing in the expressions are either calculated elsewhere within the model or are accepted as exogenous variables, the values of which are to be supplied as input. And the decision as to which variable shall be in each of these categories is an arbitrary one.

(*e*) *Division into subsystems*

Models of systems are hierarchic. Models of simpler or partial systems may be combined to form models of more complex systems, and (except in the simplest cases) models may usually be subdivided into submodels representing simpler subsystems. This hierarchic concept of systems and of their models has considerable pragmatic value.

The conceptualization of an ecosystem model is often facilitated by its division into subsystems. If the subsystems are arbitrary—if there is no objective basis for this particular method of subdivision—no overall simplification will result. But if it has a good biological basis, different processes will be operating within and between subsystems, and some of the processes will take place only within certain subsystems. Thus each subsystem will separately be a simpler problem than the system as a whole—not only in magnitude, but in complexity. This end will, in general, be attained if the subsystems into which the system is divided are internally more homogeneous than the system as a whole. If the elements of the system can be classified so that the various elements within any one particular class always have the same relationship to elements in some other particular class, generalizations about the behaviour of the class are possible, at least some of the processes operating will be common to all the elements of the class, and the total of processes operating may be much smaller than those in the ecosystem as a whole. Thus, if this class were treated as a subsystem, the submodel representing it need only take account of those processes affecting this particular class of elements. In our sagebrush ecosystem, for instance, the plants considered as a whole could constitute a subsystem, in which those processes peculiar to animals, microorganisms or the abiotic world could be ignored. And within this subsystem, one could recognize a nested subsystem consisting of the sagebrush itself, or an even more limited one consisting of the sagebrush seedlings only, with a further simplification of the subset of processes to be considered.

On a larger scale than these homogeneous subsystems, it may be worth recognizing and devoting separate consideration to subsystems which, though heterogeneous, are largely independent of one another in their operation. If submodels are to be developed first, and subsequently linked together to form a complete model, the difficulties arising at the time of linkage will be minimized if the interactions between subsystems are as few as possible. In our sagebrush ecosystem, if there are areas dominated locally by *Atriplex confertifolia*, say, it may be worth distinguishing these as spatial subsystems. Many organisms are tied to a narrowly defined locality for the whole of their lives or even for successive generations; they are thus exposed only to a local range of conditions more limited than that within the ecosystem as a whole, and can themselves have very little influence on parts of the ecosystem outside this narrow locality.

Each submodel will have its own list of exogenous and endogenous variables. But these lists will differ greatly between submodels. Variables will appear in the list for one submodel and not in that for another; an exogenous variable for one submodel will be an endogenous variable for another. Thus, when the submodels are combined to form a complete ecosystem model, the requirements of one in exogenous variables may largely be met by the output of others. For instance, a herbivore submodel may require information on the biomass of different plant species, which is provided by plant submodels, and on predator populations which is provided by a predator submodel. The requirement in exogenous variables by the model as a whole is the residuum of the exogenous variables of the component submodels after these have been matched against the endogeneous variables of other submodels.

(*f*) *Model Complexity*

A number of important decisions have to be taken in model building. One of the first is the set of desiderata already mentioned. These will at least define a minimum model. But there is no maximum and no optimum model, in a 'best-fit' sense. The degree of complexity of a model can always be increased without limit; and for any model of a given complexity there will always be one of greater complexity likely in principle to provide a closer fit and permit more accurate predictions. The degree of complexity aimed at constitutes a second important model-building decision.

There are three main roads to increasing complexity. In the first place, the state variables may be subdivided. In simpler models there is inevitably a great deal of lumping of ecosystem elements known to behave differently, but considered as one for the purpose of the model. By splitting these distinct elements, and defining state variables separately for each of them instead of for the whole undifferentiated group, one can increase complexity (and potential realism) very considerably. This splitting might, for instance, be along taxonomic lines; or it might differentiate age groups, or locations. Where in a simpler model all shrubs had been considered together, a new version might treat different species separately. A simpler model might treat the deer as forming a homogeneous population, a more complex one might distinguish stags, does, and fawns.

The second route to complexity is by taking account of a larger number of variables which may influence any given process. The number of variables which may influence each process is in principle very large; but the magnitude of their influence varies enormously. If one defines the range of values for each variable within the universe of ecosystems which it is intended to model, one could in principle determine the proportional influence of the variable, within this range, on the rate of the process (other variables being held constant at a central value). Initially, then, one would be likely to include in the model only those influential variables (exogenous and endogenous) whose effects were greatest in this sense. One could then proceed down the line, including more and more of these variables. And, as indicated above (p. 178), this might call for the inclusion of additional functions in the model. In the sagebrush system, for instance, ground-water recharge might initially be treated as a function of precipitation and evapotranspiration only; at a later stage soil infiltration would be introduced as a factor affecting it, which would in turn be determined by the lichen and algal cover of the soil surface, which themselves would then need to be modelled.

The third route to complexity is through the inclusion of more detailed mechanisms. A purely empirical model of an ecosystem can be developed without any consideration of mechanisms whatever—as with many regression models. Empirical models are generally, however, restricted in their application to the particular range of data on which they were based, whereas a model which reflects the underlying mechanisms of the processes involved is likely to have greater generality. And the more the detail of the model corresponds with the actual processes in the ecosystem, the greater the generality and precision that can be expected. For each process in the ecosystem, accordingly, it is a matter of deciding to what depth the mechanisms involved should be modelled—or, to put it in another way, what are the 'black boxes' whose internal mechanisms will be ignored, and for which only an empirical relation between inputs and outputs is to be included in the model. In our sagebrush system, the rate of photosynthesis could initially be treated as the output from a 'black box', with inputs of radiation, temperature and soil moisture; at a later stage the role of stomatal movement might be taken into account and the 'black box' would be subdivided into two, one of which (that for stomatal aperture) provided an input required by the other.

The degrees of complexity in different parts of the model will need adjustment in accordance with the purposes to which the model will be put. If certain state variables are designated as having special interest, those parts of the model which have the strongest influence on these state variables will be those where extra subdivision and detail of mechanism will best repay the effort.

(*g*) *Parameter estimation*

Once one has decided on the structure of the model, the variables to be taken into account, and their functional relationships, one needs to estimate the parameters involved. The ways in which this may be done will depend on the form of the model.

Experimental methods may sometimes be practicable. Where the variables on the right-hand side of the equations can be controlled and modified by the investigator, direct experiment can provide very satisfactory estimates of the parameters. An experimental approach may sometimes enable the full range of conditions to be covered more easily than by relying on observations without experimental manipulation. In the ecosystem context, however, few of the important variables are susceptible to experimental control, and in consequence a purely experimental approach is confined to unrealistically simple systems. A model relating the dependent variables y to cattle grazing (x_3), without regard to meteorological factors, for instance, could indeed be (and often has been) the subject of experiment. But its naïveté restricts the range of possible generalization enormously, and this largely explains the limited role of experimentation in developing and testing ecosystem models.

Where the state variables are expressed as sums of power and product terms involving independent variables and values of state variables (or their rates of change) at some previous point in time (e.g. (2)–(4) above), the usual methods of regression analysis may be used to estimate the coefficients from a set of replicate observations on samples of the ecosystem in question. If the expressions are non-linear (as in (5)–(7)), regression calculations, though much more troublesome, remain possible; and the increased number of independent variables in such a system as (8)–(10) poses no difficulty in principle.

Any model in which the state variables of interest can be expressed as an analytic function of the independent variables, which are regarded either as constant throughout the relevant period, or as constituting a step function with specified periods for the steps, or whose values can themselves be expressed as an analytic function of time, can be fitted and tested by regression methods.

Regression estimation of the parameters would require a set of observations on samples taken from the universe of ecosystems under consideration. In each ecosystem, measurements would be required of each of the variables entering into the equations—the vegetational biomass and deer populations at the beginning and end of a fixed interval, the mean temperature and cattle population, and the total precipitation and ground-water recharge over this period.

Regression methods have certain built-in assumptions. It is assumed that the variables on the right-hand side are independent ones, taking fixed and known values which are unaffected by changes in the dependent variables, and that each observed value of the dependent variable is normally and independently distributed with uniform variance around the value predicted by the model. If these assumptions are not met, the parameter estimates will not be optimal and may be biased. To fit a regression, then, implies an amplification of the underlying model, incorporating these assumptions about the residual variance.

Where some of the state variables appear on the right-hand side of certain equations, as well as on the left of one (as in (11)–(12)), a direct regression approach is not possible, for it is an assumption that the independent variables can in fact be varied independently of one another and of the dependent variables. If, however, the equations can be re-cast so that only the truly independent variables appear on the right-hand side ('path analysis'), a regression approach can be used. An added complication arises, however, in that the same parameters a_1, a_3, b_1, b_3, c_1, c_3, d_1, and e_1 will all be involved in the expressions both for y_{1t} and y_{3t}; their estimates should consequently take account of evidence from each source (see Box and Draper, 1965).

Only exceptionally can models consisting (like (13)–(15)) of a set of differential equations be treated by regression methods. A first requirement for this to be true would be that the rates of change constituting the left-hand sides of the expressions could be measured directly, and could thus be used as dependent variables. There are a few rates of change of importance in ecology to which this applies—radiation, photosynthesis, and respiration are among them—but, in the main, rates cannot be measured directly and mean rates over substantial periods have to be deduced from differences beween successive values of state variables. A second requirement is that the state variable of which the differential constitutes the left-hand side should not also appear on the right-hand side. This is exceptional in ecological systems—the rate of change of a state variable usually depends on its own current value, and also very commonly on the current value of the other state variables in the system.

Consequently, it is rare for regression methods to be applicable directly to fitting and testing an ecosystem model based on a set of differential equations. If these equations could be solved and converted into analytic expressions where each state variable was expressed as a function of the independent variables only (as in (5)–(7) above, for instance), regression methods would again come into their own. But the form of equations arising in ecosystem studies rarely makes this possible—particularly since they often involve discontinuities and threshold effects. Consequently, numerical solutions are necessary, and the regression approach falls to the ground.

Often, in a system like (13)–(15), the constants of the equations may be estimated—usually by trial and error—from a temporal sequence of observed values of the variables. This can be very laborious, however, particularly if a digital computer is used—the process can be much more expeditious with a hybrid computer. In principle, this procedure seems little different from that of fitting a regression model; however, no formal treatment has been developed for estimation of a covariance matrix for a set of parameter estimates in such a system. The usual methods of regression analysis assume that the different observations of the dependent variables are separate and uncorrelated samples from a hypothetical population of possible observations. This requirement is far from being met by successive observations of state variables in the same ecosystem.

Where the parameters of a model are estimated by *ad hoc* experiments rather than from sequential observations of the state variables in the whole system, the situation is different. The dependence of the rate of change in one variable on the factors influencing it can be studied separately, and—in principle, at least—a covariance matrix for the estimates of parameters in the expression can be obtained. In the system (13)–(15), for instance, an irrigation experiment might be set up to determine a_1, b_1, a_2, and b_2; feeding experiments with cattle and deer to determine c_1 and d_1; and experiments on deer mortality and reproduction rates to determine a_3 and b_3. The results of these experiments would be brought together into a model which would calculate changes in the three state variables.

If the various expressions in the model have been the subject of separate and independent experiments, it can be assumed that the covariance between the estimate of a parameter occurring in one expression and the estimate of a parameter in another expression will be zero, and a complete covariance matrix for the model parameters is consequently available.

Whatever method of parameter estimation is adopted, it is desirable that the data on which the estimates are based should cover the whole range—the universe—to which the predictive model is intended to apply. If the estimates are based on observations on entire ecosystems, then these ecosystems should be a random (or stratified) sample from the whole range of ecosystems of interest. If the estimates are derived from subsystems, then the values of influencing variables included should again constitute a representative sample of the values in the universe for which the predictions are intended. If this requirement is not met, the estimates are likely to be biased.

3. Model testing

Once a fully defined model of an ecosystem has been constructed, one can proceed to test its agreement with the real-life system it is intended to represent—to *validate* it, as it is sometimes termed. It must not be supposed from this term, however, that one model may be valid and another invalid.

In the first place, no quantitative model is perfect. Since simpler, a model will never be an exact representation of conditions in the real world. Almost always, then, the appropriate question is not, 'Should this model be accepted or rejected?' but 'How good are the predictions this model makes—what errors are associated with them?' Or the question may be which of two models fits more closely to reality. It will later become evident that even this question may not have an unambiguous answer. In any case, validation is never absolute. One can never even say that the predictions are unbiased—only that they, or their bias, lie within specified limits with specified confidence.

Ecosystem modelling is not usually comparable with the testing of a null hypothesis in statistically designed experimentation. It is more akin to an estimation problem—it is directed towards making quantitative statements about the universe of discourse, rather than to testing hypotheses. If two models are being compared, the 'significance' of the difference between them is not usually relevant. Even if the more complex model is not 'significantly' better, the fact that it is marginally better may make it preferable. If a factor included in a model is known to affect a process, the fact that a particular test has not shown it to be 'significant' is not a good ground for excluding it. But a statistical approach to model testing is required in that we need to know the errors of prediction, their distribution and how they differ in different parts of the universe.

If one wishes to make a statistical statement about the agreement between model and field observations, it is necessary that the two sets of figures compared—those from the model, and those from the field—should be independent of one another, or that the comparison procedure should have built-in provision to eliminate the effects of their dependence. Many standard statistical procedures use the same data for parameter estimation and testing, but are designed so that the tests and estimates are nevertheless statistically independent. It should be remembered, however, that the independence in question may itself depend on underlying distributional assumptions, and if these are not met the supposed independence of the tests and the parameter estimates tested may be illusory. This risk can always be avoided by taking independent samples from the same universe for parameter estimation and for model testing; though one may lose in efficiency by doing so, one gains in assurance.

(*a*) *The range of generalization*

In considering the adequacy of the model to make predictions, we are not usually interested in predictions for one particular concrete ecosystem—one topographically designated area of sagebrush during a specified period of time. As stated above, a model needs to have general applicability over a specified range or universe, and its fit to a particular set of data is of interest only in so far as these are representative of the universe—a sample of it. The statement of relevance of the model should indicate the precision with which its predictions will apply to *any* random sample from this universe. Sometimes this universe might be only the successive states of a single concrete ecosystem. More often it will be a class of ecosystems defined by composition, structure and/or geographical location. In the present case, we might, for instance, define our universe as consisting of all sagebrush-dominated ecosystems at altitudes between 1300 and 1660 m., between 112° and 116° W longitude and 40° and 42° N latitude. The criterion of success of our predictive model would then be the accuracy with which it could predict changes following altered grazing pressures in any arbitrary sample from these ecosystems.

Once one has a fully defined model, it may be validated by comparison with a sample of ecosystems representative of the universe to which the model is claimed to apply. For some purposes a random sample of ecosystems as they now exist—a random sample, for instance, from the set of actual sagebrush ecosystems defined in the previous paragraph—may be suitable. But often one is interested in prediction not only for ecosystems as they are now, but as they may become under certain stresses. In this case, the universe to which the model is intended to apply is only partially, and not proportionally, represented by any actual set of ecosystems from which a random sample might be taken. It may then be best to stratify the set of ecosystems available, and apportion the sample according to the interest of each stratum for one's purpose rather than their existing proportion in the landscape.

One may go further and impose modifications on some of the samples, so as to ensure coverage of the full range of ecosystems to which the model is intended to apply. This is the function of experimentation in model validation—it enables a broader range of ecosystems to be covered, and thus increases the generality which may be claimed for the model. The same purpose could often be served by seeking and waiting for the right conditions to occur in the ordinary course of events. But by taking the initiative the investigator can speed up the process and greatly increase his own convenience.

(*b*) *The use of common data for estimation and validation*

Often, the model one wishes to validate is incompletely defined. Its mathematical form may have been specified, but no (or inadequate) information may be available as to the values of the constants included in it. In such a case it is very tempting to save time and effort by using the same data for estimation of constants and for validation—that is, estimation of variance between model and reality. The extent to which this can be done will vary with the type of model.

Where the variables which one wishes to predict are expressed in the model as analytic functions of measurable variables, such joint use of the data is generally acceptable. In a designed experiment on complete ecosystems, with controlled variation in one or more of the independent variables, this can be done; and it is also possible wherever regression methods can be applied. In these cases, independent estimates are available of the error variance around the expected values, and of the error due to imperfect estimation of the parameters, so that the overall precision of the model can be deduced. It should be emphasized, however, that these error estimates will apply only to that population of ecosystems from which the sample studied was drawn. It cannot be assumed that the same precision will apply throughout the universe for which prediction is intended, particularly if the tests have only covered a portion of this universe. And if the experimental design included hierarchical features, the appropriate error term may need careful thought.

It should be remembered that the standard methods for regression analysis and analysis of variance rest on a number of assumptions, including those of normality and homoscedasticity of the error variance. If these requirements are not met, the supposed independence of the variance of parameter estimates and of the deviations from the model will no longer apply, and biases will occur.

Consequently, even though estimates of residual variation are available when a regression has been fitted or a designed experiment analysed, validation of the model by a second series of independent samples may, in some respects, be intellectually more satisfying than the usual procedures for estimating variance between the model and the observed values. Though these procedures are reasonably robust in cases where the assumptions as to population structure are not met, there is always room for suspicion that the deviations are sufficient to falsify the conclusions. If a second sample of ecosystems is taken, and, for each of them, predicted values of the dependent variable are calculated based on the parameter estimates developed from the first sample, then comparison of these with the observed values will provide an estimate of bias and of variance in the predicted values which does not depend in any way on the assumptions underlying the regression analysis or analysis of variance.

(*c*) *Errors of estimation and model imperfections*

In regression analysis, the uncertainties may be divided into the errors in estimating the regression coefficients and the variance around regression (which includes both random variation in the values of y for any given x, and model inadequacies); likewise, in a model of any arbitrary type one can also distinguish variance due to model inadequacies from that due to imperfect fitting. As already indicated a covariance matrix for the model parameters can usually be deduced—except in cases where the fitting is by trial and error. To proceed from this to deduction of a covariance matrix for the model output—the new values of state variables after an interval of time—is a task which has never been attempted in the general case, and may well prove baffling. A Monte Carlo approach may, however, be used; the model is run repeatedly with random samples from the fiducial distributions of parameters, and thus an empirical distribution of output values is obtained for comparison with those observed. Provided no observations used in this comparison enter into the calculations of parameter estimates (or were correlated with other observations which did so), then the conclusions will be unexceptionable. If, however, some of the data used in estimation of parameters were associated with those used for comparison purposes, assessment of the agreement or disagreement between model and reality, in a statistical sense, may become far more complicated.

However, the errors in fitting a model—in estimating the parameters—are rarely of interest in isolation from imperfections of the model itself. The variance of model output which depends on error in parameter estimates will never be more than a part of the variance between model output and reality; it forms a lower limit only. It is the larger quantity—the total variance between predictions and reality—which is of prime concern, and the contribution from error in parameter estimates is of interest only in that it points to the potential value of improving these estimates, as against a possible change in model structure. For general validation of the model—in order, that is, to assess the precision of the estimate it might give for an ecosystem arbitrarily selected from the universe of reference—figures for the total variance between predicted and observed values are needed, and these cannot be obtained from the model alone. Comparisons with replicate samples of field data are required.

It should be remembered that precision is unlikely to be uniform over the whole universe to which the model is intended to apply. Bias may also vary over the universe, and may even differ in sign. Variation in precision for the different types of prediction in different parts of the universe may indeed not be unwelcome—the precision should be greater in those parts where predictions are regarded as more important. It also can provide valuable clues to deficiencies in the model, and guide its further development and improvement.

(*d*) *Tests on subsystems*

Where a model is built from functions describing a number of partial processes, each of which has been studied independently, it may be useful to perform validation tests on subsystems—physical constructs incorporating interactions between some of the processes that have been studied separately, but still at a lower level of complexity than the system as a whole. If the interactions can be ignored in the whole system, it should usually be possible also to ignore them in a subsystem—and in the simpler context the sources of unacceptable discrepancies should be more readily identifiable. Consequently, in developing models of complex systems, validation of subsystem models against artificially simplified real-world systems may play an important part, though the real test of the model comes only when the comparison involves the full complexity of the universe of reference.

(*e*) *The data required for validation*

An ecosystem model will generally require input in the form of a vector of values of state variables at a starting time, together with records of exogenous variables, including meteorological factors, impinging on the system from that date onwards. Some of these may take the form of step functions—like the predation pressures on ungulates when the hunting season opens. With this input data, the model will enable the changes in the state variables to be calculated over any specified period. The requirements for a field study intended to validate a computer model are consequently an inventory of the values of state variables at a fixed point in time—an inventory which must include all those taken into account in the model; and a continuous or near-continuous record of exogenous variables involved in it. Without this, the model cannot operate. And it is evident that at least one observation of one state variable on one subsequent date will be needed to compare with the performance of the model. It need be no more; but clearly the more variables are observed and the more frequently they are recorded, the more complete will the validation be, and the better the chance of making intelligent guesses at the origin of discrepancies, so that the model may be improved.

As regards the initial inventory, the criterion of simultaneity must be emphasized. If some of the inventory is performed in March, some in April, the model will be able to make use of the April data only by making assumptions about the values then of the state variables that were measured only in March, or vice versa. The model itself will not provide means of extrapolating forward or backward, given only part of the data—unless, perhaps, by an iterative procedure. Of course, it is often hardly practicable to make observations on all state variables on the same day, so that the ideal of simultaneity can only be approximated. In this case, one will have to use general knowledge of the system—not the model itself—to extrapolate or interpolate for some of the variables, which will be an additional cause of uncertainty in the validation process.

Error from this source will be minimized if the variables whose initial values are approximated in this way are not changing fast. This argues that a suitable time for the initial inventory will be when the ecosystem is relatively quiescent—when few or no state variables are undergoing rapid change. Then observations on the stable variables can be spread over a longer time period, and a special effort can be made to concentrate observations on the more labile variables at a fixed time, to which the others can then be extrapolated or interpolated.

(*f*) *Use of existing data*

In the literature, and in the records of field stations, there are many cases where a study area has been observed repeatedly year after year, and the question is sometimes asked whether such series of observations cannot be used to validate models. In such cases, unfortunately, it is rare for the records to include all aspects of the ecosystem; they may include meteorology, soils and plants, but no animals. Or they may include larger vertebrates but none of the smaller animals. Consequently, a full initial inventory as a starting-point for the model is not available—unless the model is simple enough only to require such state variables as happen to have been measured. However, for certain types of model existing series of records, even if incomplete, can provide useful validation; this is true *if the model can be run backwards*. In this case, it would be possible to take a complete inventory of the ecosystem now, run the model backwards (making use of the meteorological and other exogenous factors recorded), and compare the values of particular state variables on some earlier date with those actually observed.

It is always an assumption of a deterministic model that a subsequent state can be predicted from an adequately specified antecedent state; and even with a stochastic model it can be predicted with a known penumbra of uncertainty. In order that a model should be run backwards it is necessary that a given subsequent state of the system, adequately specified, should follow only from one particular calculable antecedent state—with or without definable stochastic uncertainty. There may, however, be circumstances where a given present state of an ecosystem may correspond with a variety of previous states, and in this case the model cannot be operated backwards. This would happen, for instance, if a natural calamity destroyed all members of a particular species which were not in certain protected localities. Where, however, no such discontinuities occur, it may well be worth studying the possibility of using incomplete past data, supplemented by a full inventory at the present time, for validation purposes.

(*g*) *Comparison of models*

As stated above, validation of a model is never absolute. It should be expressed in terms of the precision of estimation of one or more state variables after a specified period, in an ecosystem constituting a random sample from the universe to which the model is stated to apply—or the variation in precision in different parts of this universe. Models may differ greatly in precision, and it may be possible to say that one represents the ecosystem more precisely or less precisely; but to say that a model is or is not a valid representation of an ecosystem would imply an accepted criterion of precision, against which the actual precision of the model predictions were being compared. Generally, we have no such absolute criterion. Rather than tests of single models in isolation, we are much more often concerned with tests of the *relative* precision of two or more alternative models. Comparison of models is in fact the way in which progressive model development and improvement takes place; the new model is compared with the old, and is retained only if it proves to be more precise.

The principle of continuity implies that there will normally be progressive divergence between the true values of variables in an ecosystem and those predicted by a model, however good. The value of each variable at a moment in time is serially correlated with the values of the same variable before and after, and the correlation approaches unity as the time interval decreases. Consequently, any continuous model will give virtually perfect prediction of the values of state variables after a very short interval, and the longer the interval over which prediction is attempted the more rigorous the test of the model.

When only a single prediction is at issue, a comparison of models seems straightforward enough; one could presumably choose the model giving a minimum mean-square deviation of predicted from observed values, when predictions were made over numerous samples from the universe specified. In making these comparisons, it would be necessary to specify the interval of time over which prediction was to be made, as well as the state variable in question and the universe of discourse, if the conclusion was to be unambiguous.

But an ecosystem model will usually predict changes in numerous state variables, and will predict them with widely varying precision. Where one is comparing alternative models, it cannot consequently be expected that one model would give better predictions for all the variables. The relative merits of the alternative models will also differ in different parts of the universe covered, and with the time interval over which the prediction is made. Thus it is to be expected that the choice of model would depend on which variable one is considering, on the time interval, and on the region of the universe in question. As mentioned earlier, the universe need not—and perhaps should not—be taken as including only ecosystems actually existing, and those into which they may change by succession or other natural modification. The practical value of a model will be greatly enhanced if the universe to which it applies includes also such modified ecosystems as might result from human manipulation of those actually existing.

It is clear, then, that a comparison of models making multiple predictions calls for differential weighting of the various predictions. The appropriate weighting depends on the purposes for which the model is to be used, and seems inescapably subjective. Unless the goal can be equated with prediction of a single variable only (or a single definable function of several variables), the weighting of different variables will be arbitrary. And since, as indicated, the universe though definable is not enumerable and has no common measure, the weighting of its various regions is also arbitrary. Once this arbitrary weighting function has been selected, however, its application to selection of models can be objective and unambiguous.

Since a model cannot be simultaneously optimal for different predictions, it follows that the best strategy may call for different models for different purposes. The extra precision obtainable must then be weighed against the extra modelling effort involved. The latter, may, however, be considerably reduced if the models are built on a modular principle, so that different subsets of models can be combined according to the needs of the situation.

(*h*) *Choice of variables for validation*

While the initial inventory of an ecosystem intended for validation of a model should cover completely the input information needed by the model, subsequent observations for comparison with model output need only be a selection of the variables included in the model. So far, the question of how these variables should be selected has been avoided.

Clearly, the more measurements there are, the better the validation. Every additional measurement that can be predicted by the model, whether a state variable, a rate of change, or some function of them, can make a contribution and increase one's knowledge as to how precise the various predictions given by the model are, and where the gaps between model and reality are widest. But the resources available for validation are in practice limited, and the problem therefore becomes one of optimization: given these limited resources, how should they be expended to give the best possible information on the precision of the model?

In solving this problem, many diverse considerations need to be taken into account. First and foremost, we may mention the desiderata whose prediction is the prime purpose of the model. If the desiderata themselves can easily be measured, these values will provide the best possible validation of the model. Otherwise, other values more easily measured which are closely related to them may be substituted.

It has been pointed out above that, for model comparison, a weighting for the different desiderata is required. In selecting an optimum set of variables for monitoring, too, the subjective weights accorded to the desiderata should be brought into the calculations.

Another point to be taken into account is the correlation matrix of the variables under consideration. Recording numerous highly correlated variables will involve great redundancy and wasted effort. If a set of variables with minimal mutual correlation can be selected, the return in terms of unit effort expended will be maximal.

Redundancy of correlated measurements applies strongly to repeated observations of the same variable. The serial correlation of such observations will be very high, though the time scale over which it operates will vary greatly from one variable to another—with perennial plants, correlations may be perceptible over many years, whereas for some insect populations they may hardly extend from one week to the next. By choosing for repeated monitoring a sub-set of variables in which serial correlation is low, observational effort will give better validation return.

Another consideration in selecting variables to be monitored is the ease of making the observations at a given accuracy. It may well be better to choose an easy measurement less closely related to the desiderata than one very closely related to them which requires a considerable expenditure of effort. This factor may often lead to the measurement of a derived quantity rather than a state variable. For instance, soil nitrification rate or carbon dioxide production may be measured more easily and accurately than the biomass of soil-inhabiting micro-organisms; and it may be possible to use the proportional change in a trapping rate for small mammals, under specified conditions, for testing a model without converting the figures to absolute population density.

Yet another factor in the choice of variables to monitor is the sensitivity with which the different variables reflect differences in model structure. Clearly, the more the predicted value of a variable differs from model to model, the more useful is a real-life value for that variable in guiding the choice of models.

The way in which these various factors should be combined in order to optimize the distribution of effort in validation measurements would take us beyond the scope of this paper. In principle, the answer will clearly be different for each pair of models one might wish to compare. But since, in general, model development and validation studies will proceed side by side, a temporary answer must be developed which, though not necessarily optimal for any particular model comparisons, will approach optimality for a wide range of comparisons. Only when model development has reached an advanced stage and the range of choice is quite narrow can the optimization procedure be expected to become really effective.

Reference

BOX G.E.P. & DRAPER N.R. (1965) The Bayesian estimation of common parameters from several responses. *Biometrika* 52, 355–65.

Population models as a basis for pest control

G.R.CONWAY and G.MURDIE *Department of Zoology and Applied Entomology, Imperial College of Science and Technology, Silwood Park, Ascot, Berkshire*

Summary

Population models are of value as aids to pest control at a number of different levels. This is illustrated with examples drawn from models of component processes, i.e. reproduction and predator-prey interactions, and from population models of mosquitoes and red bollworms.

Before the Second World War there was good reason to expect that pest control would develop as a sound, ecologically-based applied science. In the intervening years, good, basic research has been done, and at times there has been an infusion of ideas from the pure sciences. However, there have been relatively few attempts to create a general and workable theory. As a result, pest control has been dominated by an empirical approach which has relied heavily on cure-all solutions. The discovery of DDT in 1944 heralded an era of pesticide discovery and use in which ecological considerations took second place. Then, in the middle 'fifties, as awareness grew of the drawbacks of pesticide use, biological control emerged as the new panacea. Subsequently, other solutions have been pressed in turn, the most recent being the various techniques aimed at pest eradication through sterile matings.

Nevertheless, in recent years there has been an attempt to form a middle ground. This began with the concept of integrated control, first applied to combining chemical and biological control techniques and later expanded to include the evaluation and integration of all possible control techniques for a given pest problem (Smith & Reynolds 1960, Smith & van den Bosch 1967). Where it has been applied, this has met with much success (Pickett *et al.* 1958, Stern *et al.* 1959, Wood 1971). But it is still largely an empirical approach. As the experience of one of us (GRC) in Malaysia typifies, integrated control packages usually emerge as a result of a laborious trial and error process (Conway 1971 *a* & *b*).

The problem has been the sheer complexity of pest situations. Even the simplest of pest problems involves a complex system in which at least the interrelation of a number of key variables has to be understood before the effect of various control actions can be predicted. This is perhaps an obvious point, but as the simplistic nature of post-war pest control practices indicate, it has been largely ignored, deliberately in many instances, but for the most part because of the lack of appropriate techniques and tools for handling complexity of this kind. It is only in the last ten years that entomologists have begun to come to grips with this problem.

Progress has been made along three lines. First, there has been an increasing effort to gather and sift information on pest situations within a more logical framework. Techniques such as mark-release-recapture, life table construction and key factor analysis have come to play an important part in unravelling what has been called the 'life system' of a pest (Clark *et al.* 1967, Morris 1963, Southwood 1966). Second, there has been a growing body of knowledge on predator-prey relations based on laboratory experiments which has contributed to the theory of biological control (Burnett 1960, Flanders & Badgley 1963, Huffaker *et al.* 1963). Finally, attempts have been made to build mathematical models of pest situations using the techniques of systems analysis and modern computing methods (Cuellar 1969 *a* & *b*, Hughes & Gilbert 1968, Jones 1965, Hassell & Varley 1969, Watt 1961, 1963 & 1964).

Modelling has always been implicit in pest control. Faced with a pest situation, the pest control worker invariably creates a model of it either in his mind or in the form of a laboratory population or by taking a sample plot of field. He then simulates control on this model producing a new state which he compares with the actual pest situation that he desires. The advantage of mathematical models in this context is that they are more ordered, and more easily lend themselves to manipulation and simulation. All models are, of course, abstractions of the real world, and no model is ever true in any absolute sense. Different models are only relatively better or worse at representing events and entities in the real world. We can characterize models in terms of their generality, realism and accuracy. Generality is a measure of the breadth of situations to which a model applies; realism is a function of how closely the language and structure of the model mimic the actual entities and processes in the real world; and accuracy is a measure of the degree of prediction. Ideally, perhaps, a good model combines each of these characteristics to an equally high degree. But it does not mean, as some workers imply, that this is the yardstick by which all models should be judged. Mathematical modelling is not an end in itself and criticism should depend on the objectives of the model and on the context in which it is formulated. This is particularly true of an applied science such as pest control where the aim is to solve a real-time problem in the real world. Then, the nature of the problem and the balance between results and costs of modelling in terms of time and money have to be considered.

In pest control, we are concerned foremost with accuracy. The primary test of a pest model is that it will lead to a control strategy which will give, within certain limits, predictable results. For example, Morris (1963) and Watt (1968) found that they could relate the fluctuations in pest numbers to environmental variables such as midsummer temperature. Models of this kind can form the basis of effective control but they are rarely consistent in the long term. Moreover, when there is a failure in prediction, an explanation cannot be sought from the model itself, since the causal mechanisms relating the environmental variables to the pest numbers are not included. Predictive pest models thus require a certain minimal degree of realism built into them if they are to be of more than transient use. However, if we are to produce such models speedily and economically, we also require models of a quite different kind, in which a high premium is placed on generality. These can contribute to the general theory of pest control but, more important, serve as guides to the form of the more specific predictive models. In this paper, we will describe a number of models of both kinds which we have been developing over the past four years.

A teaching model

The first of these is a very simple model of a hypothetical insect population and is intended primarily as a teaching tool (Conway 1970). If we take a population occupying a uniform discrete habitat, then we can represent its changes in numbers of days in a flow diagram (Fig. 1). In the model, development rate is taken to be constant with the population classified according to the stage in the life cycle and to age. Migration is ignored, but survival and reproduction are regarded as simple density-dependent processes. Thus, each day the average mortality for each stage is computed as a function of density in the following manner:

$$\text{AVMORT} = 1 - e^{-k\text{DENS}}$$

where AVMORT is the average mortality for the stage, DENS is its density, and k is a mortality coefficient.

A stochastic element is then introduced by assuming that the actual mortality is normally distributed about the average mortality. A separate sub-program takes the average mortality and its standard deviation and, using a random number generator, allots an actual mortality from this normal distribution. Reproduction is treated in a similar manner, average fecundity

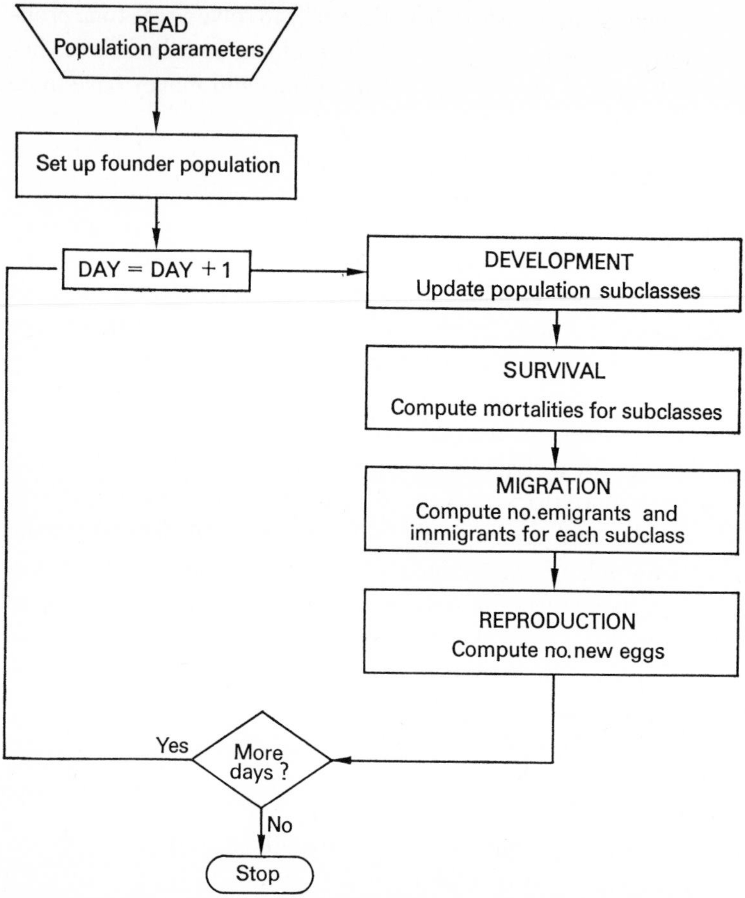

Figure 1. Flow diagram of the changes in numbers of a simple insect population.

declining exponentially with density.

When the standard deviations used in computing the actual fecundities or mortalities are very small, the result is a very tightly regulated population. As the standard deviations are increased, so the density dependence weakens, and a much more violently fluctuating population is produced. On this basic model, a number of different pest control techniques can be simulated. In Fig. 2, the effect is shown of seven releases, at two-day intervals, of 250,000 sterile males on a tightly regulated population. In Fig. 3, the same release program is made against a strongly fluctuating population. As can be seen, eradication occurs in the latter case but not in the former, illustrating the greater vulnerability of less tightly regulated populations to this kind of technique. Combinations of different techniques can also be simulated. Figure 4 shows the effect on the tightly regulated population of combining

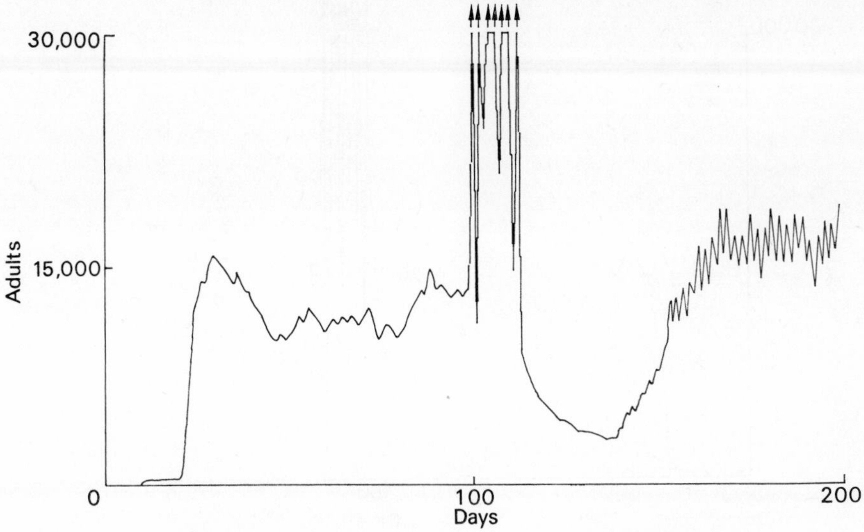

Figure 2. A hypothetical insect population under strong density dependent regulation with 250,000 sterile males released on day 100 and on six subsequent occasions at 2-day intervals.

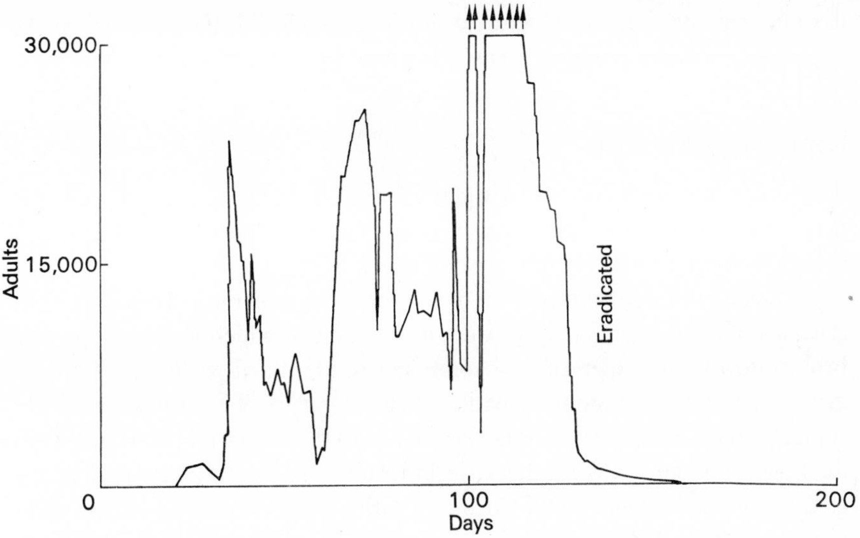

Figure 3. A hypothetical insect population under weak density dependent regulation with sterile males released as in Fig. 2.

the same program of sterile male release with an application of an insecticide on day 100 which produces an initial 90 per cent kill of larvae and pupae, followed by a residual kill decreasing each day by 5 per cent of the previous

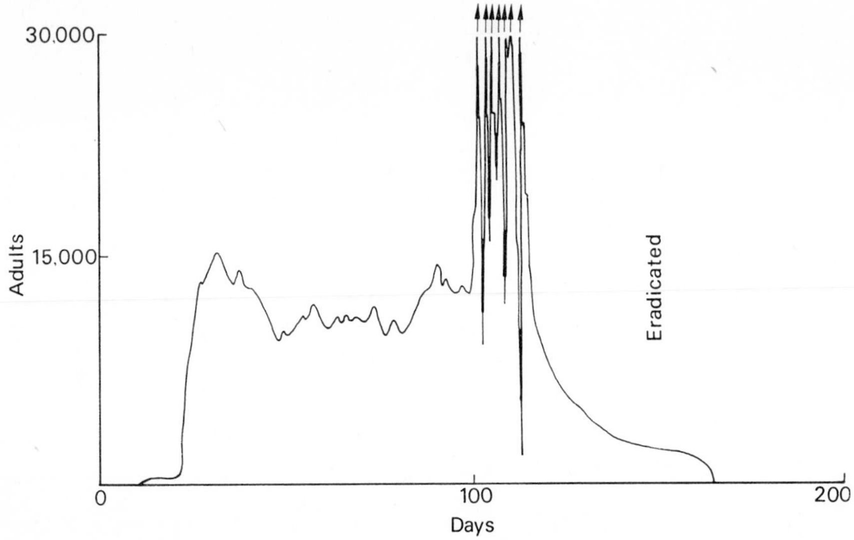

Figure 4. A population as in Fig. 3 but with a single insecticide application on day 100 producing 90 per cent larval and pupal mortality with 95 per cent residual kill in addition to the sterile male release.

days percentage. This produces eradication, which did not occur when the sterile male release program was used alone.

Reproduction model

The next two models are similarly general models, but they aim at a more detailed understanding of some of the component population processes, which in the teaching model were only treated superficially.

The first of these is an attempt to represent the dynamics of reproduction. A series of component analyses were carried out in which reproduction was broken down into a hierarchy of components, identified partly on deductive grounds, and partly from information in the literature. The analysis was continued as far as relevant data was available or to the extent that reasonable deductions could be made. Mathematical models were then built for each of the components and then assembled into a large computer model of the daily changes in an insect population, the other major components, development, survival and migration, being represented in the simplest manner possible. The whole model was then used as a basis for simulation studies aimed at understanding the different ways in which various pest control strategies affect population size through the reproduction component (Conway 1969).

The simulation we report here was concerned with exploring the implications when fecundity per female decreases exponentially with density. Watt

(1960) has suggested that in such a situation insecticide applications could, in fact, induce pest outbreaks.

Fecundity was represented as follows:

$$\text{FEC} = \text{BFEC}\ e^{-k\text{DENS}}$$

where FEC = no. of eggs laid, BFEC = the basic fecundity,

DENS = the density, k = density coefficient of fecundity

BFEC was put equal to 100 and k to ·1. It was assumed that reproduction was assexual and the ovarian maturation period was set at 5 days. This component was then assembled as part of a model describing a hypothetical population inhabiting an area of 100,000 units. Individuals in the population were assumed to pass through a life cycle consisting of two days in the egg stage, ten days as juveniles and then living as adults up to a maximum of twenty days. Mortality was set at ·01 per stage per day and there was assumed to be no migration. A founder population of 500 new eggs was established and studies made of the subsequent daily population changes. Figure 5 shows the kind of regular population oscillations which are set up in these circumstances. Simulations were then carried out in which the application of insecticides at various phases in the oscillations were mimicked. The applications were assumed to give 90 per cent kill of the adults on the first day of application with a residual kill decreasing by 10 per cent each subsequent day. As can be seen, insecticides applied during or after the peak of the oscillation (Figs. 6, 7) instead of dampening the fluctuations, produce a massive outbreak.

When fecundity per female declines exponentially in this way, there is a critical density at which the total fecundity is maximal (Fig. 8). In the uncontrolled population, the density changes fairly rapidly and little time is spent at the peak egg laying density. However, following insecticide application, the population falls well below the critical density, but then recovers slowly as the insecticide becomes less effective. In consequence, considerable time is spent in the region of peak egg laying and an outbreak occurs. Outbreaks of pests following insecticide applications are known to occur and the model is valuable in indicating one mechanism that might be responsible.

Predation model

The second model looks at another of the major population processes, survival, and at one component of this, predation (Murdie 1970). It is concerned with effects of spatial heterogeneity of prey distribution on the success of predators having different search behaviour patterns.

Prey populations of varied density were distributed either at random or in clumps in a matrix of 1000 × 1000 locations in the computer core. Two types

(5)

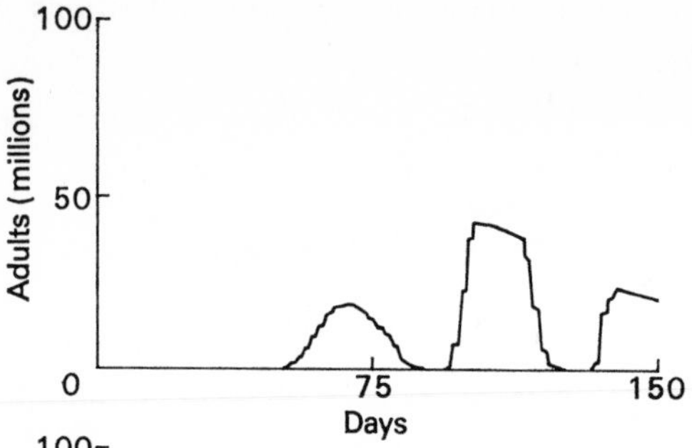

(6)

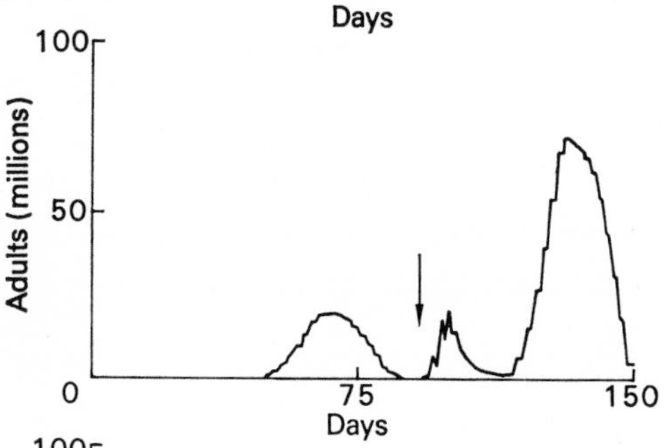

(7)

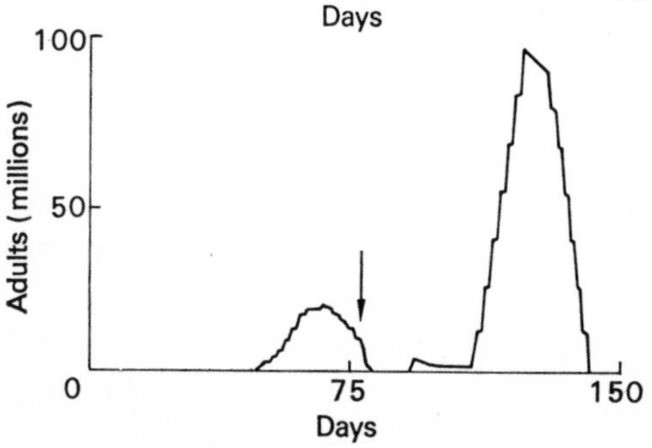

Figure 5. Simulation of an insect population in which fecundity per female is a declining exponential function of density (see text).
Figure 6. Population as in Fig. 5 with an insecticide application on day 90 producing a 90 per cent adult kill with 90 per cent residual.
Figure 7. As in Fig. 6 but with an insecticide application on day 79.

of predator movement were then simulated. In the first or 'blundering idiot' model, the predator follows a simple random walk which is unaffected by successful prey encounters. In the second or 'adaptive' model (Fig. 9), the predator similarly follows a random walk, but with a low probability of turning until a prey is encountered. There is then an obligatory increase in turning so that search is restricted to the region of prey encounter. After a defined period in which there are no further encounters, the original random walk is resumed. In each simulation, a single predator was introduced into the quadrat and permitted 200 consecutive movements. A total of 200 simula-

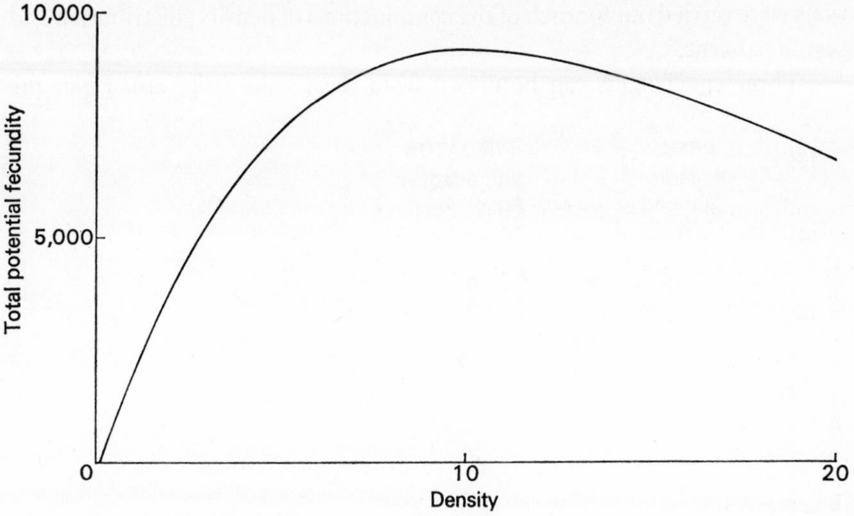

Figure 8. Graph of total fecundity as a function of density when fecundity per female declines exponentially (see text).

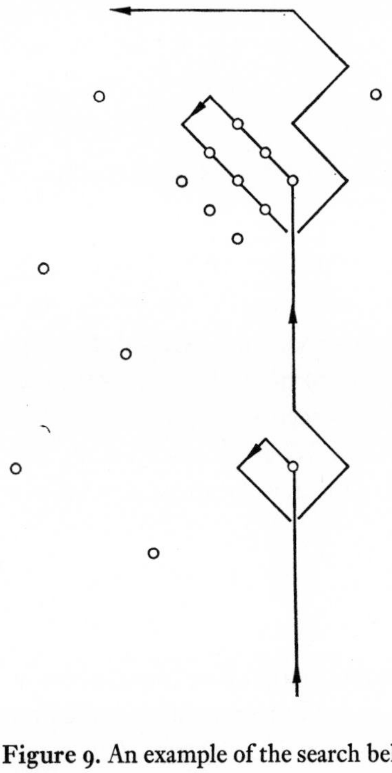

Figure 9. An example of the search behaviour pattern of the hypothetical 'adaptive' predator.

tions were carried out for each of the combinations of density, distribution and search patterns.

Three conclusions can be drawn from the results (Fig. 10). First, the

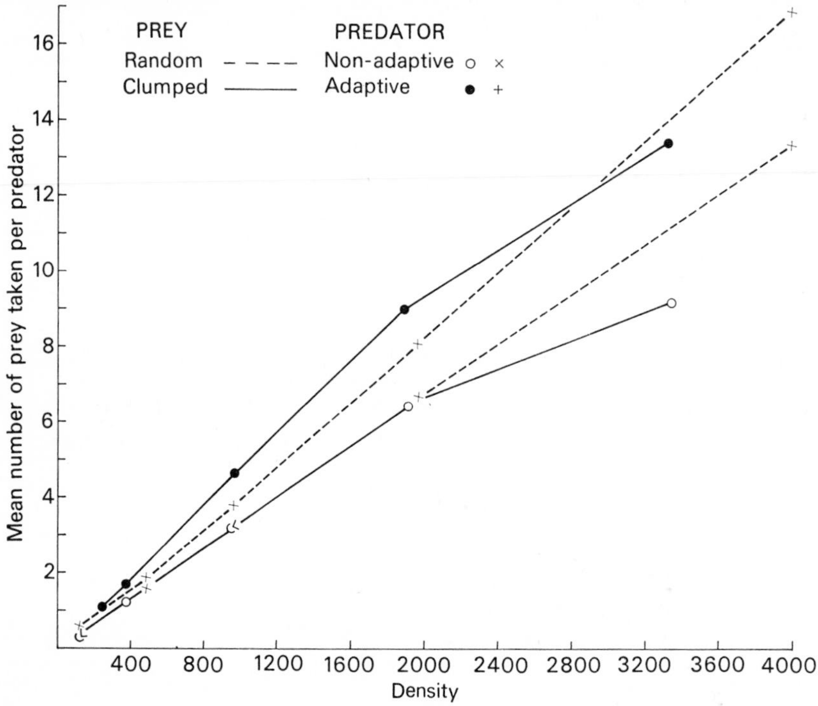

Figure 10. Graph of the relationship between predator success and prey density with differing search behaviour models and prey distributions.

'blundering idiot' predator has a lower success rate than the 'adaptive' one, under all circumstances. Second, the adaptive behaviour of the second predator is particularly effective at all densities below 2000 per quadrat. Finally, aggregation by prey confers some protection at the highest density studied. The potential value of this model lies in its relevance to biological control of pests. Successful introductions of predators could well depend on finding species with a searching behaviour that matches the density and characteristic distribution of the pest to be controlled.

Cotton red bollworm model

The two final models represent attempts to mimic actual pest populations in the field, in the hope of providing practical guides to pest control strategies. The first of these is concerned with the cotton red bollworm (*Diparopsis castanea* Hmps.) in Central Africa. It arises from an investigation into the use of sex attractants and sterile male release as whole or partial replacements for the routine spraying programme traditionally aimed at control of the first instar bollworm larvae (Murdie & Campion 1971).

The system can be represented as in Fig. 11. Planting and uprooting of the

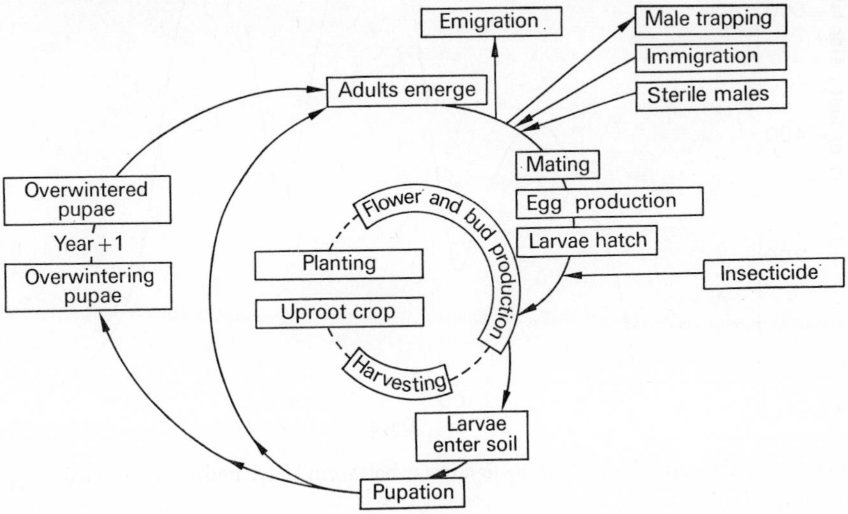

Figure 11. The dynamics of the red bollworm–cotton system.

crop provide the end points of the simulations. The development of cotton flowers and buds has been represented by a simple logistic sub-model based on data from Munro (1971). Input parameters for bollworm development, survival, adult emergence and reproduction have been taken from various sources (Pearson 1958, Matthews 1966, Tunstall 1968). Simplifying assumptions have been made where knowledge is lacking, for example, of mating success in the field, of the relative competitiveness of sterile males and the exact proportions of pupae entering diapause. Figure 12 shows the pattern and rate of boll attack by first instar larvae when no control is practised.

The standard programme of weekly applications of a larval insecticide with short persistence is simulated in Fig. 13. Since egg laying is continuous, there is a continual potential for damage which the insecticide, because of its short persistence, can only partially control. Larvae which have entered the bolls are protected from subsequent sprays and are the source of later generations. At the end of the season, the population, although reduced, is still sufficiently large to produce overwintering pupae.

As an alternative to insecticide control, simulations indicate that extinction of the population can be achieved by repeated releases of sterile males such that a ratio of 20 sterile to one wild male is maintained (Fig. 14). This

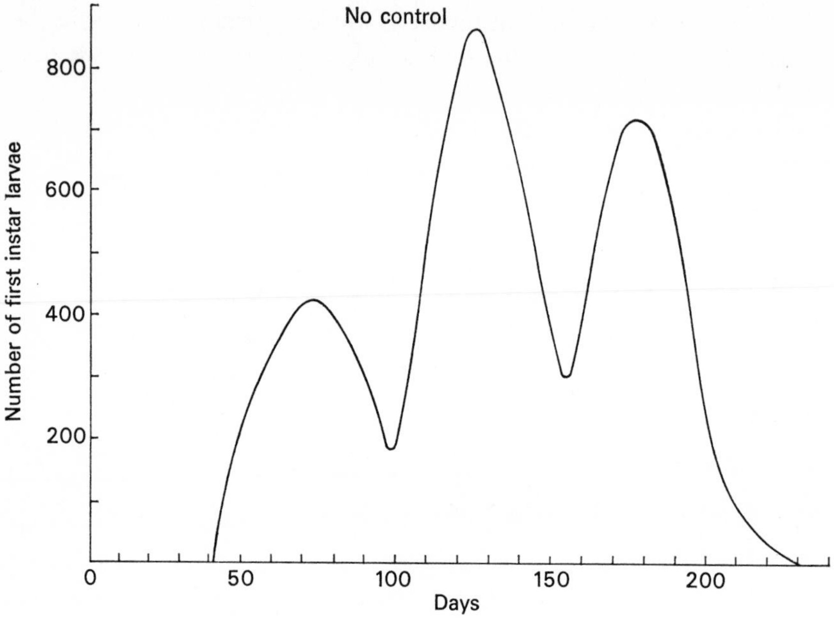

Figure 12. The rate of boll attack by first instar bollworm larvae under uncontrolled conditions.

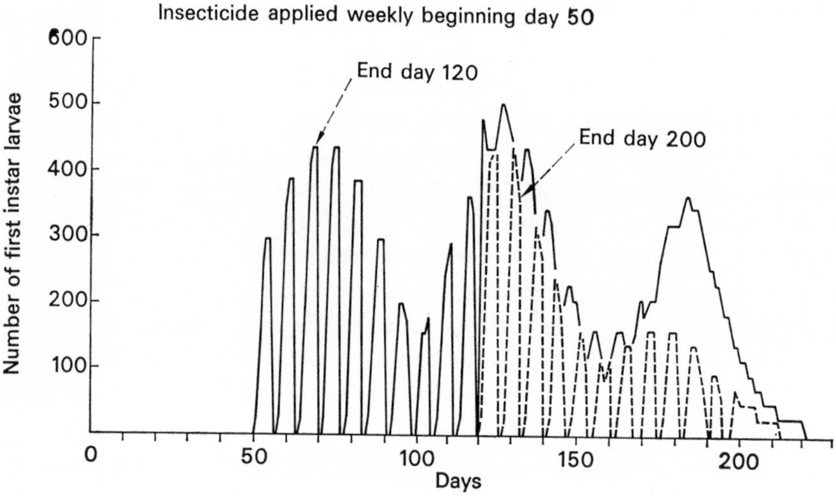

Figure 13. Effect of weekly applications of a short-persistence insecticide continued for 70 and 150 days on the rate of boll attack by first instar bollworm larvae.

represents a total release of 40,000 males per acre. However, it is also apparent that control equivalent to that obtained by insecticide treatment can be obtained with a ratio of only 2:1 and this may be economically feasible (Table 1).

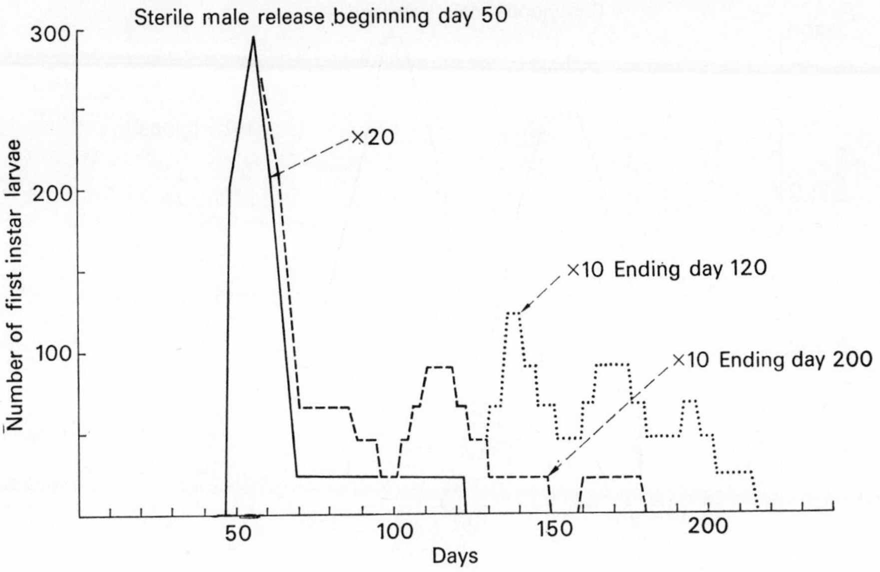

Figure 14. Effect of daily releases of sterile males to maintain ratios of 20 and 10 : 1 wild males on the rate of boll attack by first instar bollworm larvae.

Table 1. Summary of boll damage and diapause pupae production following simulation of sterile male release and insecticide control of *Diparopsis castanea*. Insecticides applied weekly from day 50 to day 200 and the sterile males released daily over the same period.

Control	Damaged bolls per acre	Percent damage	No diapause pupae per acre
No control	78,901	100	32,079
Insecticide	18,080	23	4,703
Sterile males			
2 × wild males	17,912	23	3,948
10 × wild males	7,046	9	453
20 × wild males	5,807	7	210

The simulation of control by use of sex attractant traps are less promising. Figure 15 shows that a satisfactory level of control can only be achieved when 80 per cent of the emerging males are trapped. This will undoubtedly be difficult to achieve in the field.

Aedes aegypti—yellow fever model

The final model refers to the mosquito *Aedes aegypti* and arises out of a short-term consultancy of one of us (GRC) to the World Health Organization East

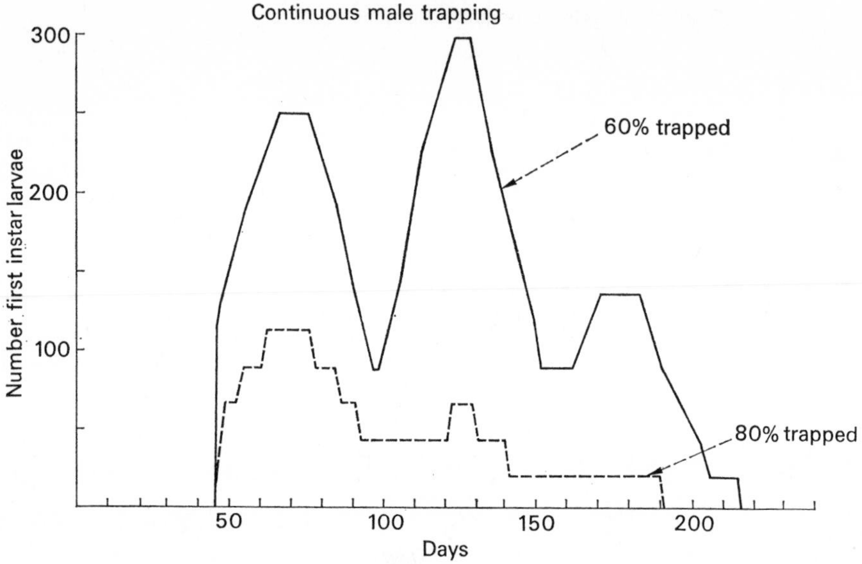

Figure 15. Effect of trapping 60 per cent and 80 per cent of the emerging red bollworm adult males on the rate of boll attack by first instar bollworm larvae.

Africa *Aedes* Research Unit at Dar-es-Salaam in Tanzania. *A.aegypti* is an important vector of yellow fever, chikungunya and a number of other viruses affecting man in Africa. As part of a broad strategy of preventing or suppressing epidemics of these diseases, measures of epidemic risk are required for different localities at different times of the year. The total system, of which the viruses and mosquitoes are part, is exceedingly complex, including not only men but a number of species of monkeys and bush-babies. However, it has been recognized for a number of years that, in the urban man/mosquito/man cycle, the key parameters are five in number (MacDonald 1961). They are:

m the *Aedes* density in relation to man,
a the average number of men bitten by one mosquito in one day,
p the probability of a mosquito surviving through one day,
n the time taken for incubation of the virus in the mosquito,
r reciprocal of the period of infectivity in man.

The values of n and r are known from studies of previous epidemics. By means of a mark-release-recapture experiment (McClelland & Conway 1971) and by a blood meal analysis, it was possible to obtain measurements of the parameters m, a and p for a site in Dar-es-Salaam, Tanzania, in the summer of 1970. Hitherto, the values for these parameters, if they could be obtained, were put together in a simple algebraic equation to give a value for the basic reproduction rate for yellow fever:

$$\frac{ma^2p^n}{-r\log_e p}$$

From it, can be obtained a measure of the daily reproductive rate or vectorial capacity in terms of the number of days between each new case (Table 2).

Table 2. Vectorial capacities for populations of *Aedes aegypti* with daily mortalities from ·5 to ·05 using the deterministic equation (see text).

Mortality	One new case of yellow fever every
·5	106 days
·45	39 days
·4	15 days
·35	6 days
·3	2·6 days
·25	1·1 days
·2	·5 days
·15	·2 days
·1	·08 days
·05	·02 days

On this occasion, however, it was decided to develop a computer model which would give a rather more realistic representation of the transmission of the virus in the system. The general flow diagram of the model is shown in Fig. 16. Both the mosquito and human populations are stored in terms of age classes which are further divided into subclasses for infected and infective individuals. Mortalities and virus transmissions are then allotted according to the probabilities given using a random number generator. A number of simulations were carried out on the model based on the parameter values obtained in the experimental work at Dar-es-Salaam, i.e. a stable mosquito population of about 1,000, a human population of 50 and a four day feeding cycle with an extra feed on the first day. The simulations consisted of 10 runs with different sets of random numbers for each of differing mortalities varying from ·5 to ·05 in steps of ·05. The results are shown in Table 3.

The feature that comes out most clearly from these results is the on/off characteristic of the epidemic situation being described here. A single infected human being entering the population may result in the whole population becoming infected or in a few or no other infections. Very rarely is there a situation in which an intermediate number of the population succumb. This feature is not revealed when the parameters are dealt with in the algebraic fashion previously referred to (Table 2). Part of the reason for

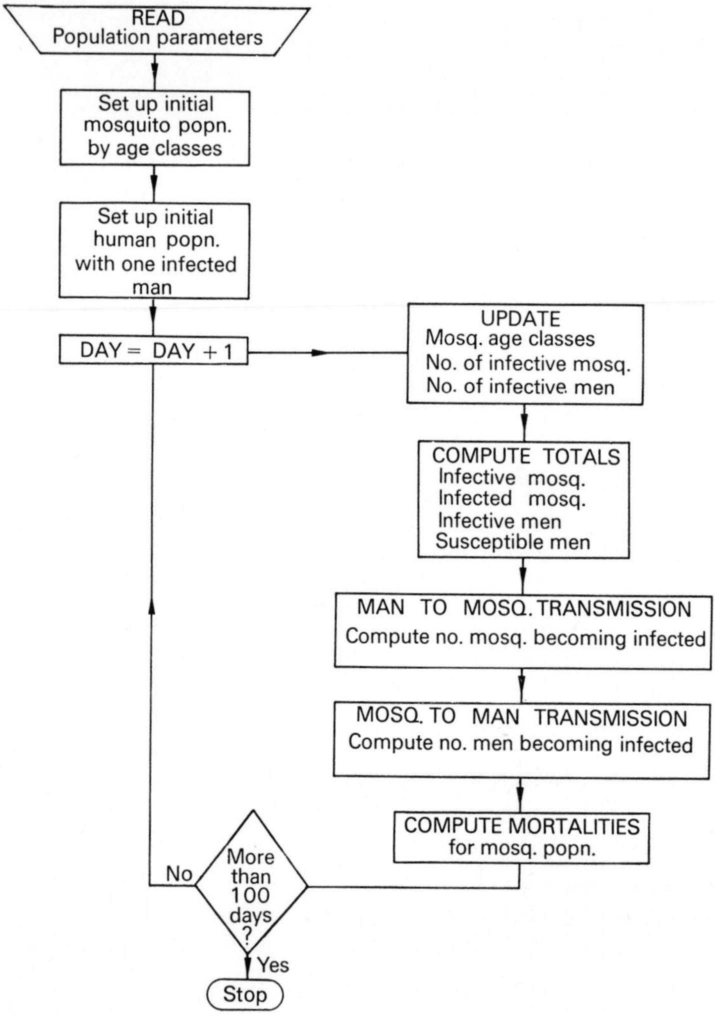

Figure 16. Flow diagram of the *Aedes aegypti*—yellow fever model.

this lies in the representation of the man-biting rate. As the mark-recapture experiment in Dar-es-Salaam revealed, the mosquito bites on two consecutive days and then does not bite again until three days later, after oviposition. In the algebraic calculations this has to be represented as a biting rate of ·5 per day. However, in the computer model the discrete nature of the biting cycle can be preserved.

Secondly, the simulation studies reveal the very narrow ranges within which critical parameter values are to be found. For example, the difference between a mortality of ·3 and ·2 is the difference between and almost certain epidemic occurrence and a fairly rare one. The importance of this lies in the

Table 3. Output from simulations on the *Aedes aegypti*—yellow fever model. The number of human cases of yellow fever arising from a single case after 100 days. Mosquito mortalities ranging from ·5 to ·05; 10 runs each with different sets of random numbers.

Mortality	·5	·45	·4	·35	·3	·25	·2	·15	·1	·05
Runs 1	1	1	1	14	26	44	50	50	50	50
2	2	1	2	1	1	48	50	50	50	50
3	1	1	1	6	43	50	50	50	50	50
4	1	1	1	1	1	44	50	50	50	50
5	1	1	1	1	1	1	50	50	50	50
6	1	5	1	2	1	43	50	50	50	50
7	1	1	1	1	1	49	50	50	50	50
8	1	1	1	1	3	49	50	50	50	50
9	1	1	4	4	20	49	50	50	50	50
10	1	1	1	1	1	1	50	50	50	50

demand that it creates for highly accurate field measurement of population parameters. In the population that was studied in Dar-es-Salaam, despite the very high recapture rate in the mark-recapture experiment, the estimates of survival rate had standard errors of 50 per cent or more. Quite clearly, it is impossible on these grounds to make accurate predictions of epidemic potential. If the full capability of computer models as guides to pest control strategy is to be realized, significant progress will have to be made in developing field techniques which will give better estimates of ecological parameters.

Acknowledgements

Support for this work came from the Office of Resources and Environment in the Ford Foundation under grants to the Institute of Ecology, University of California, Davis, and to the Environmental Resource Management Research Unit at Imperial College. We are also indebted to the World Health Organization for supporting the work on *Aedes aegypti* and to the Overseas Development Administration and the Cotton Research Corporation for assisting the investigations related to the bollworm model. Computing facilities have been provided by the University of California, Davis, and the University of London Computer Centres. We are grateful to Miss Denise Richards for assistance in carrying out the programming and simulations.

References

Burnett T. (1960) An insect host-parasite population. *Canad. J. Zool.* 38, 57–75.

Clark L.R., Geier P.W. & Hughes R.D. (1967) *The ecology of insect populations in theory and practice*. London.

CONWAY G.R. (1969) A basic model of insect reproduction and its implications for pest control. Unpublished Ph.D. Thesis, University of California, Davis.

CONWAY G.R. (1970) Computer simulation as an aid to developing strategies for anopheline control. *Misc. Pub. Ent. Soc. Amer.* 7, 181–93.

CONWAY G.R. (1971*a*) Ecological aspects of pest control in Malaysia. *The Careless Technology: The Ecology of International Development.* (Ed. by T. Farvar & J. Milton, Chicago.)

CONWAY G.R. (1971*b*) Pests of cocoa (*Theobroma cacao* L.) in Sabah and their control, with a list of the cocoa fauna. *Bull. Dept. Agriculture, State of Sabah, Malaysia* (in press).

CUELLAR C.B. (1969*a*) A theoretical model of the dynamics of an *Anopheles gambiae* population under challenge with eggs giving rise to sterile males. *Bull. Wld.Hlth. Org.* **40**, 205–12.

CUELLAR C.B. (1969*b*) The critical level of interference in species eradication of mosquitoes. *Bull. Wld.Hlth. Org.* **40**, 213–19.

FLANDERS S.E. & BADGLEY M.E. (1963) Prey-predator interactions in self-balanced laboratory populations. *Hilgardia* **35**, 145–83.

HASSELL M.P. & VARLEY G.C. (1969) New inductive population model for insect parasites and its bearing on biological control. *Nature* **223**, 1133–7.

HUFFAKER C.B., SHEA K.P. & HERMAN S.G. (1963) Experimental studies on predation: Complex dispersion and levels of food in acarine predator–prey interactions. *Hilgardia* **34**, 305–30.

HUGHES R.D. & GILBERT N. (1968) A model of an aphid population—a general statement. *J. Animal Ecol.* **37**, 553–63.

JONES F.G.W. (1965) The population dynamics and population genetics of the potato cyst–nematode *Heterodera rostochiensis* Woll. on susceptible and resistant potatoes. *Rep. Rothamsted exp. Sta.* 1965 pp. 301–16.

MCCLELLAND G.A.H. & CONWAY G.R. (1971) The frequency of blood feeding in the mosquito *Aedes aegypti*. *Nature* **232**:485–6.

MACDONALD G. (1961) Epidemiological models in studies of vector-borne diseases. *Publ. Hlth. Rep.* **76**, 753–64.

MATTHEWS G.A. (1966) Investigations of the chemical control of insect pests of cotton in central Africa. II Tests of insecticides with larvae and adults. *Bull. ent. Res.* **57**, 77–91.

MORRIS R.F. (1963) (ed.) The dynamics of epidemic spruce budworm populations. *Mem. ent. Soc. Can.* **31**, 1–332.

MUNRO J.M. (1971) The analysis of earliness in cotton. *Cotton Growing Review*, **48**, 28–41.

MURDIE G. (1970) Simulation of the effects of predator/parasite models on prey/host spatial distribution. *Statistical Ecology Vol* **1** (Ed. by G. P. Patil, E.C. Pielou and W.E. Water) pp. 215–233. Pennsylvania State Univ. Press.

MURDIE G. & CAMPION D.G. (1971) The evaluation of control methods for red bollworm using a computer population model. *Proc. Conf. Cotton Insect Control, Blantyre, Malawi* 1971 (in press).

PEARSON E.O. (1958) *The insect pests of cotton in tropical Africa*. London.

PICKETT A.D., PUTMAN W.L. & LE ROUX E.J. (1958) Progress in harmonizing biological and chemical control of orchard pests in Eastern Canada. *Proc. XTH Int. Cong. Ent. Montreal* **3**, 169–74.

SMITH R.F. & VAN DEN BOSCH R. (1967) Integrated control. *Pest Control, Biological, Physical and Selected Chemical Methods* (Ed. by W.W. Kilgore and R.L. Doutt) pp. 295–340.

SMITH R.F. & REYNOLDS H.T. (1966) Principles, definitions and scope of integrated pest control. *Proc. F.A.O. Symp. Integrated Pest Control*, 1, 11–17.

SOUTHWOOD T.R.E. (1966) *Ecological Methods*, London.

STERN V.M., SMITH R.F., VAN DEN BOSCH R. & HAGEN K.S. (1959) The integration of chemical and biological control of the spotted alfalfa aphid. *Hilgardia* 29, 81–154.

TUNSTALL J.P. (1968) Pupal development and moth emergence of the red bollworm (*Diparopsis castanea* Hmps) in Malawi and Rhodesia. *Bull. ent. Res.* 58, 233–54.

WATT K.E.F. (1960) The effect of population density on fecundity in insects. *Canad. Ent.* 92, 674–95.

WATT K.E.F. (1961) Mathematical models for use in insect pest control. *Canad. Ent. Suppl.* 19, 62 pp.

WATT K.E.F. (1963) Dynamic programming, 'Look ahead programming' and the strategy of insect pest control. *Canad. Ent.* 95, 525–36.

WATT K.E.F. (1964) The use of mathematics and computers to determine optimal strategy and factors for a given insect pest control problem. *Canad. Ent.* 96, 202–20.

WATT K.E.F. (1968) *Ecology and resource management*. New York.

WOOD B.J. (1971) Development of integrated control programme for pests of tropical perennial crops in Malaysia. *Biological Control* (Ed. C.B. Huffaker) pp. 422–57, New York.

'Accuracy without precision'—an introduction to the analogue computer

J.K.DENMEAD *Department of Systems Engineering, University of Lancaster*

Summary

An analogue computer can often provide a relatively cheap and very convenient method of obtaining results from mathematical models formulated as differential equations. The paper attempts to convey an impression of the nature and scope of analogue computation and its recent development, hybrid computation. It is hoped to mount a demonstration using a small computer.

1. Introduction

The term analogue computer may validly be used to describe a wide variety of calculating devices which have in common the property that they solve mathematical problems by the process of modelling. Such computers consist of a model (or analogue) whose describing equations are of the same form as the set of equations one wishes to solve and one obtains a solution by measuring, on this model, physical quantities whose magnitudes are proportional to the variables in the original set of equations.

The particular type of analogue computer dealt with here is what is usually called a general purpose electronic analogue computer. The term 'general purpose', when applied to an analogue computer, may be taken as indicating a very flexible device suitable for use in a number of different fields with a common mathematical background.

A general purpose electronic analogue computer is a collection of electrical circuit elements (principally electronic direct-coupled voltage amplifiers, fixed and variable resistors and capacitors) which may be connected together in a variety of ways by leads to form complete circuits. The interconnections of the circuit elements are so arranged that the equations describing the magnitudes of the voltages existing at various points in the circuits so formed are analogous to the set of equations one is solving. In other words, the analogue is an electrical one and the measurements taken are measurements of voltage. It is in the field of the solution of differential equations (ordinary or partial, linear or non-linear) that the electronic analogue computer finds its main application.

This paper is intended to impart familiarity with the essentials of analogue computing to an audience which it is assumed will have a knowledge of the significance of differential equations and an appreciation of the importance of their accurate solution. No attempt will be made to explain hardware in great detail and virtually no electrical knowledge will be assumed beyond perhaps a nodding acquaintance with Ohm's Law. The basic concepts of that comparatively recent phenomenon, the hybrid computer, will also be briefly discussed and an attempt made to compare analogue, hybrid and digital methods of dealing with problems involving the solution of differential equations. On occasions complete generality will be sacrificed in the interests of clarity and it is hoped that any analogue enthusiasts will recognize the reasons for such lapses and pardon them.

In what follows, the term 'analogue computer' may be taken as meaning a general purpose electronic analogue computer.

2. Equipment

The principal components of a modern analogue computer are briefly described below.

2.1. *Reference voltage source and voltmeter*

An analogue computer must be provided with three fixed voltages—a positive reference voltage (usually either 100v or 10v), an equal negative voltage and zero (earth). The positive and negative reference voltages are generally referred to as plus and minus Reference, and written $+R$ and $-R$. These voltages are maintained constant to an accuracy of the order of one part in $10^3 - 10^5$ depending on the computer and all other voltages are measured in 'machine units', i.e. as fractions of reference voltage relative to earth, rather than in absolute volts. A voltmeter (or ratiometer device) for carrying out such measurements is another essential component of the computer.

2.2. *Potentiometers*

These are used for performing the simple but frequently required operation of multiplying a voltage by a constant less than one. A potentiometer is a resistance with a tapping point whose position is set by the computer user. An electrical diagram is:

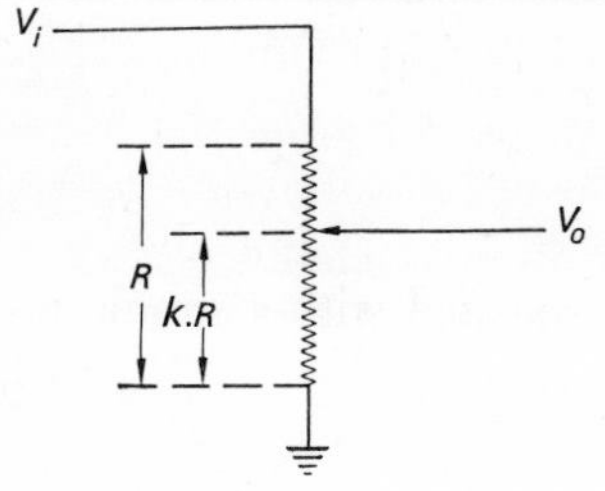

The relation between input and output voltages is:

$$V_o = k.V_i$$

The programming symbol used for a potentiometer is:

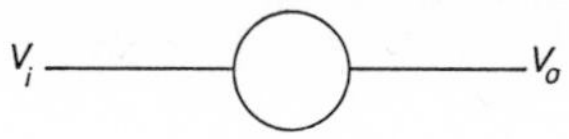

i.e. it is represented as a two-terminal device since the user does not normally need access to the earthed terminal. The resolution of potentiometers used in analogue computers is generally of the order of one part in 10^4.

2.3. *Operational amplifiers*

These are the most important components of an analogue computer and are used in a variety of ways. An operational amplifier is an electronic direct-coupled amplifier with a high voltage amplification (of the order of 10^5 to 10^9 depending on the accuracy of the computer), a high input resistance and a low output resistance. The programming symbol is:

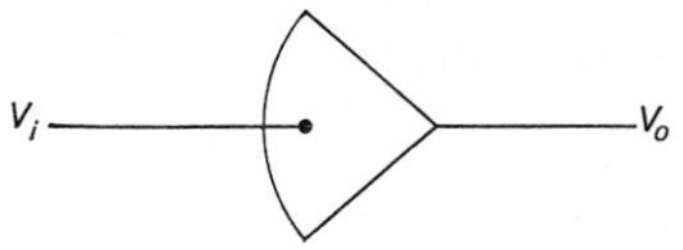

and the input-output voltage relation is:

$$V_o = -\mu.V_i$$

where μ is the voltage amplification factor (or gain).

The linear relation between input and output voltages applies only over an output voltage range not much greater than $+R$ to $-R$. This fact is one of the chief inconveniences of the analogue machine and introduces the necessity for scaling the equations to be solved. This will be discussed later.

The operational amplifier is seldom used alone in a practical computing circuit. It usually has resistors and/or capacitors connected externally to it to enable useful operations to be performed. The two fundamental operations are summation and integration.

2.4. *Summers*

A summer consists of an operational amplifier associated with a simple resistor network:

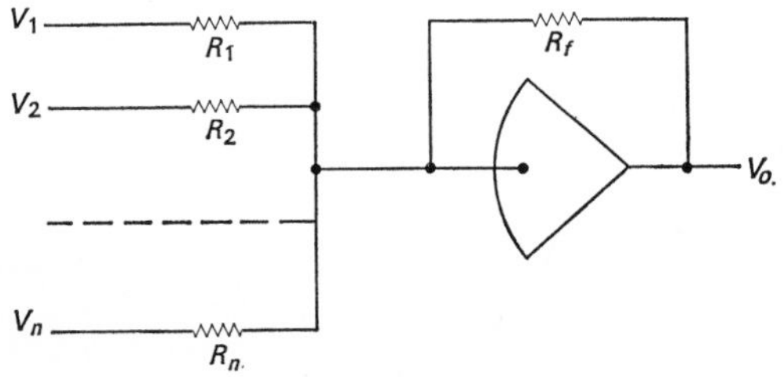

The input-output voltage relation of such a combination is:

$$V_o = R_f \sum_{i=1}^{n} V_i/R_i$$

In practice a restricted range of sizes of input resistors (R_i) and feedback resistor (R_f) is used. If R_f and all the R_i have the same value then the relation simplifies to

$$V_o = -\sum_{i=1}^{n} V_i$$

i.e. this computing element performs summation, albeit with a change of sign. This sign change, sometimes but by no means always, embarrassing, is a feature of quite a number of analogue computing operations.

The programming symbol for a summer is:

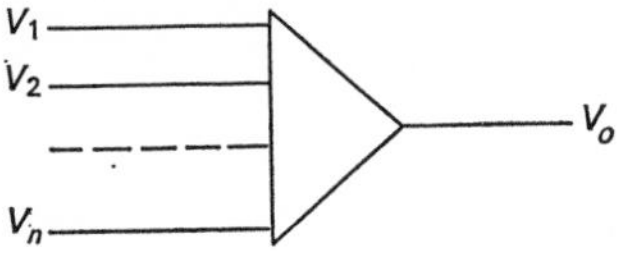

The accuracy of summation is normally of the order of one part in 10^4.

2.5. *Integrators*

An operational amplifier may be used as an integrator by associating a resistance-capacitance network with it, thus:

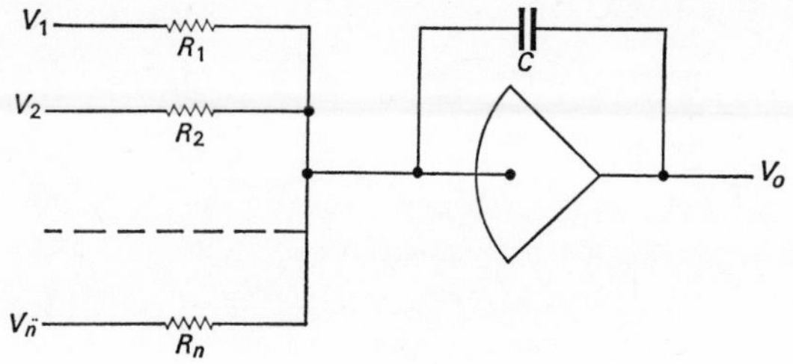

The input/output voltage relation is:

$$V_o = -\, 1/C \int \sum_{i=1}^{n} V_i/R_i \, dt.$$

It is normal in a modern computer for a range of values of the terms $1/R_i.C$ to be made available to the user by arranging all the R_i to be equal and providing a choice of capacitors. R_iC is referred to as the integrator 'time constant', τ—the product of resistance and capacitance has the dimension of time. The above expression then simplifies to:

$$V_o = -\frac{1}{\tau} \int \sum_{i=1}^{n} V_i \, dt$$

The smaller the time constants of the integrators used in a particular computing circuit the faster will be the time scale in which the solution to the equations being modelled will be obtained. Thus the user has control over the speed of his solution. Factors influencing his choice will be discussed later. Values of τ commonly provided range from 10 seconds to 1 millisecond.

The programming symbol for an integrator is

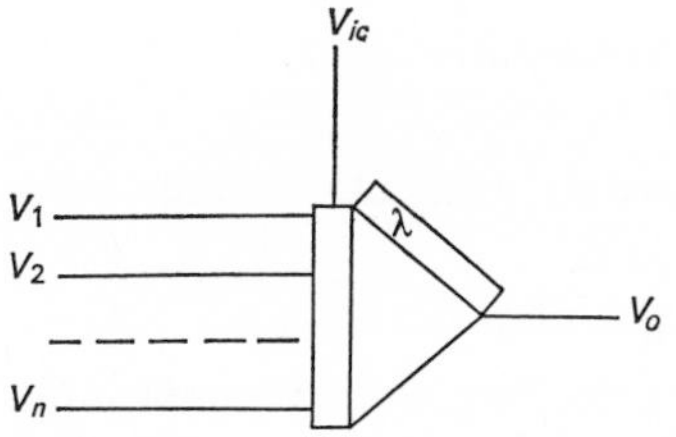

The voltage V_{ic} represents the initial value (i.e. the value at zero time) of $-\,V_o$ and a facility for imposing such a value is necessary so that the familiar constant of integration may be introduced. A more accurate statement of the integrator function is:

$$V_o(t) = -\left\{ V_{ic} + \frac{1}{\tau} \int_0^t \sum_{i=1}^{n} V_i(t) \, dt \right\}$$

i.e.

$$dV_o/dt = -\frac{1}{T}\sum_{i=1}^{n} V_i$$

with

$$V_o = -V_{ic} \text{ at } t = 0$$

Note that the integrator combines summation and integration and introduces a sign-change into the result.

The accuracy of integration is normally of the order of one part in 10^4.

2.6. *Mode control*

To enable the computer to be used effectively some organization of the modes of operation of the various components is necessary. This is achieved usually by providing the operator with a set of push buttons so that he may cause the computer to carry out operations in the sequence he desires. The principal modes of operation he can select are 'Potentiometer Set' (PS), 'Initial Condition' (IC), 'Operate' (OP) and 'Hold' (H).

The PS mode is provided to enable potentiometers to be set to their desired values. In this mode positive reference voltage is applied to the potentiometers in turn and their settings adjusted until the output voltage obtained (measured on the computer's voltmeter) is equal to the desired setting.

In the IC mode the integrator outputs take up their desired initial values (usually set as the outputs of potentiometers) and computer time is prevented from advancing by disconnecting the input resistors of integrators from the operational amplifiers.

The provision of this mode allows the operators to use a standard starting condition (or set of such conditions) from which integration may subsequently proceed.

In the OP mode the integrator input resistors are connected to the operational amplifiers and problem solution takes place.

The H mode, which may be invoked at any time the operator desires, 'freezes' the solution at this time simply by disconnecting the integrator input resistors from the operational amplifiers.

2.7. *Multipliers*

Whereas a potentiometer will multiply a voltage by a constant factor, a special multiplier is required to multiply one time-varying voltage by another. The most common of such devices used today is the 'quarter-square' multiplier which, using a fairly simple electrical circuit, produces the product of two voltages V_1 and V_2 by mechanizing the identity.

$$V_1 V_2 = 1/4 \{(V_1 + V_2)^2 - (V_1 - V_2)^2\}$$

A quarter-square multiplier may or may not introduce a sign-change and has an accuracy of the order of a few parts in 10^4.

Most modern multipliers will accept either sign of V_1 or V_2 and may be used for division as well as for multiplication.

A simplified programming symbol for a quarter-square multiplier is

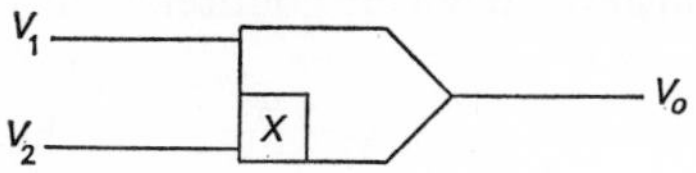

2.8. *Function generators*

A function generator enables an output voltage which is an arbitrary (analytic or non-analytic) function of an input voltage to be produced. The normal type is the diode function generator (DFG) which approximates to the desired function curve by a series of linear segments. It is fairly commonplace for up to 20 such segments to be available on a DFG. Many computers possess fixed function generators to provide commonly used mathematical functions such as cos, sin, square, square root, log, etc.

A programming symbol for a function generator is

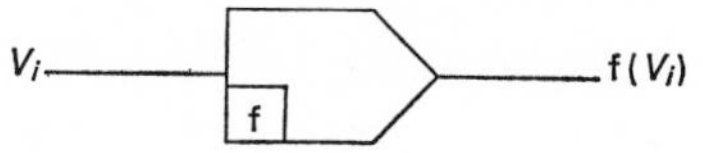

2.9. *Switches*

A selection of switches is generally provided to facilitate ease of use of the computer. By means of manually operated switches the user may switch in or out terms in his equations either prior to obtaining a particular solution or during the actual course of the solution. It is frequently desired that the solution itself should decide the point at which switching operations occur and this need is catered for by providing switches designed to be power-operated under the control of logic elements (see below). These latter switches are usually quite rapid in operation—switching times of the order of a microsecond are common.

2.10. *Logic*

The ability to make decisions and to change the course of computation as a result thereof is a common requirement. A variety of logic elements is provided in the modern analogue computer enabling quite complex decisions to be made. This logic is digital in nature but parallel in operation as distinct from the serial facilities provided with digital computers—i.e. many decisions may be made simultaneously.

Digital input to the logic elements is provided by analogue/digital comparators which provide a logic signal which is either 'true' or 'false' depending on whether the sum of a number of analogue voltages is positive or negative. A wide variety of logic elements can be provided, including AND/NAND gates, bistable (digital memory) devices, monostable devices, counters, shift registers, timers, etc.

Output from the digital logic section to the analogue elements is via the logically operated switches mentioned above which provide a path for analogue voltages or not, depending on whether their controlling logic signal is 'true' or 'false'.

Logic may also be used to control the modes of integrators (either every integrator in the machine or individual ones as desired). In this way quite complex sequences of computing operations may be performed without operator intervention.

2.11. *Patch panels*

To facilitate interconnection of the various computing elements a patch panel is used consisting of a matrix of terminal holes to which the inputs and outputs of all the elements in the machine are brought. Leads connected (or patched) between these holes then enable the desired circuits to be made. On many computers the patch panels are removable so that they may be stored off the machine if required. An analogue 'program', consisting of a patched-up panel plus a list of potentiometer settings, can thus be preserved for as long as needed.

2.12. *Output equipment*

The traditional way of obtaining results from an analogue computer is the plotting of one variable against another. Equipment capable of performing this function is readily available and the display of results in graphical form is the most common requirement when solving differential equations.

A rectilinear plotting table is the most generally useful device for producing a permanent graphical record of results. A pen writing on a sheet of squared paper is the recording medium. The paper is held on to a rectangular table by some means such as a vacuum system, its grid being accurately aligned with the table axes. A carriage holding the pen is driven by a control system along an arm parallel to the Y-axis of the table, its displacement from one end of the arm being kept proportional to the analogue computer voltage representing the Y co-ordinate of the curve to be plotted. The whole arm can move parallel to the X-axis of the plotter and is controlled so that its displacement from some reference point on the X-axis of the table is proportional to the X co-ordinate of the function plotted. Thus any variable may be plotted against any other. An accuracy of 0·1 per cent of full scale may be obtained but, because of the inertia of the moving parts, the speed of response attainable is usually limited—full-scale travel in 1 second is probably a fairly representative maximum plotting speed.

Another useful recording device is the chart recorder. This has the advantages over the rectilinear plotter that it is simple to record several variables simultaneously on the same chart and that it is possible to attain high speeds of response. On the other hand its use is restricted to the plotting of variables against time since the X movement is provided by driving the chart paper at constant speed past the recording heads. It is also generally true that high writing speeds are obtained at the expense of accuracy.

A device of a somewhat different kind is the oscilloscope, whose main usefulness is in displaying results rather than recording them. Its characteristics are very high speed response and limited accuracy. To enable an oscilloscope to be used successfully one needs to generate the computer solution on a fast time scale and to repeat the solution process possibly many times per second, so that a seemingly stationary trace is displayed on the screen. The 'repetitive operation' mode of working the computer provides this facility. In this mode the integrators are cycled rapidly between the 'initial condition' and 'operate' states and, if sufficiently small values of time constants are chosen for them, the desired solution rates may be obtained.

In spite of its limited accuracy an oscilloscope can provide the computer user with a high degree of 'man-machine rapport'. It is rarely possible to determine in advance the exact conditions under which solutions to a particular problem are to be obtained. More commonly one wishes to select from a range of possible solutions a few which have particular characteristics. In computing terms this requires alterations of various kinds to the computing circuit and observation of the effects of these alterations. The high speed of the oscilloscope enables the results of changes to be observed and evaluated as they are made. Hence a large number of possibilities may be reviewed in a short time and unacceptable results rejected. The solution may then be slowed down to permit the use of a plotter for taking accurate permanent records of selected results. Thus the oscilloscope and plotter are complementary devices. If great recording accuracy is not important, an alternative to plotting is photographing the oscilloscope trace.

3. Programming

The traditional field of application of the analogue computer is the solution of sets of simultaneous differential equations with one independent variable. The computer's independent variable is always time since the integrators integrate with respect to time. The problem independent variable itself need not be time but it will be represented on the computer by time.

The aim of analogue computer programming is to prepare a circuit diagram showing what computing elements will be needed to solve a problem and how they must be interconnected plus lists of settings of various elements (principally potentiometers). As a minimum prerequisite the programmer must know the equations it is desired to solve, he must have estimates of the maximum values all variables are likely to take during the solution and he must know the initial values of those variables which are defined by differential equations.

In the interests of clarity the programming method will be explained by reference to a simple worked example. The problem is to provide solutions to the differential equations describing the population, p_f, of fish in a bay which contains a limited supply of food and which is also inhabited by a population, p_s, of sharks who exist by preying on the fish.

The fish population balance has the form:

$$d/dt\,(p_f) - (k_o - k_1 \cdot p_f)\,p_f + k_2 \cdot p_f \cdot p_s = 0$$

where k_o is the rate of natural increase of the fish population due to the excess of births over deaths under conditions of unlimited food supply,

$k_1 \cdot p_f$ is the diminution of this natural expansion rate attributable to a limited supply of food,

k_2 is the death rate of the fish population caused by encounters with sharks,

t is time in seconds.

The shark population balance has the form:

$$d/dt\,(p_s) - k_3 \cdot p_f + k_4 \cdot p_s - k_5 \cdot p_f \cdot p_s = 0$$

where k_3 is the rate at which sharks are attracted into the bay because of the presence of fish,

k_4 is the rate at which sharks leave the bay because the presence of other sharks cramps their style,

k_5 is the shark birth rate—the number of sharks born is assumed to be proportional to the number of encounters between sharks and fish (i.e. well-fed sharks are healthy).

3.1. *Obtaining the analogue circuit*

The first step is to write the differential equations with the highest order derivative standing by itself on the left-hand side. Differential equations of order, n, greater than 1, must be reduced to a set of n first order differential equations. Any algebraic equations should be written in a form explicit in one variable. (If this latter condition cannot be achieved, special techniques are available to deal with algebraic equations—these will not be discussed here.) In our case the equations will become:

$$d/dt\,(p_f) = k_o \cdot p_f - k_1 \cdot p_f^2 - k_2 \cdot p_f \cdot p_s$$

$$d/dt\,(p_s) = k_3 \cdot p_f - k_4 \cdot p_s + k_5 \cdot p_f \cdot p_s$$

Next, the programmer assumes that he has available to him voltages representing all the terms on the right-hand side of his set of differential equations. He may then draw the first part of his circuit which will consist of integrators, one producing as output each variable defined by a differential equation. Voltages representing each term in the expressions for the derivatives are shown at the inputs to the integrators. At this stage our circuit looks like this:

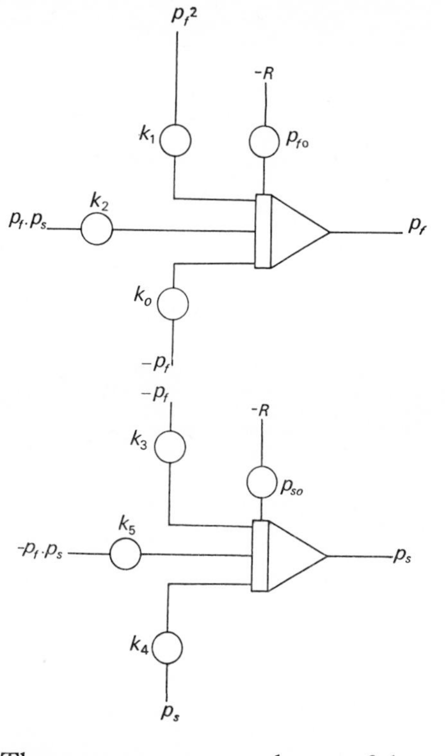

The next steps are to make use of the variables generated by the integrators to produce the input voltages which were initially assumed to be available and then to connect these to the appropriate integrator inputs, thus 'completing the loops' and validating the original assumption that this could be done. This commonly used 'analogue fiddle' has quite a respectable ancestry. It was proposed by Lord Kelvin in 1876, as a method of solving continuously a second order differential equation using mechanical integrators of a type invented in the previous year by his brother, Professor James Thomson.

The analogue circuit for our example after the first of these steps is:

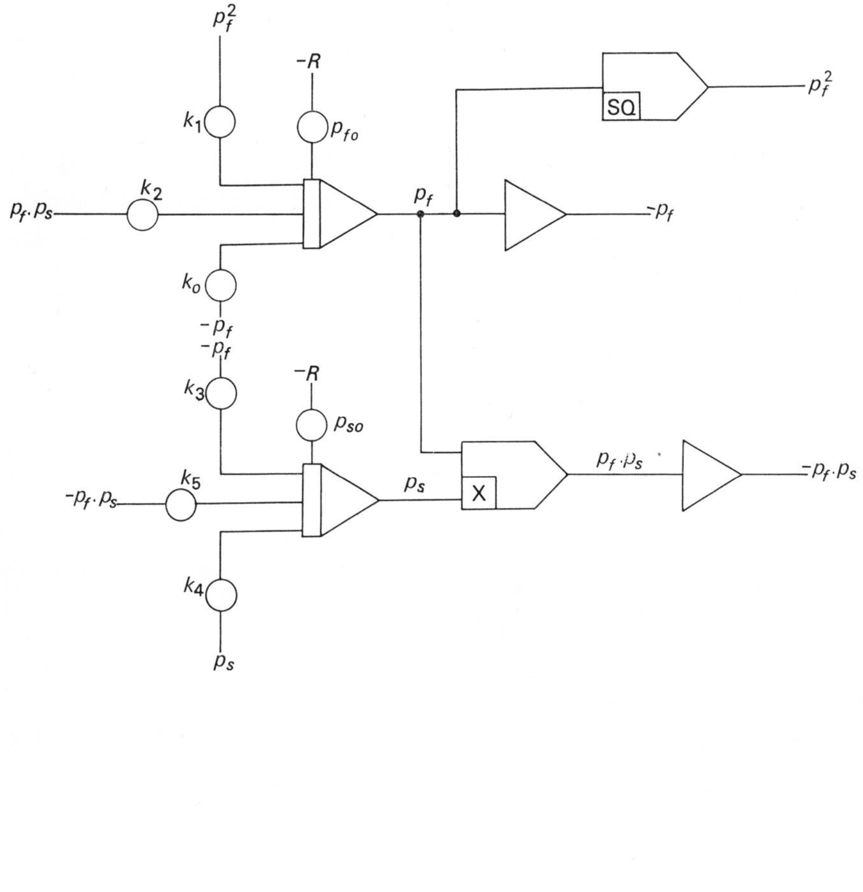

After all the connections have been made the final circuit is:

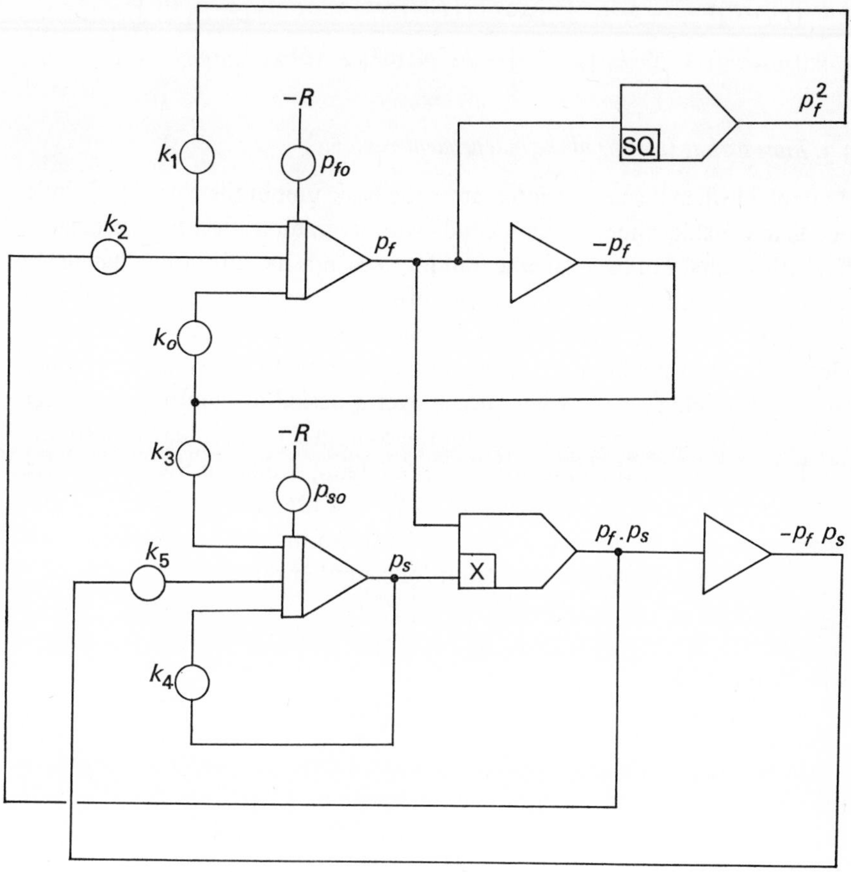

3.2. *Scaling of the variables*

Because the permitted range of voltage swing of the operational amplifiers is ± 1 machine unit, the problem variables will in general need to be scaled. The process consists of carrying out simple transformations on the equations to provide a new set containing scaled variables whose excursions during solution will lie within the above range. The operation consists simply of replacing every variable by a new one which is

$$\frac{\text{Variable}}{\text{Maximum numerical value of variable}}$$

and then adjusting the constant coefficients in the equations accordingly.

In our problem we shall take the maximum values of p_f and p_s as 10^5 and 10^2 respectively. The maximum value of the product $p_f \cdot p_s$ is, therefore,

10^7 and the maximum value of p_f^2 is 10^{10}. Thus our scaled equations are:

$$d/dt([p_f/10^5]) = k_o[p_f/10^5] - 10^5 \cdot k_1[p_f^2/10^{10}] - 10^2 \cdot k_2[p_f \cdot p_s/10^7]$$

$$d/dt([p_s/10^2]) = 10^3 \cdot k_3 \cdot [p_f/10^5] - k_4[p_s/10^2] + 10^5 \cdot k_5 \cdot [p_f \cdot p_s/10^7]$$

3.3. *Time scaling (scaling of the independent variable)*

As implied when discussing integrators the basic unit of the computer's independent variable, time, is the second—an integrator with a time constant τ fed with a voltage of one machine unit for τ seconds has its output changed by one machine unit. By a suitable choice of τ a problem solution taking place in real time (if time is the problem independent variable) may be obtained. However, this is not always convenient for the user. In our case, for instance, we are examining events which occur over a period of months but we certainly do not wish to wait this long for the computer to churn out a solution.

The way out of the difficulty lies in applying scaling to the independent variable of the problem. To accomplish this we use a new variable T—computer time—and we let one unit of the problem independent variable, t, be represented on the computer by β units of computer time,

i.e.
$$t = T/\beta$$

which means that

$$d/dt = \beta \cdot d/dT$$

In our case, assuming we wish to use a plotter to record the computer results, it will be convenient to make β equal to 4×10^{-7}, i.e. one second on the computer will represent $2{\cdot}5 \times 10^6$ seconds (approximately one month) of problem time.

Changing d/dt in our equations to $\beta \cdot d/dT$ we have their final form, which is:

$$d/dT([p_f/10^5]) = 1/\beta\{k_o \cdot [p_f/10^5] - 10^5 \cdot k_1 \cdot [p_f^2/10^{10}] - 10^2 \cdot k_2[p_f \cdot p_s/10^7]\}$$

$$d/dT([p_s/10^2]) = 1/\beta\{10^3 \cdot k_3 \cdot [p_f/10^5] - k_4 \cdot [p_s/10^2] + 10^5 \cdot k_5 \cdot [p_f \cdot p_s/10^7]\}$$

We may now calculate numerical values for the various constant coefficients in the above equations, which will be potentiometer settings on the computer. These values, if inconveniently large or small, may need to be modified by an appropriate choice of integrator time constants. If an oscilloscope is to be employed to display results in addition to the plotter, an alternative smaller value of β may also be chosen and corresponding τ values selected. We then assign computer addresses to the various elements in our diagram and prepare a list of the potentiometer settings. Once this is done the formal programming process is virtually complete. To implement the solution on the computer it is only necessary to connect the patch panel following our diagram, to set the potentiometers and to carry out a check in the initial condition mode that all the computer elements are functioning correctly and that all the connections have been made in the way which we intended.

It will be seen that the programming process is essentially quite a simple and logical one. It is, however, tedious and with large problems the possibility of human error is finite. Because of this considerable effort has been put in recent years into the writing of digital computer programs to carry out analogue computer programming automatically. Several successful programs have been produced.

4. Hybrid computation

The last ten years has seen the rise of a very significant extension of analogue computation in the form of the hybrid computer. This is a combination of a digital and an analogue computer engineered to exploit the best features of each machine so as to produce a system more powerful for dealing with problems involving differential equations than either of the two principal components in isolation. The greatest single impetus to the development of modern hybrid systems undoubtedly came from the United States Government's investment in its missile and space exploration programme which demanded simulation facilities previously unavailable. It was in the aerospace industry that hybrid computation was first generally applied but it is now rapidly becoming accepted and used for the solution of a wide range of problems in other areas.

The main idea behind hybrid computation is that an analogue computer, although capable of producing very fast and accurate solutions of differential equations, has a limited arithmetic and organizing capability. A digital computer, on the other hand, is very strong in these weak areas of the analogue machine but is not outstanding when it comes to integrating differential equations rapidly. Broadly speaking a hybrid system attempts to overcome these difficulties by effectively making an analogue computer a differential equation solving peripheral device for a digital computer. Organization is in the hands of the digital machine, whilst arithmetic is shared between the two machines. In practice, it is not possible to set rigid boundaries to the areas of use of the two machines as a great deal depends on the problem being solved and the characteristics of the machines themselves.

An interface system is required to match the characteristics of the two machines and, although it is not true that any analogue computer may be used with any digital computer to form a viable hybrid combination, a well-designed interface can remove many of the difficulties associated with such a marriage. The cost of the interface is generally not negligible compared with that of the component computers of the hybrid system. An interface will usually have:

- A monitoring and control linkage enabling the digital computer to perform many of the tasks carried out by the operator of a conventional analogue computer.
- High-speed data conversion channels to pass numerical information to, or from, either machine.
- A logic linkage to transmit status signals and commands between the machines.

Application areas for hybrid computation are rapidly becoming more and more numerous. A few possibilities will be described here.

Firstly, the use of a digital computer as an analogue computer operator can result in large savings in time and reductions of human errors. The ability to describe in advance a series of runs and to have results stored permanently in digital form and/or analysed on-line is particularly valuable for very large problems where a human operator cannot really keep an adequate watch on all parts of the analogue model during the course of a solution. The installation of a hybrid system may sometimes be justified from arguments based simply on the savings which this super-operator capability makes possible.

The use of the digital machine to perform some arithmetic has already been mentioned. Another important application in the same category is function generation. It is, of course, very simple to generate complex (i.e. complicated) non-linear analytic and logical functions digitally. Non-analytic functions can also be generated by looking up a table of the function stored in an array and, if necessary, interpolating between the stored points. Functions of more than one variable may be generated by the table look-up technique and, as this is usually a difficult operation on an analogue computer, it is a particularly useful application.

The digital machine may be used not only to output fixed preset functions, but also to store complete solutions obtained from one run and play them back as inputs for another run. This ability forms the basis of several hybrid methods which have been developed for solving partial differential equations more satisfactorily than may be done on either analogue or digital machine alone. It is also a very satisfactory method of simulating transport delay phenomena. Either fixed or variable delay may be dealt with in this way.

One of the earliest applications, and one of great usefulness in the right circumstances, is the use of the digital machine to integrate some of the differential equations of a problem in parallel with the analogue computer. It sometimes happens that the values of particular variables defined by differential equations need to be evaluated more accurately than is possible with analogue integrators. If the precision of the digital machine is sufficient and if the rates of change of the variables are small, digital integration may provide a good solution. Whilst on this point, and to emphasize the fact that there are no hard and fast rules to the hybrid game, at least one hybrid computer has been built (for solving hydrological problems) in which the digital machine carried out all the integrations whilst the analogue machine was used for arithmetical matrix operations.

As a final example, hybrid computation lends itself very well to a certain class of optimization problems. Techniques for selecting optimum system parameters depend on multiple evaluations of a performance criterion or objective function for the system with different values of these parameters. When the objective function evaluation consists of the determination of the value of an arithmetic function for given parameter values, such optimization may be carried out successfully on a digital computer. However, when the value of the objective function depends on the solution of a complex set of non-linear differential equations, the time required for digital solution can easily become prohibitive.

An analogue computer solution of the equations may be possible in a time which is shorter than that required for digital solution by a factor of 10 or 100 or more, but the analogue machine is incapable of deciding, in all but the simplest cases, what new values to assign to parameters as a result of its previous runs.

In situations like this the use of a hybrid computer can be a very attractive proposition. Sophisticated numerical optimization routines may be used on the digital machine and the function evaluations performed on the analogue. A variant of this technique, known as parameter estimation, may be used to adjust parameters in a model set up on the analogue machine in such a way as to match the transient response of the model to a curve obtained experimentally from an actual process (the word is used here in a broad sense). If a sufficiently good match can be obtained, final model parameters will be an estimate of the unknown process parameters.

Hybrid computer programming involves not only traditional analogue and digital techniques, but also a realization of all the implications of interactions between the machines. Because of this it is an extremely stimulating and challenging occupation. A hybrid computer is a real-time system in that each machine is required to produce results for the other to process on a time schedule which must be strictly adhered to if meaningful results are to be obtained. Thus, hybrid computer programming is somewhat more difficult than programming either of the individual machines but the potential benefits to be obtained can be substantial.

5. Analogue/hybrid v. digital

It is, of course, an undoubted fact that solutions to sets of ordinary differential equations can be produced by numerical integration techniques and that such solutions, obtained from a digital computer, can be as accurate as those obtained from an analogue computer, or more so. Nevertheless, the process can be rather more laborious than some of its advocates will admit. Whilst the digital computer is undoubtedly a very powerful tool for the solution of a vast range of mathematical problems, the way in which it is applied in this particular area is somewhat unnatural. The solution is produced as a series of still pictures separated by small time intervals and as a consequence numerical difficulties arise.

These difficulties can be overcome by the choice of an appropriate integration algorithm, and a judicious selection of basic time interval but to achieve accurate results this time step must be very small compared with the total solution time. Consequently a great deal of repetitive calculation (and a correspondingly long computing time) is required to obtain answers. A further time penalty is introduced by the fact that the digital computer is a serial device. Calculations must be done one at a time, so that the more equations one has to solve, the longer it takes.

The analogue computer, on the other hand, has no such problems. The integration process is continuous—there is no time step and no stability problems exist. Even the modest tolerances commonly employed for computing components can usually provide solutions as accurate, for practical purposes, as those obtained from a digital computer operating with far greater precision in its number representation—'accuracy without precision'. Moreover the analogue computer is a parallel machine. All the equations are solved simultaneously so that there is no time penalty associated with large sets of equations. Of course, this benefit is not obtained without cost—the amount of computing equipment needed increases with problem size and it is not possible to trade off speed against size.

Analogue integration, as well as being stable, is rapid. It is doubtful whether even the fastest digital machines working today can approach the solution times possible on a modern analogue computer for any but the most trivial of problems. This fact is often overlooked. Comparisons are made between the digital machines of today and the analogue machines of ten years ago, ignoring the advances in analogue computer technology in that time.

Curiously enough, the analogue computer can also sometimes score over the digital machine in algebraic calculations. Because of the serial nature of digital arithmetic care must be taken that a logical order is chosen when the equations for solution are presented to it. This remark applies to any subsidiary algebraic equations which need to be solved before the derivatives of the variables can be evaluated. This order is not always obvious and can require a fair amount of preliminary analysis. It can happen, of course, that the calculation of the derivatives requires the solution of a set of simultaneous non-linear algebraic equations. In this case an iterative procedure will have to be invoked and, with all but the simplest of integration methods, the set will need to be solved several times per time step. This again lengthens the solution. With an analogue computer, the solution of simultaneous algebraic equations is a straightforward process which introduces no time penalty.

A final point worth making in favour of analogue computing is concerned with the somewhat nebulous but very real quality, accessibility. The ability to fiddle with a simulation and to see immediately the results of variations of parameters and/or configuration is a valuable property of the analogue method which is not at the present time generally available with the digital method. Since the exploration of many possibilities in a short time is such a simple matter with an analogue machine, an analogue solution to a problem will generally represent less of a compromise than one obtained digitally. The wider use of multi-access systems will go some way towards redressing the balance in favour of the digital machine. However, only a very large speed increase, achieved either by changes in basic machine characteristics or (less likely) by the development of more efficient numerical methods and the extensive use of graphical output will make the digital machine truly comparable with a modern analogue computer in this respect.

In spite of its difficulties, however, the digital simulation of continuous systems has some definite advantages over analogue simulation. Perhaps the most significant of these is availability. Many organizations with some interest in the field are in possession of a general purpose digital computer and cannot justify the purchase of an analogue computer on economic grounds. In such circumstances the simulation practitioner will want to use the digital machine regardless of its disadvantages.

There are also technical advantages. A digital machine with floating point arithmetic does not need the equations put to it to be pre-scaled by the user. The accuracy obtainable from a digital machine can, in the right circumstances, be far greater than is technically possible with an analogue machine. When using a digital computer it is almost always possible to trade off speed against size. The machine size constraint is much softer with a digital computer than with an analogue computer so that the former is potentially capable of solving larger sets of equations than the latter. The ability to store results and programs in digital form can be very useful. The implementation of complex non-linear functions of problem variables and the extensive use of programmed logic become relatively simple matters. Setting up a problem on the machine requires no effort from the user—it is done automatically by the operating system software. Finally, comprehensive documentation can be provided by the computer itself as a matter of course.

Although, as indicated previously, hybrid computation can considerably widen the horizons of the analogue computer in many directions, in some ways it can have a restricting influence. Some of the flexibility of a conventional analogue computer often has to be sacrificed in the interests of efficiency and the comparatively slow speed of the digital machine often lengthens solution time. However, if these facts are recognized and can be tolerated, tremendous advantages can be obtained.

Finally, a word about one possible future development. The existence of the analogue patch panel, whilst providing flexibility, also imposes restrictions. A stored program in the form of a patched-up panel is a vastly more expensive item than a deck of computer cards. Also, unless meticulous attention is paid to documentation, the ease with which patching changes may be made can easily lead one to make mistakes and accept incorrect results—even analogue computer users are human and can err. A hybrid system containing a digitally controlled switching matrix to interconnect the analogue elements could remove both these difficulties. Until fairly recently the economics of doing this have been heavily weighted against it—manufacturers have estimated that the provision of an automatic patching system could double the cost of a hybrid installation. However, there are now signs that advances in technology may make it a viable proposition within the not too distant future. A breakthrough in this direction combined with the use of 'analogue compilers' of the type mentioned earlier could go a long way towards making analogue computation an even more powerful and widely used technique in the future than it has been in the past.

Bibliography

1. *Analog Computer Techniques*, C.L.Johnson (McGraw-Hill, 1956).
2. *Analog Simulation*, W.J.Karplus (McGraw-Hill, 1958).
3. *Principles of Analog Simulation*, G.Smith & R.Wood (McGraw-Hill, 1959).
4. *Analog Computation in Engineering Design*, A.E.Rogers & J.W.Connolly (McGraw-Hill, 1960).

The above four are 'golden oldies', written when valves were preferred to transistors, but they contain much that is still worth knowing.

5. *Analogue Computing for Beginners*, K.E.Key (Chapman and Hall, 1965).
6. *Systematic Analogue Computer Programming*, A.S.Charlesworth & J.A.Fletcher (Pitman, 1967).

These two contain basic material. The second avoids the descriptions of amplifier circuits which earlier writers apparently felt obliged to present to the reader.

7. *Hybrid Computation*, G.A.Bekey & W.J.Karplus (Wiley, 1968).

This book covers a great deal of ground and is written very readably.

8. *Simulation*, J.McLeod (ed.) (McGraw-Hill, 1968).

Descriptions of a number of applications, drawn from different fields of analogue, hybrid and digital simulation.

9. Useful Periodicals:

I.E.E.E. Transactions on Computers.

Proceedings of the International Association for Analogue Computation (*A.I.C.A.*).

Simulation.

Parameter sensitivity and interdependence in hydrological models

D.T.PLINSTON *Institute of Hydrology, Wallingford, Berkshire*

Summary

The importance of parameter sensitivity and of the interdependence of parameters is illustrated by reference to simple models used to predict run-off volumes from rainfall. Numerical values of these factors are derived and their use in improving the efficiency of optimization and in the choice of parameters is discussed.

Introduction

Advances in computer technology have allowed the rapid development of sophisticated conceptual models of the hydrological processes governing the generation of runoff from rainfall on a catchment area. Such models are usually made up of two more or less distinct parts. The first part keeps a running water balance so that, with the knowledge of the current 'state' of the catchment, the input rainfall can be distributed between the various outputs—runoff, transpiration and addition to soil moisture storage—according to empirical relationships representing the more complex division which takes place in a natural basin. The second part of the model takes the generated storm runoff volumes and distributes them in time, usually by routing through one or more hypothetical storages, to give an outflow hydrograph. In many models, these storages are arranged to be equivalent to the classical, though subjective, division of the hydrograph into surface, subsurface and groundwater flow.

The constants in the various empirical relations, together with any limits constraining certain operations within the model, are collectively known as the model parameters. It is assumed that these parameters can be adjusted to values which give a 'best fit' between predicted and observed runoff hydrographs. The choice of a criterion of 'best fit' is clearly subjective and will depend on the purpose of the model. The process of parameter adjustment or optimization can be made automatic and objective provided that the number of parameters to be optimized is not too large and that there is not a large measure of interdependence between parameters. Of these two conditions, it is unfortunate that the second is often a consequence of the first. The problem of interdependence and the related factor of parameter sensitivity can be studied by examining the behaviour of, say, a simple sum of squares error function following changes in parameter values away from their optimum values. Improvement in the efficiency of optimization might be expected when parameter scales are chosen so that the respective sensitivities are approximately equal; an improvement in the precision of parameter estimates might follow from the removal or addition of parameters so that serious cases of parameter interdependence are avoided.

This paper approaches these problems through studies of relatively simple models used to predict storm runoff volumes from the 19 km^2 catchment area of the River Ray at Grendon Underwood.

The models

The two models described briefly here have been reported in full elsewhere.[1,2] The catchment area suffers a significant soil moisture deficit in some years and both models attempt to allow for the effect of this deficit on evaporation and runoff. As far as possible, the models have been constructed so that, by the successive elimination of various sections of the models, the effects of each parameter can be considered separately.

The first model, termed the *Layer* model, considers the catchment as a spatially homogeneous vertical stack of soil layers, each holding 25 mm of water at saturation. Evaporation is allowed at the potential rate from the top layer and is then reduced by a parameter C (< 1) successively from layer to layer as the soil moisture is depleted. A parameter Z denotes the total soil moisture storage available for evaporation. The Penman open water evaporation, which is estimated from meteorological data, is reduced to potential evaporation from the catchment by a parameter T. Runoff occurs when the catchment is saturated, although a fraction H of the excess rainfall can be allowed to become runoff directly.

The second model, termed the *Area* model, is based on Penman's study of the Stour catchment.[3] It is assumed that vegetation transpires at the potential rate until wilting point is reached throughout the rooting depth. Thereafter, transpiration is reduced to one tenth of the potential rate. Three distinct areas are defined on the basis of root constant. A known area (measured) of trees has a root constant D_T; an area A of grass remote from the river has a root constant D; and the remaining riparian area (obtained by difference) is assumed to have an unlimited supply of water for transpiration. In practice, for this catchment, the root constant for the tree area never optimized to a value which affected transpiration and thus the tree area may be considered as lumped with the riparian area. A direct runoff factor H can be included in the same way as in the layer model.

The data comprised 3-hourly values of rainfall, open water evaporation and runoff and for the study of storm runoff volumes the record for the whole period (approximately 4 years) was divided into 204 periods, each containing one or more significant rainfall events. When computing the observed runoff for each period, allowance was made for the runoff in storage at the beginning and end of each period from an examination of the recession curves. The 3-hourly computed runoff volumes were summed over the same periods and an error function or residual variance was taken as the sum of squares of differences between the observed and computed runoff volumes over the 204 periods.

The optimization procedure adopted is based on that developed by Rosenbrock.[4] In this method, the initial search directions are those defined by the parameter directions. When all the directions have been searched once and a provisional optimum found, a new set of orthogonal search directions is defined, one of which is the vector joining the initial and final points of the first iteration. Further searches and realignment of search directions continues until the maximum change in parameter values over the final iteration is within a predefined limit. In the optimization procedure, the parameter values are normalized by a sine squared transformation which also allows the parameter range and thus the parameter sensitivity to be controlled.

Sensitivity and interdependence

Having chosen an objective criterion of fit, in this case a sum of squares function, it is possible to examine the shape of the function surface in the n-dimensional parameter space. A cross-section through this space on a plane defined by any 2 parameters can be drawn as a contour map of the function (F^2) as shown in Fig. 1. The ideal model would give a series of concentric circles, Fig. 1a, indicating equal sensitivity in each parameter direction, that is, the function value would be changed by the same amount following a constant change in either parameter value. In this case, there is no interdependence between parameter estimates, that is, any error in the estimate of parameter X would have no effect on the precision of estimate of parameter Y. If the parameter sensitivities are different, the resulting map would show a pattern of ellipses, Fig. 1b. Clearly, a linear scale change could transform the pattern of ellipses to one of circles. The case where the parameter estimates are interdependent is shown in Fig. 1c. The axes of the ellipses are inclined to the parameter directions and a simple scale change is insufficient to transform the pattern to one with circular contours without rotation to new parameter directions.

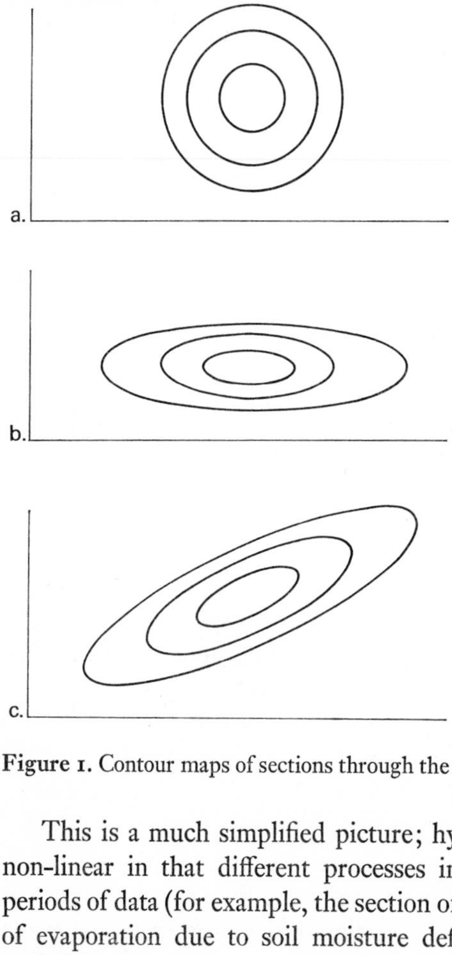

Figure 1. Contour maps of sections through the F^2 surface.

This is a much simplified picture; hydrological models are often highly non-linear in that different processes in the models operate on different periods of data (for example, the section of the model governing the reduction of evaporation due to soil moisture deficit rarely operates during winter periods) and switching or threshold parameters often lead to discontinuities in the F^2 surface. A complete picture of the behaviour of the model requires the mapping of the surface on all the planes defined by pairs of parameters. This is invariably a time consuming and expensive procedure which is best avoided if more simple estimates of sensitivity and interdependence can be obtained. The simpler the model the more likely is this to be possible.

(a) Sensitivity

Sensitivity is perhaps best represented by the second derivative of F^2 with respect to each parameter. When measured at the optimum, this is inversely proportional to the radius of curvature of the F^2 surface in the parameter directions. Alternative methods, such as the increase in F^2 for a given percentage departure from optimum parameter values, have been suggested.[5] However, it is desirable to be able to compare both the sensitivities of different parameters on a single catchment and the sensitivity of a given parameter from catchment to catchment. Thus, sensitivity should be linked to parameter scale rather than to a percentage of the optimum values obtained in each case. Irrespective of the definition adopted, discontinuities and irregularities in the F^2 surface are a hazard which must be treated with caution.

(*b*) *Interdependence*

Provided that the F^2 surface can be represented reasonably by a second order uni-modal function, a useful measure of interdependence might be obtained by analogy with the coefficient of correlation from a bivariate probability function. If x and y were not parameter estimates but random variates, they would have a bivariate normal distribution of the form,

$$p(x,y) = \frac{1}{2\pi\, \sigma_x \sigma_y \sqrt{1-\rho^2}} \exp\left[-\frac{1}{2(1-\rho^2)}\left(\frac{x^2}{\sigma_x{}^2} - \frac{2\rho x}{\sigma_x}\cdot\frac{y}{\sigma_y} + \frac{y^2}{\sigma_y{}^2}\right)\right] \qquad 1.$$

with x and y having a mean of zero, standard deviations σ_x and σ_y, and a coefficient of correlation ρ.

The contours of equal probability associated with combinations of x and y would be defined by $z =$ constant where,

$$z = \frac{x^2}{\sigma_x{}^2} - 2\rho\frac{x}{\sigma_x}\cdot\frac{y}{\sigma_y} + \frac{y^2}{\sigma_y{}^2} \qquad 2.$$

A second order approximation to the F^2 surface, such as the elliptic paraboloid, has contours given by $F^2 =$ constant where,

$$F^2 = \frac{a^2}{2}(x\cos\theta + y\sin\theta)^2 + \frac{b^2}{2}(y\cos\theta - x\sin\theta)^2 \qquad 3.$$

with x and y having optimum values of zero and where θ is the angle between the parameter axes and the axes of the ellipse.

Equations 2 and 3 are of the same form which suggests that a measure of interdependence between parameter estimates could be taken as,

$$\rho = -\frac{\sin\theta.\cos\theta.(a^2 - b^2)}{(a^2\cos\theta^2 + b^2\sin^2\theta)^{\frac{1}{2}}(a^2\sin^2\theta + b^2\cos^2\theta)^{\frac{1}{2}}} \qquad 4.$$

A convenient method of computation is to set up a 5×5 matrix of F^2 values centred on the optimum and to compute the second derivatives of F^2

in the axial and diagonal directions. These derivatives can be expressed in terms of a^2, b^2 and θ as follows for the assumed second order surface defined by equation 3.

$$D_1 = \frac{\delta_2(F^2)}{\delta x^2} = a^2\cos^2\theta + b^2\sin^2\theta \qquad 5.$$

$$D_2 = \frac{\delta_2(F^2)}{\delta y^2} = a^2\sin^2\theta + b^2\cos^2\theta \qquad 6.$$

$$D_3 = \frac{\delta_2(F^2)}{\delta p^2} = \frac{a^2}{2}(\sin\theta + \cos\theta)^2 + \frac{b^2}{2}(\sin\theta - \cos\theta)^2 \qquad 7.$$

$$D_4 = \frac{\delta_2(F^2)}{\delta q^2} = \frac{a^2}{2}(\sin\theta - \cos\theta)^2 + \frac{b^2}{2}(\sin\theta + \cos\theta)^2 \qquad 8.$$

where p and q are the diagonal directions of the matrix.
Substitution for a^2, b^2 and θ in equation 4 gives,

$$\rho = \frac{D_4 - D_3}{2(D_1, D_2)^{\frac{1}{2}}} \qquad 9.$$

It can be verified that equation 9 is independent of any scaling effects introduced to balance the parameter sensitivities.

As in the case of the computation of sensitivity, care must be taken to ensure that the assumptions inherent in the computation of ρ are not seriously violated. This can be done quite simply by computing values of sensitivity and ρ over a progressively wider region of the F^2 surface. Consistent values would indicate a good approximation to a second order surface. Inconsistent values would indicate that mapping is desirable before further progress could be made.

Results from the models

Two interesting results illustrating the effects of parameter interdependence arose during the testing of the *Layer* model, and in both cases they involved the parameters which limited the evaporation during periods of soil moisture deficit. Figure 2 shows the F^2 map on the plane defined by parameters C and T. The obvious feature is the long diagonally oriented valley along which there is little change in F^2 for different combinations of C and T. All other parameters were held at inoperative values. The obvious interdependence of C and T is confirmed by the value of −0·95 obtained for ρ. It is therefore unreasonable to expect precise fitting of both these parameters and a realistic course of action is to set one parameter at a physically realistic value so that the other can be optimized with a better precision. Although, in this case, the

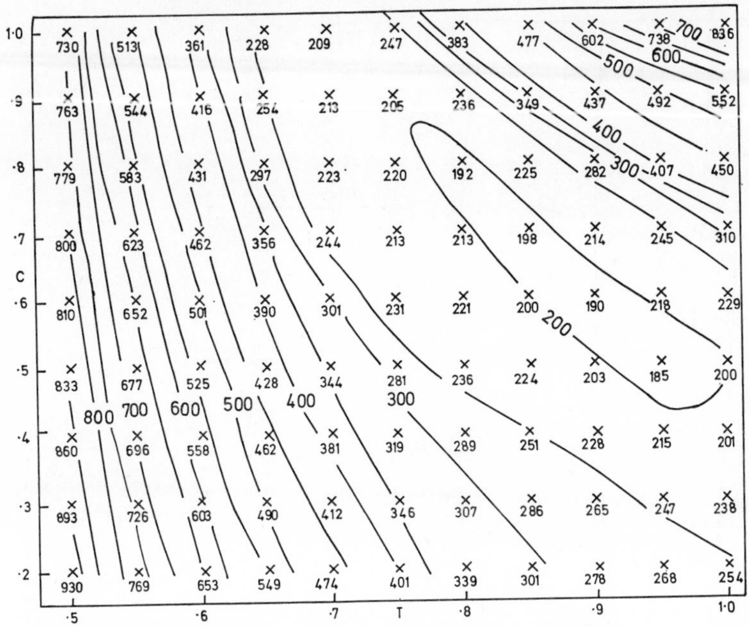

Figure 2. Dependence of F^2 on C and T—*Layer* model (Mandeville).

F^2 surface looks to be a good approximation to a second order surface, closer examination of selected cross-sections revealed isolated irregularities superimposed on the main relief. However, data from a fairly wide region was used in the computation of ρ so that the effect of these irregularities was minimized.

A quite different picture emerged when the model was fitted to data from the Brosna catchment at Ferbane, Ireland.[1] Figure 3 shows the F^2 map for the plane defined by parameters C and T. In this case, C optimized at a virtually inoperative value and was extremely insensitive. On the other hand, T was well defined. This result is not surprising as the Brosna catchment has a higher rainfall than the Ray catchment and consequently the soil moisture deficit does not normally reach the level at which evaporation would be reduced significantly from the potential rate. Thus, the section of the model which controls this reduction could be removed in this application.

The second case, illustrated in Fig. 4, concerns interdependence between parameters C and Z although, in this case, the interdependence is such that only one of the parameters can be optimized at once, the other becoming inoperative. When Z exceeds the maximum computed soil moisture deficit, it becomes inoperative (zero sensitivity) and the effective soil water storage is controlled by parameter C. Conversely, when Z is low, C is ill-defined as it makes little difference, in terms of computed runoff volumes, whether the small amount of water available for transpiration is used at a fast or a slow

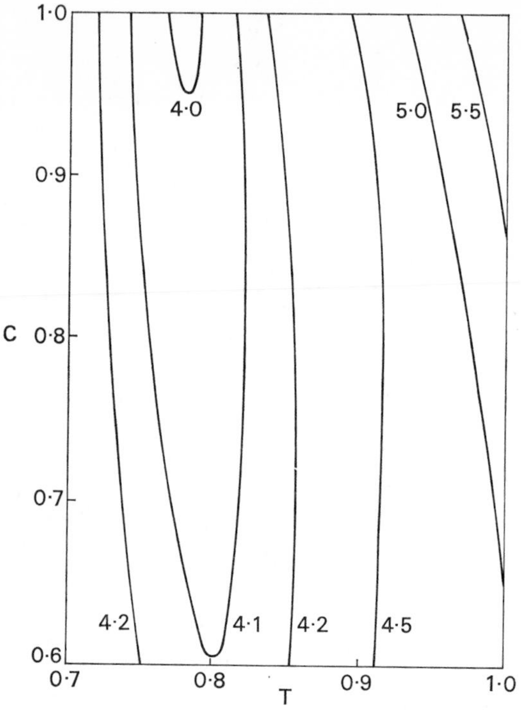

Figure 3. Dependence of F^2 on C and T—*Layer* model Brosna catchment (O'Connell).

rate. In this case, further tests showed that a model with C only was only slightly superior to one containing Z only on the basis of explained variance and that either parameter could be eliminated without significant reduction in the explained variance. Obviously the F^2 surface is not at all well fitted by a second order surface and the computation of ρ would be meaningless. In this case, a map was essential as was indicated by the obvious insensitivity of the parameters at different times during the optimization search.

The dependence of F^2 on the parameters A and D of the *Area* model is illustrated in Fig. 5; the parameter H was inoperative in this case. The optimum is not well defined and there is evidence of a local optimum at high values of D. It would seem that the poor definition of the optimum is due to the low sensitivity of both parameters near the optimum rather than to interdependence between them. A value for ρ of 0·30 indicates only a moderate degree of interdependence. When parameter H was introduced, the interdependence between A and D was reduced ($\rho = 0{\cdot}23$) as was the sensitivity of parameter A. There was no significant interdependence between H and either A or D.

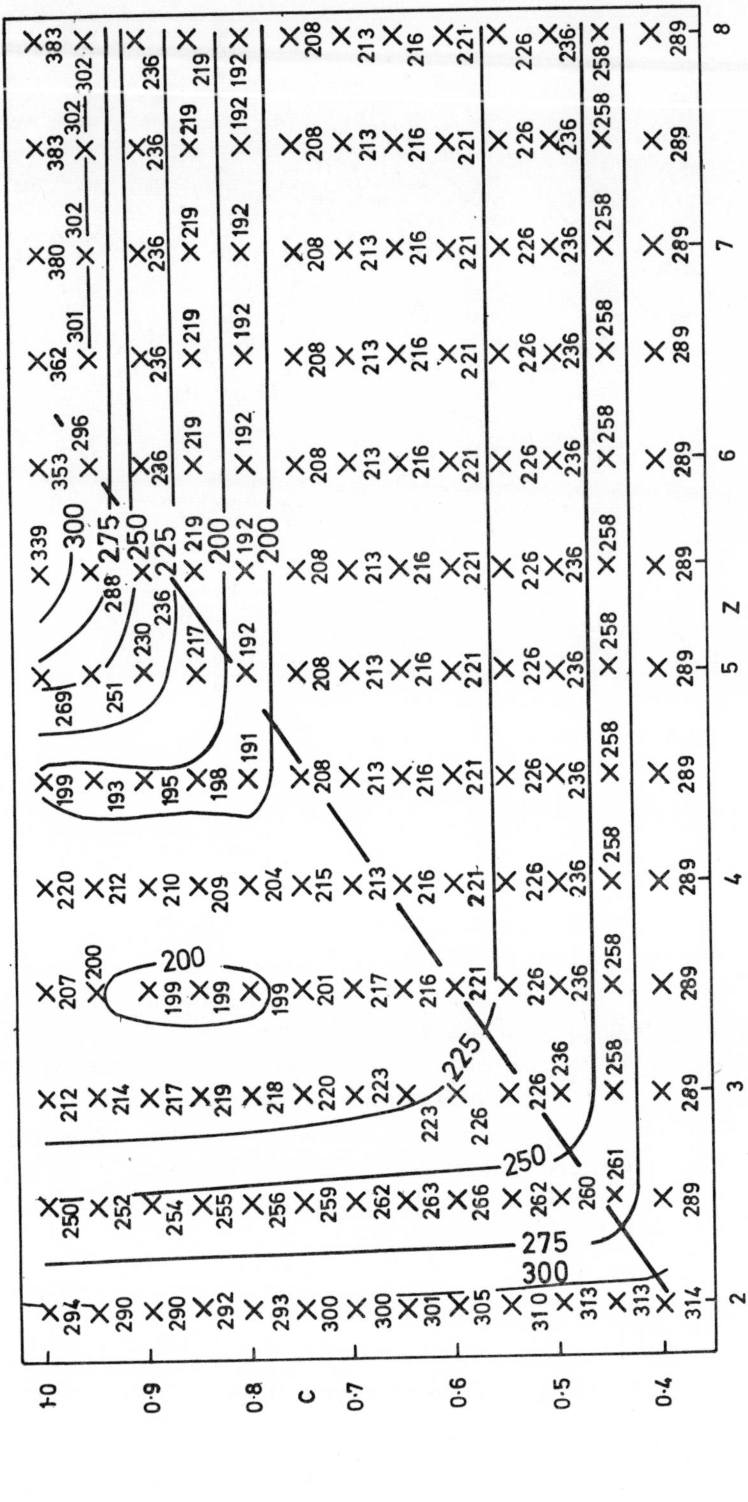

Figure 4. Dependence of F^2 on C and Z—*Layer* model (Mandeville).

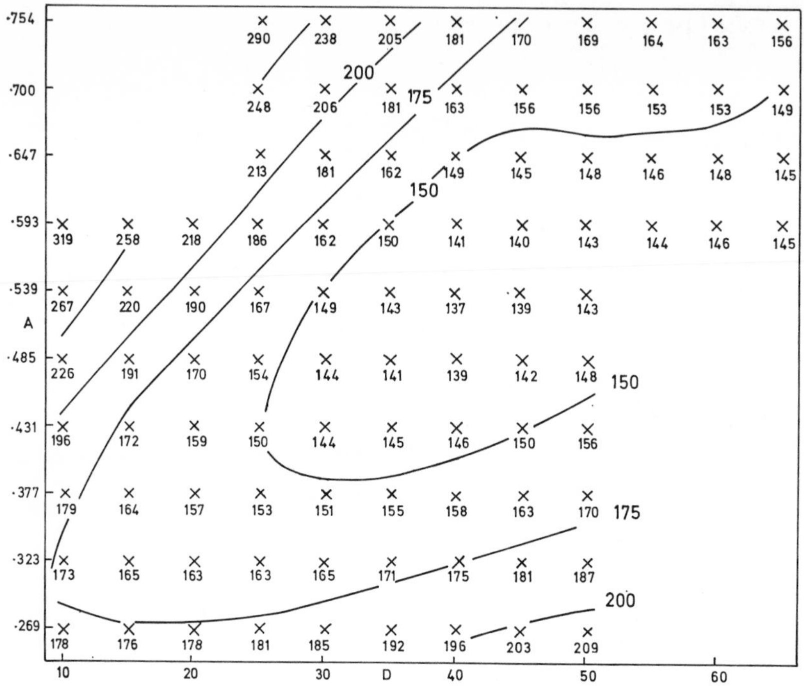

Figure 5. Dependence of F^2 on A and D—*Area* model (Mandeville).

Conclusions

These examples, albeit on relatively simple models, have shown that much useful information concerning the behaviour of the model can be gained from a study of the surface of a function representing the goodness of fit of the model. Complete mapping of sections of this surface is not always essential; estimates of parameter sensitivity and interdependence based on fewer computations of the function can give a valuable indication of the effectiveness of the model parameters. In some cases, the results can be explained by physical reasoning and can suggest changes in the structure of the model or can indicate when certain sections of the model may be made inoperative without any significant reduction in the goodness of fit.

Acknowledgements

Figures 2 to 5 of this paper appeared originally in the papers referenced 1 and 2 below. The author gratefully acknowledges permission to use them here.

References

1. O'CONNELL P.E., NASH J.E. & FARRELL J.P. (1970) River flow forecasting through conceptual models: Part II—the Brosna catchment at Ferbane. *J. Hydrol.* **10**, 317–29.
2. MANDEVILLE A.N., O'CONNELL P.E., SUTCLIFFE J.V. & NASH J.E. (1970) River flow forecasting through conceptual models: Part III—the Ray catchment at Grendon Underwood. *J. Hydrol.* **11**, 109–28.
3. PENMAN H.L. (1950) The water balance of the Stour catchment area. *J. Inst. Water Engrs.* **4**, 457–69.
4. ROSENBROCK H.H. (1960) An automatic method of finding the greatest or least value of a function. *Computer Journal* **3**. (3) 175–84.
5. DAWDY D.R. & O'DONNELL T. (1965) Mathematical models of catchment behaviour. *Proc. ASCE HY4*, 123–37.

The use of computer simulation in conservation management

C. MILNER *The Nature Conservancy, Edinburgh*

Summary

Conservation, however it is defined, involves the management of ecosystems and collections of ecosystems to optimize pre-determined variables or numbers of variables. Upland ecosystems, although simplified by climate and history, are nevertheless too complex for conventional agricultural experimentation to be appropriate and more useful techniques are necessary to allow their efficient management. The Nature Conservancy in Scotland has begun to explore the use of simulation as a management tool which is central to a continuing data acquisition operation within a major National Nature Reserve, St Kilda.

This paper reports on the early use of this strategy and explores its advantages for assisting in this intrinsically difficult management problem.

Introduction

Conservation is a difficult concept. The major problem is to achieve an operational definition or statement of aims which does not involve value judgments in its formulation but which recognizes the subjective nature of its implementation in particular circumstances.

There are however very many statements of definitions and aims which cover a wide spectrum of opinion. For example, Margalef (1968) implies that any exploitation or harvest from an ecosystem, or any interference with the normal evolving processes and feedback mechanisms within the system, is incompatible with conservation. Holdgate (1970), however, attaches a considerable importance to the management of natural resources as part of a conservation ethic and his statements of aims have wise use of such resources as an important feature. They do, however, illustrate a fundamental difficulty in all such definitions as a guide for policy as, dependent on the weight given to important emotive words such as 'unnecessary' and 'irreversible', which are not defined further, the statements may be impossible to reconcile internally.

Elton (1958) suggests that conservation is the retention and extension of ecological variability at the ecosystem, region, national and world scale. His view is that ecological variability preserves the possibility of varying subsequent utilization strategy. The reconstitution of natural or quasi-natural plant and animal communities is not excluded, nor is the use of non-native biota, or the management of plant and animal communities, if variability is thereby increased. Such increase in diversity by management has as its main constraint that no decrease in ecological stability ensues. This statement of conservation aims appeals intuitively mainly in the higher probability of precise definition and measurement to produce a coherent body of ecological knowledge capable of objective interpretation. This must be done before value judgments are made using humanistic criteria.

With the exception of Margalef's idealistic statement of aims, management features in them all, and I suggest that the management content of conservation is one of its most important features. Conservation ecologists are therefore committed to a considerable management responsibility at many levels. This paper describes the early stages in the development of a logical structure for the management of upland ecosystems in Scotland, using computer simulation as its basis within a general computer-based data acquisition strategy. The pilot scheme has been centred on a single National Nature Reserve at this stage of its development.

Management objectives

The intuitive broad aims of management must be put into a logical and numerically precise form, before a basis for conservation management and management itself can be undertaken. Harvesting of animals or plants from within the ecosystem biota must be expressed numerically as a rate or total, and the consequences of the operation predicted with known confidence. That is, not only do we need clear statements of numbers and rates arrived at intuitively, but also the ability to predict the numerical consequences of a particular harvest rate on the population harvested, and other relevant ecosystem variables. In circumstances where there is the chance of the extinction of a very small population, these predictions and confidence levels are of particular importance.

The numerical statements must also be given spatial dimensions within which to operate. For example the management policy to be followed if the same age class structure of Scots pine (*Pinus sylvestris*) was required on every National Nature Reserve would be qualitatively and quantitatively different from that required to produce skewed distributions of age classes in particular Reserves. Management would also be different if the problem of the conservation of Scots pine (*Pinus sylvestris*) were considered within a regional context rather than an individual Reserve context. Although problems of this type can be dealt with intuitively, the complexity of ecosystems is such that intuitive methods fail to give management strategies of sufficient sensitivity for a sophisticated conservation management operation.

A statement of objectives, therefore, does not in itself help the study of the complex interactions occurring, even in the simplest ecosystem. If time is introduced as a major variable, then the dimension of complexity is considerably increased. In the face of such complexity, it is possible to abandon prediction and manage the ecosystem or ecosystems as a land-holding exercise, changing the management empirically in the face of pressures from owners, tenants or other interested parties. Although such pressures are legitimate, reaction to them is often undesirable, and one aim of a modelling operation is to provide a better objective management plan, decided on ecological rather than on an empirical land management basis. Such ecological management should have predefined conservation aims known to be capable of achievement within a relevant time scale. In this management situation, predictive models, and possibly decision models, have been shown to be of value in many disciplines, and the described strategy was built around a predictive model, simulated over time on a digital computer.

Management of upland nature reserves in the interest of conservation, as previously discussed, may be said to have four main objectives.

(1) To maintain a vegetation cover sufficient to prevent soil removal by weather catastrophes occurring with a greater probability than ·001 per year.

(2) To maintain varied ecosystems both in biotic, structural and trophic level diversity.

(3) To increase variety, where lacking, by shifting communities and ecosystems from sub-climax or plagio-climax states towards climax states.

(4) To preserve individual species of high interest, rarity and vulnerability, irrespective of their functional position in an ecosystem.

To accomplish these objectives in upland Britain, the most usual and effective agent is the grazing herbivore or fire, the use of which has often maintained less than conservation excellence and which has been reviled by no less an authority than Fraser Darling (1947). However, variation in sheep and cattle numbers and burning, alone or together with grazing, can effect changes in the upland vegetation complex (King & Nicholson 1964) which may increase both structural and floristic diversity in moorland ecosystems, if used positively. In the example described below, the St Kilda National Nature Reserve is atypical in being grazed by one large herbivore, the Soay sheep, and in a truncated range of vegetation structural types. It does, however, provide an easily defined universe, and for this reason was chosen to test the relevance of computer simulation as a management tool.

The St Kilda ecosystem

The St Kilda National Nature Reserve is an archipelago with four main islands and several isolated stacks. The model described is of the largest of these, Hirta, which is approximately 645 ha in extent. This reserve has five major advantages for the development of a model to be used in management.
(1) It is an island and the important biological processes are restricted spatially. The input to the system of sea bird excreta and spray, whilst important, are probably insignificant in the model developed.
(2) The island has few vegetation types.
(3) There is only one major vertebrate herbivore, the Soay sheep, and no predators on this population. The Soay sheep population is unmanaged by man.
(4) The island has been the focus for much research activity since 1958, mainly on the population characteristics of the Soay sheep. The model has drawn freely on data from this research activity organized by J.M.Boyd into the Soay Sheep Team. In particular, data and ideas from P.A.Jewell, P.Grubb, D.Gwynne, I.A.Cheyne, R.G.Gunn and R.N.Campbell have been used.
(5) The island is a major National Nature Reserve of considerable importance to conservation. The management of the island in a logical way is therefore highly desirable.

The island system, or important aspects of it, are therefore simple enough to reduce the interactions to a reasonable number. This has made the modelling simpler, whilst retaining many of the major characteristics of the uplands which makes the test of the strategy realistic.

Detailed description of the physical features of the island are given in Williamson & Boyd (1960), MacGregor (1960) and Boyd, Doney, Gunn & Jewell (1964). Further descriptions are in Boyd, Jewell & Milner (in press) and no detailed description is needed here. However, some concept of the island will indicate its value as a pilot site.

The physical environment

Hirta has a maximum altitude of 419 m (1397 ft) O.D. at the summit of Conachair, whose north east face, sheer to the sea, is the highest sea cliff in Britain. The island has cliffs to sea level around much of it and only the Village Glen facing south east provides opportunity for landing. It is in this Glen that the Village, evacuated in 1930, is located and much of the research effort has been centred. Village Glen is an amphitheatre with screes and crags and unmistakable signs of glaciation and late snow beds. An opposing glen, facing north west, Gleann Mor, is, however, smooth and wide with few crags and no scree. There are many large ledges on the seacliffs utilized by the Soay sheep, which are broken by buttresses and talus slopes. There are also hundreds of small stone structures throughout the island used by the St Kildans for storage and called *cleitean.* These provide considerable shelter for the sheep which, together with the broken topography, reduces the effects of high mean wind speed which is a major feature of the island climate. Although no long term weather records from Hirta are available, its major characteristics can be inferred from data collected at Benbecula 50 miles east, where it has been assumed the weather at sea level is similar although, because of the mountainous topography, there will be much variability over the island. South westerly winds are most important and the mean annual wind speed is over 20 m.p.h. The mean daily temperatures at Benbecula range from 6°C in January to 13°C in August with daily ranges of less than 5°C summer and winter. Rainfall is about 114 cms per annum at sea level. On Hirta, there is perhaps 25 cm more on the summit ridge. The frequent orographic cloud cover of Hirta considerably reduces insolation at all times of the year and probably reduces both the daily and annual temperature range. Further discussion of the physical environment of the island is in Campbell (in press Boyd, Jewell & Milner).

The vegetation

Most of the island is grassland or dwarf shrub heath accessible to sheep and providing a variable food source throughout the year. The vegetation has been described by Turrill (1927), Petch (1933), Poore & Robertson (1949) and McVean (1961) and a recent description occurs in Gwynne & Milner (in press Boyd, Jewell & Milner).

The plant associations of Hirta may be split into five arbitrary groups.

Calluna heath
Nardus grassland
Mixed grassland
Biotic grassland
Maritime grassland

This series follows an increasing nutrient gradient from Calluna heath and Nardus grassland to Maritime grassland and associated increasing biotic effect of sheep and birds. There is however considerable variability in this series and some of the associations which floristically would be in the mixed grassland group are sufficiently base rich to be considered in the biotic grassland group. This simply indicates the difficulties of vegetation classification into arbitrary hierarchies. The broad distribution of the five groups of associations are shown in Fig. 1.

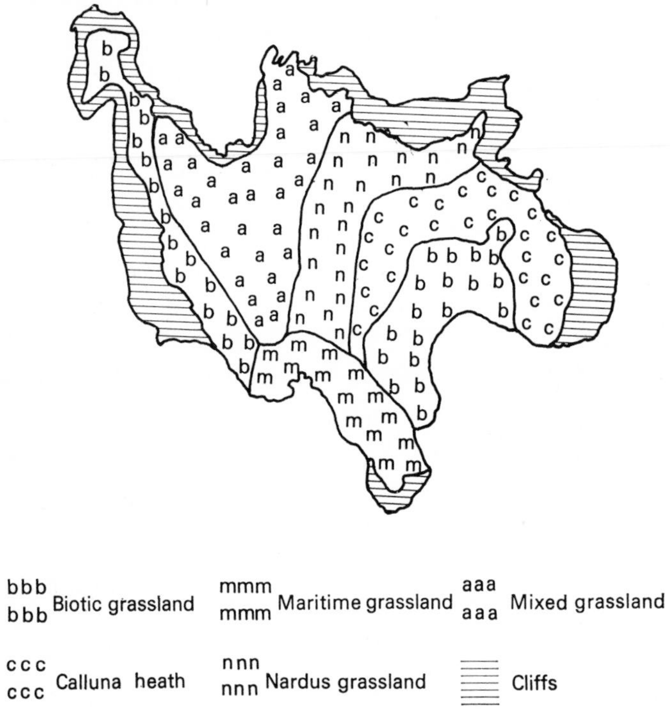

Figure 1. The vegetation of Hirta (highly simplified).

The groups of plant associations are dominated by six main grass species in various combinations and by *Calluna vulgaris*. The main grass species are *Nardus stricta*, *Agrostis tenuis*, *Festuca ovina*, *Festuca rubra*, *Holcus lanatus and Poa pratensis*. This species series also follows an increasing nutrient and animal activity gradient in broad terms although not in palatability terms as both *Festuca rubra* and *Poa pratensis* are selected whereas *Holcus lanatus* and *Agrostis tenuis* are rejected. The biological characteristics of the main communities can therefore be effectively described in terms of the growth patterns and interrelationships of these grass species and *Calluna vulgaris*, which together produce a high proportion of the plant biomass on Hirta. There have been changes in the relative proportions of the species in the years since the first account of the vegetation and changes in the relative proportions of the different communities. *Calluna* heath for example has extended its boundaries considerably. This species shift is one of the features to be modelled and is important in conservation terms as one of the processes whose rate may be increased or decreased to promote diversity.

The Soay sheep

The population of the Soay sheep, which is the only large herbivore on the island of Hirta, has increased to its present size from 107 animals released on the island in 1932. Their fluctuations since then are shown in Fig. 2. Since

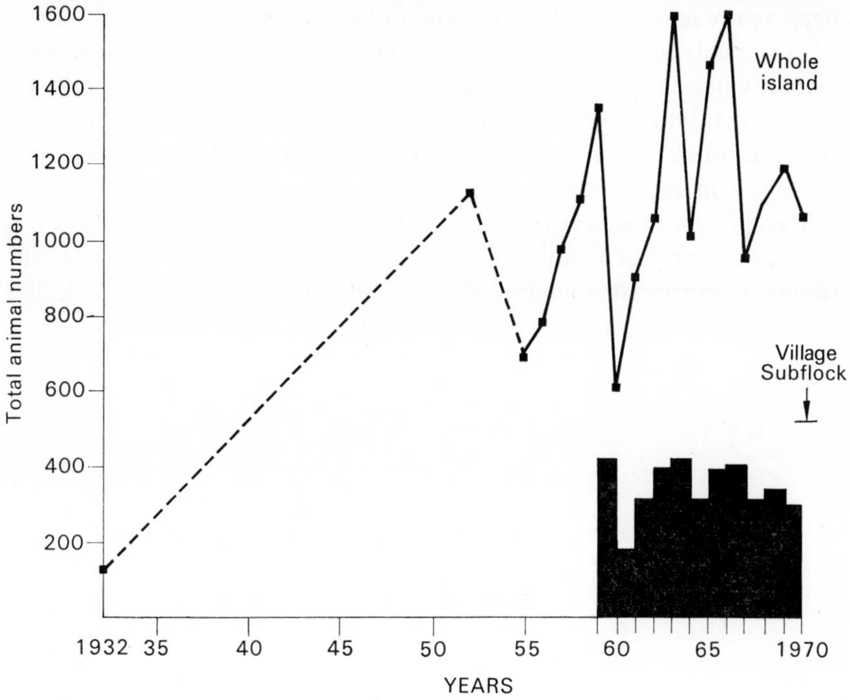

Figure 2. Population fluctuation of Soay sheep on Hirta, St Kilda.

1958, a considerable research effort has resulted in a very accurate description of the major characteristics of the population and of the biology of the sheep and a high proportion (80 per cent +) of lambs born in the Village have been individually marked. A high percentage of carcases dying over winter have been recovered and identified owing to an unusual feature of this population in using *cleitean* for shelter where they frequently die. This is dealt with in greater detail by Boyd, *et al.* (1964) and Grubb & Boyd (in press Boyd, Jewell & Milner). Although the total numbers of animals have increased from 107 to the present 1006 the increase has not been smooth. Since 1958 the pattern has been one of population build up to a high level followed by increased mortality resulting in a sudden decrease in numbers in one year (in early spring). Following such a reduced population, natality has been high for several years, increasing the population to numbers similar to those at which high mortality occurred previously, when high death rates have again reduced numbers. The periodicity of high death rates has been variable but of the order of 3 to 4 years. Some degree of density dependent mortality exists although it is not a simple relationship. A weather effect is apparent since almost all the deaths and births occur during the months of March and April, deaths being particularly marked in years in which the mean temperature in March and April is lower than average.

Fecundity and death are also age dependent and there is differential mortality between males and females. This aspect is discussed in Grubb (in press Boyd, Jewell & Milner) and is not discussed in detail here. However, two features are important for the model described. The ratio of males to females is 1 or approximately so immediately after neo natal mortality. This shifts rapidly with increasing age and at 3 years the ratio is approximately 0·5, decreasing to 0·3 at 5 years. Fecundity in the females rises asymptotically (fecundity expressed as number of lambs born per ewe) and does not decline with age, most non pregnant ewes being one or two years old (see Fig. 3 from

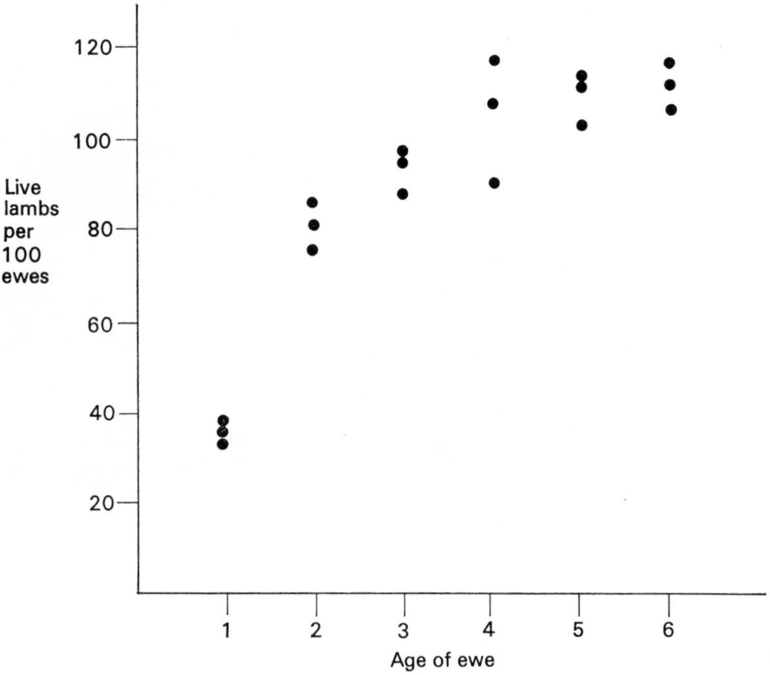

Figure 3. Age specific fecundity of Soay ewes, St Kilda.

data in Grubb & Jewell, in press Boyd, Jewell & Milner).

Death is associated with a low body weight (Fig. 4) and on post mortem there are very low fat deposits both subcutaneously and pericardially. No

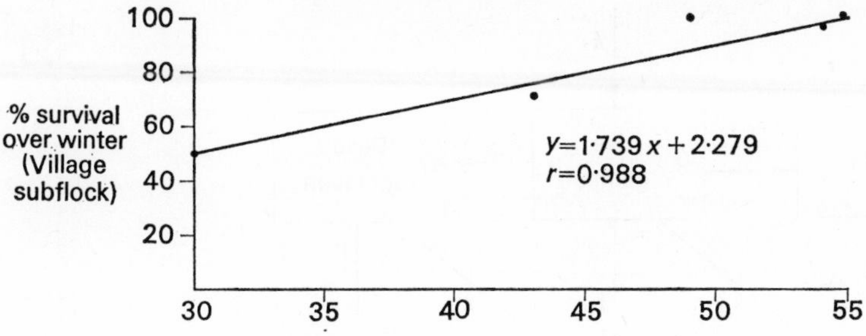

Figure 4. Per cent survival of adult Soay ewes on St Kilda (five year data).

evidence of pathogens or excessive parasitism was noted in over 80 post mortems, and it appears likely that death is caused by a lowered intake of digestible energy and increased energy requirements causing fat reserves to be utilized ultimately resulting in death by starvation. Body weight changes are considerable over the year and between years.

The model

There are as many models of an ecological system as there are ecological viewpoints of that system (Schultz 1969) and the model proposed for Hirta is therefore only one of the large number possible. In this sense it is not an ecosystem model; nor is it desirable, in my view, that it should be so. The view of the system of interest is that concerned with the principles laid down on page 251 and therefore concerns the sheep population and the vegetation on which they feed, as the important components. Such a model has the advantage of relevance to management, and an important practical advantage in the low range of response times of the system or rather that, at the level of study adopted, they have similar response and relaxation times. The main compartments and movements of matter and energy are shown in Fig. 5.

Compartmental models of this type, in which compartments represent space-time units occupying a volume of space for a finite time and its associated states, and the arrows represent the physical movement of matter between compartments such as to alter their states, have been increasingly used in ecology. Olsen (1964) Gore & Olsen (1967) and Kelly *et. al.* (1969) use models of this type of explore the interrelations between compartments of a system in which the rates of movement are simulated by differential equations solved over time, the change of state of the compartments being visualized as the results of the iterations involved in solving the system of differential equations.

Such relatively simple models with a set of equations concerned only with intrinsic factors of the system and with steady state situations have been developed by Bledsoe & Jameson (1969), Goodall (1967, 1968) and Van Dyne

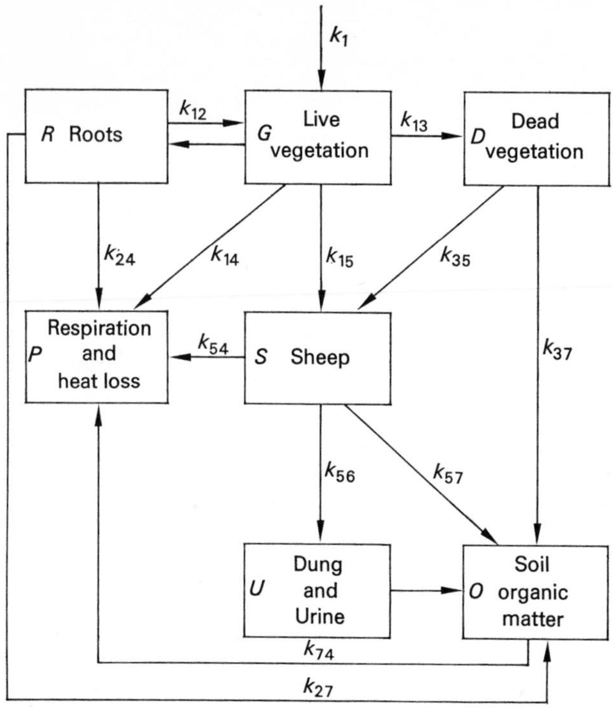

Figure 5. Compartmental model of sheep/vegetation system, Hirta, St. Kilda.

(1969) into more realistic systems of differential and difference equations in which the rates of movement and the states of the compartments are affected by extrinsic variables and by interactions within the system not necessarily capable of expression as a physical movement of material. That is, a rate or state factor may be influenced by a state factor or rate not directly linked physically. An obvious example is competition for light or nutrients between different species of the same plant communities.

The Hirta model has been of this second type for various reasons. The simple closed models are of considerable value in exploring the detailed features of a system in which the main controlling factors are intrinsic. In the sheep/grass system, however, extrinsic variables such as climate are important. The growth of plant species is weather dependent, as is the energy output by sheep. In order to achieve a usable management model, therefore, those variables associated with climate must be included. The disadvantage of including such variables is that the functions describing their effects on the rates of movement (*k* values) are incompletely known and must be deduced from ecological and biological first principles. It is considered that this is worthwhile, despite the disadvantages, for two main reasons.

(1) A model is produced in which the predicted output varies correctly

despite its possibly low prediction of actual values. Any major discrepancy in the directional qualities of the prediction indicates a major fault in the structure of the model, and, if this occurs, it is not sufficient to modify the variables; the relationships between components must be re-examined.

(2) The model allows the tolerance of the variables to error and change to be explored, resulting in priorities in measuring such variables being correctly assigned. Such sensitivity analyses are among the more important tactical uses of the model and are discussed later.

In the isomorphic model shown in Fig. 5, therefore, the compartments represent biomass or energy without reference to time and the arrows represent rates of movement of biomass or energy between compartments. G, for example, represents a total dry weight of green grass at time t. S represents a total weight of sheep at the same time t, k_{15} is the rate of movement of G (green grass) to S (sheep), i.e. weight of green herbage grazed per unit time interval. In this model, however, K_{15} is not necessarily considered as occurring over a very small time unit, i.e. a continuous process, but can be considered as a discrete process involving difference equations rather than differential equations. Each of the transfer rates is also a function of the weights (or numbers) in the linked compartments, of the weights or numbers in other compartments not directly linked, and of extrinsic variables such as tempera-

Table 1. Transfers in St Kilda model (*k* values as in Fig. 5).

Transfer	Description	Function
k_1	Solar radiation input	Not considered separately, see subroutine PROD
k_{12}	Translocation	$f.\ (t, T, G, R)$
k_{13}	Death of vegetation	$f(t, T, G, D)$
k_{14}	Respiration of above ground	Not considered separately, see subroutine PROD
k_{15}, k_{35} (bracketed)	Grazing	$f(G, S, P, k_{15}+k_{35}/k_{56})$
k_{24}	Respiration of roots	Not considered separately, see subroutine PROD
k_{37}	Litter fall	$f(D)$
k_{54}	Maintenance of sheep	$f(S, T, k_{15}, k_{56})$
k_{56}	Defaecation	$f(S, k_{15}, k_{35}, k_{54})$
k_{57}	Death of sheep	$f(S, t)$
k_{27}	Death of roots	$f(R, T)$
k_{74}	Decay of soil organic matter	$f(O, T)$
k_{67}	Decay of dung and urine	$f(U, T)$

where G, R, D, S, O, U and k_{xy} are as in Fig. 5, T is temperature (°C) and t is time in appropriate units.

ture, solar radiation and other meteorological variables. The main transfers considered in the model are given in Table 1, together with the functional relationships considered.

In practical terms, the model is described by a series of Fortran statements assorted into a main program controlling the subroutines which simulate the subprocesses of the system such as plant growth, grazing and sheep population change.

The basic structure of the program is shown in Fig. 6 in the form of a

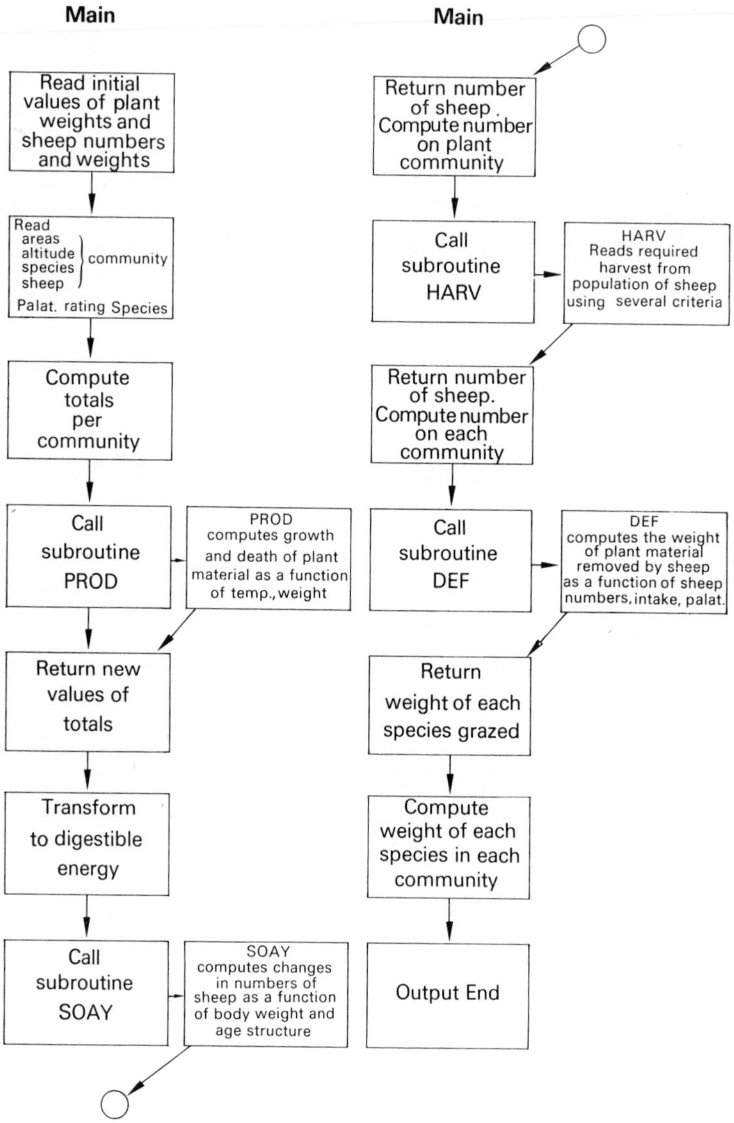

Figure 6. Flow diagram HIRTA.

flow chart. The time scale for the present program is such that one pass through the program simulates one actual day. This allows sufficient resolution to investigate inter-year variations although there are problems of computing time when intra-year variations over a reasonable time span are to be considered. The main variables of the system are set as appropriate and descriptive variables of the various plant communities and species read in. The island is divided into m plant communities made up of various combinations of n plant species. (m has a maximum value of 5 and n a maximum of 6.) The total weight of green herbage of each species is computed for each community and the total dead, i.e. the sum of the dead of each species in each community. The submodel calculating increase or decrease in green and dead weight of each species (PROD) is then utilized to change the weight of green and dead. Change in the live root compartment is also calculated and new values for community totals computed. This change is then used to calculate the total weight of digestible energy as a function of time from empirical data on energy and digestibility obtained in the appropriate community. The submodel of the Soay sheep population (Soay) is then used to calculate new numbers of sheep. The sheep are divided on sex and age class (r). The facility is available to remove or harvest sheep from a predetermined sex or age class or combination of these (HARV). The sub-model concerned with the grazing of the various species is then used to calculate the amount of each species removed from each community (DEF) and the total weights remaining are returned to output if appropriate. The next time interval is then simulated and, the main program and sub-programs are utilized to give changes in the various compartments of interest for a predefined length of time (i.e. 1 yr or more).

Subroutine PROD

The primary production subroutine has several unsatisfactory features, related mainly to a lack of data on some of the functional relationships, in particular those affecting the partitioning of assimilates between root and shoot. The model does not adequately incorporate solar radiation as a variable but utilizes empirical observations on the changes in biomass over a 12 month period. The main features of the subroutine are given in Fig. 7.

Rate of increase in green biomass for each separate species in each community is calculated as

$$\dot{G}_n = p_n\,(G_n * d_n * G') \qquad \dot{G}_m = \Sigma \dot{G}_n \qquad \left\{ \begin{array}{lll} n = 1\text{—}6 & \text{species} \\ m = 1\text{—}5 & \text{communities} \\ T \;\; 20^\circ\text{C} & \end{array} \right.$$

where p_n and d_n are coefficients $G_n < Gc\ g$ for

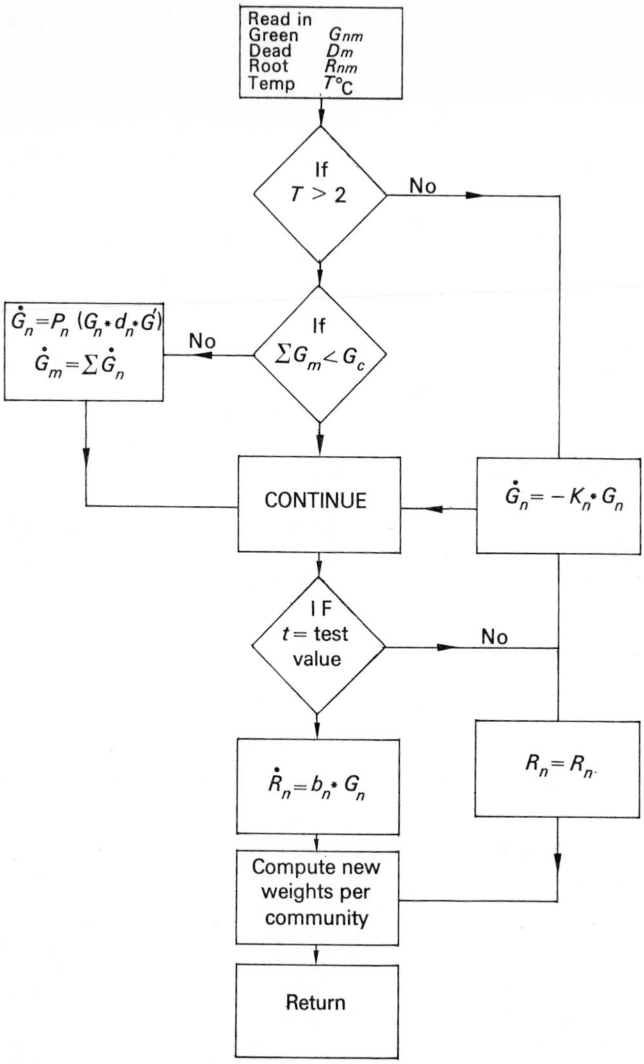

Figure 7. Subroutine PROD.

each species (G_c is the weight at which the solar radiation does not illuminate all leaves sufficiently for photosynthesis) and

$$p_n = p_n * \text{Sin}\ (cT - a)$$

modified by a sinusoidal function with O values at −6°C and 43°C (Fig. 8.1) (T varies with time and with the mean altitude of the community above the sea level datum), and where G' is a weighted sum of the species other than the

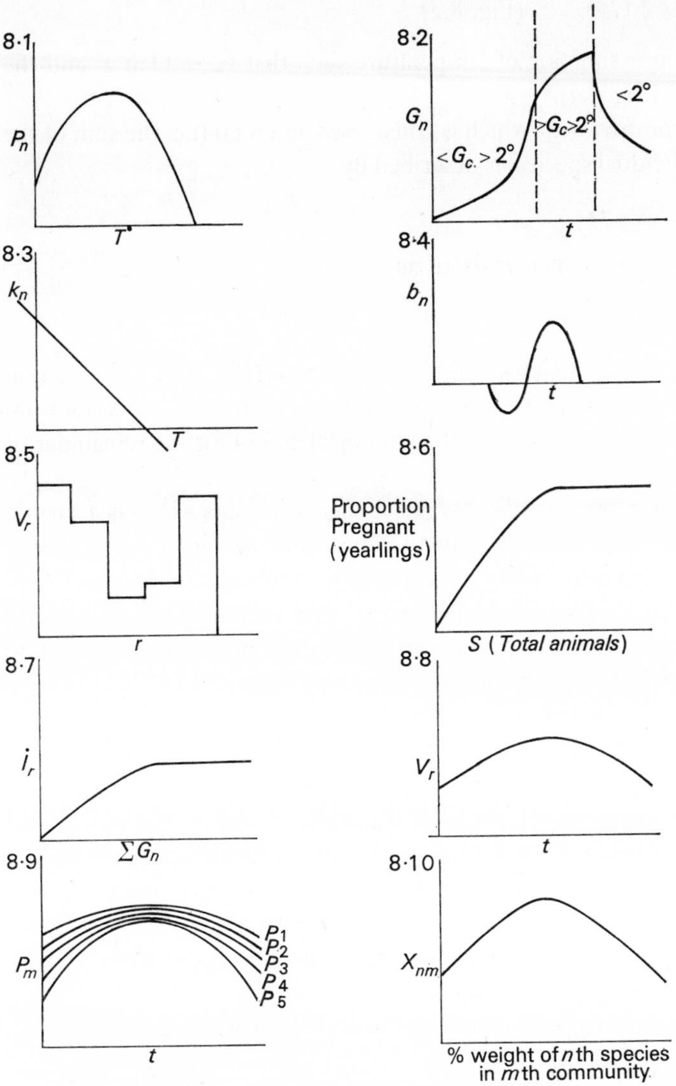

Figure 8. Form of important curves used in 'Hirta'.

one being calculated such that $O > d_n *G' > 1$ allowing for a competition factor between plant species.

The slope of the growth curve for each species reverses at a variable green weight (*Gc*) to approach a variable maximum (*G* max). This switch is necessary to allow for a lowered total photosynthetic efficiency when the level to which light penetrates is higher than the soil surface and only a proportion of the leaves are illuminated (Fig. 8.2). Below 2°C transfer to the dead compartment occurs and the weight of green decreases such that

$\dot{G}_n = - k_n * G_n$ (Fig. 8.2)

where k_n is a linear function of temperature such that $k_n = O$ at 2° and the slope is negative (Fig. 8.3).

The dead compartment which is considered as a total (i.e. the sum of the inputs from individual species) is described by

$$\dot{D} = \Sigma(\dot{G}_n * - 1)$$

The growth of roots is calculated as

$$\dot{R}_n = b_n * G_n$$

such that b_n is a modified sinusoidal function of time (Fig. 8.4). This function is such that there is a negative value for $\dot{R}n$ from $t = 90$ to $t = 120$ (spring) and positive from $t = 120$ to $t = 280$ (summer) and O for the remainder of the year.

This function represents the storage of energy in roots which is available in spring for additional and early growth. This function is imperfectly known and considerable experimental work is required to obtain the required variables. The feedback effect from one year to succeeding years is as yet not known sufficiently well and without empirical adjustments to the death rate the amplitude of the recurrent curve becomes very large.

Subroutine Soay

This subroutine calculates the intake of digestible energy by the various age classes of sheep, calculates the main energy losses by the sheep resulting in a positive or negative energy balance and the resulting gain or loss in weight. The actual live body weight is then used to predict the death rate of a particular cohort of sheep and the population size adjusted as appropriate. A flow chart is given in Fig. 9.

Intake of organic matter is related to the available herbage such that

$$\dot{I}_r = a\ \Sigma G_n / 1 + b * \Sigma G_n$$ $r = 1$—5 in males see (Fig. 8.7)

$r = 1$—10 females

r is age class

an asymptotic relationship where $\dot{I}_r$ is intake per minute per kg weight. The actual intake per day for each age class (I_r), is given by

$$I_r = \dot{I}_r * A * W_r$$

where A is the number of minutes spent grazing per day which is a function of time (see Fig. 8.8). W_r is the mean body weight of a particular cohort.

The digestible energy intake (D) is calculated as

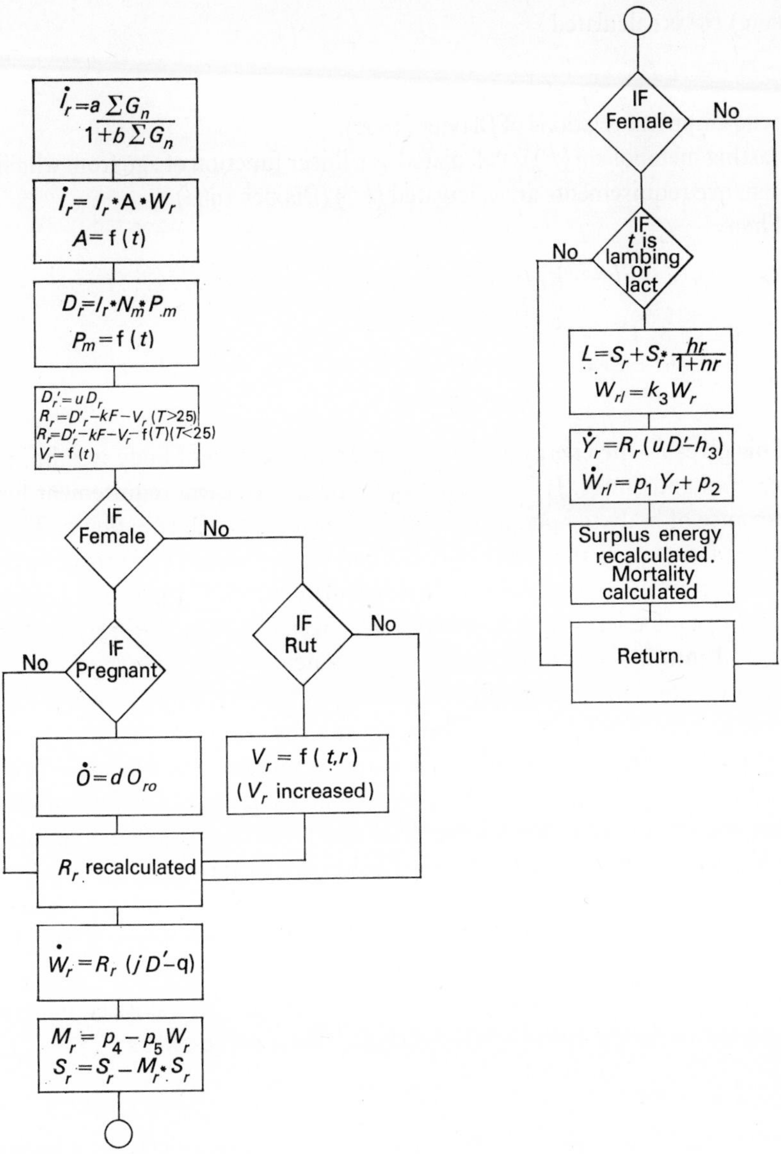

Figure 9. Subroutine SOAY.

$$D_r = I_r * N_m * P_m$$ [r is as before and m = 1—5 number of separate communities.]

where I_r is as before and N_m is the gross energy (measured) of a plant community, and P_m is the digestibility of that energy determined by 'in vitro' techniques over time (see Fig. 8.9).

Metabolizable energy (Digestible energy—energy losses as urine and methane) D_r^1 is calculated

as $D_r^1 = u\,D$

using the empirical relations of Blaxter (1962).

Fasting metabolism (F) is calculated as a linear function of age from which maintenance requirements are calculated (kF) (Blaxter 1962).

Then

$$R_r = D_r^1 - kF - V_r \text{ (if } T>25) \quad \ldots\ldots\ldots 1$$
$$(r \text{ as before})$$

$$\text{or } R_r = D_r^1 - kF - V_r - f(T)\ (T<25) \quad \ldots\ldots\ldots 2$$
$$(r \text{ as before})$$

This gives R_r the energy available for the production of body weight increase, wool and lambs. If T is below 25°C there is an extra requirement for maintenance of body temperature which is shown as $f(T)$ in equation 2. This is a linear function in the limits encountered on the island. V_r is the energy output for movement and activity. It is calculated from Clapperton's (1961) measurement of energy requirements for movement (e.g. during grazing) partitioned into horizontal and vertical components. The amount of movement is a function of time on Hirta and data has been collected allowing this function to be derived Fig. 8.8. The male cohorts show a considerably increased activity within the limits of the rut, and this activity is increased dependent on the age of the ram. The activity is at a maximum for the first two and last cohort, and minimal for the third and fourth cohorts (see Fig. 8.5). This difference is the result of observed behaviour at the rut when younger and old animals have less mating success than medium aged animals and spend a higher proportion of time searching for oestrus females and in agonistic activity against other males.

All the females of cohorts greater than 1 year have an increased energy requirement in spring due to foetal demands; this is expressed as

$$\dot{O}_r = d * O_{ro}$$

where $\dot{O}_r$ is rate of increase in energy requirement of foetus and O_{ro} is an initial empirical variable which is a function of the age class of the ewe. d is a constant. O_r is included in equations 1 or 2 as a negative and R_r recalculated.

The first cohort are divided into pregnant and non pregnant as a function of the total number of animals present with a maximum of ·6 pregnant (see Fig. 8.6). These are then considered as separate cohorts with the non pregnant cohort having no energy demands for foetal demands or lactation.

Weight increment is then calculated for all classes as

$$\dot{W}_r = R_r\,(jD^1 - q) \ldots\ldots\ldots 3$$

The mortality for each age class (cohort) is calculated as a linear function of live body weight determined empirically from observation on Hirta (see Fig. 4).

$$M_r = p_4 - p_5 W_r$$

where M_r is mortality and P_4, P_5 constants and W_r is body weight from which the number of animals remaining is calculated

$$S_r = S_r - M_r * S_r$$

The number of lambs born (L) is calculated as a function of the age of the surviving females

$$L = S_r + S_r \left(\frac{h_r}{1 + mr}\right) \quad r = 1\text{—}10 \qquad \text{(see Fig. 3)}$$

S_r is the number of females of each cohort.

& h and m are constants.

L approaching an asymptote of 1·2 per ewe at ages greater than 6 yrs.

Lamb birth weight is computed as an empirical linear function of the mean weight of ewe passing through the origin. This is obtained from data collected on the island.

$W_{rl} = k_3 W_r$ ($r = 1$—10) W_r is the mean body weight of the rth cohort, the lambs of each ewe cohort being treated independently.

Neonatal mortality of lambs is calculated as a linear function of birth weight for each lamb—ewe cohort group.

Lactation is related to the 'surplus' of energy (R_r) previously calculated and metabolizable energy intake (Blaxter 1962).

$\dot{Y}_r = R_r (uD^1 - h_3)$ where R and D^1 are as before and u and h_3 constants dependent on age of ewe.

R_r and W_r are then recalculated as in equation 1, 2 or 3.

Lamb weight increase over the arbitrarily chosen first 20 days is given by

$\dot{W}_r = p_1 Y + p_2$ [$r = 1$] (P_1 & P_2 are constants)

After 20 days the lambs are presumed to be new first cohort. The mortality of each age class is again calculated including lambs.

Total numbers and liveweight are then returned to the main program as is the intake of organic matter from the mth community.

Subroutine DEF

The removal of organic matter from a particular community ($I_r * S_m$) has been computed as a total and no account taken of the relative acceptability of each of the species within the community. This subroutine partitions the intake of green material between species in the mth community and computes faecal return to the mth community. It provides facilities (not yet utilized) to input the nutrient return to communities. The number of animals on each community which is used by earlier subroutines has been computed as a function of time only (empirically) for each community read in as a monthly mean in the main program.

The defoliation of each species is then computed as

$$Z_{nm} = X_{mn} (I_r * S_m)$$

where I_r is intake per animal, S_m is the number of animals on the community and Z_{nm} is the percentage removed from the nth species in the mth community. X_{nm} is a palatability factor of the nth species in the mth community which is a function of the percentage weight of the species present in the mth community, such that X_{nm} rises to a high value at a mean weight of the species for all communities dropping to a low value above this, i.e. a parabolic function (see Fig. 8.10).

The weight of each species removed from each community is calculated, and the weight remaining for each species in each community and a total for the community is calculated and returned to the main program.

The weight of faeces deposited on each community (F_m) is calculated from

$$F_m = I_r * S_m * 100 - /P100$$

where I_r is the intake by the cohort mean animal and S_m is the total number of animals on a community and an exponential decay rate is introduced

$$\dot{F}_m = -k F_m$$

From $\dot{F}_m$ it is possible to compute the return of major nutrients to the system using mean concentration of nutrients in dung organic matter. This is an important feedback mechanism to plant growth which has not so far been incorporated into the model, but data is now available allowing a better calculation of $\dot{F}$ and hence nutrient return.

This subroutine has the important output of the total weight of each species in each community. Although over the time period the model has been run no trend has developed, it is in this subroutine that species compositional changes will be calculated. Further work is in hand to make the model more realistic and a separate subroutine is projected.

Subroutine HARV

This subroutine requires little description. The percentage of each age class to be harvested on any date is read and computed. Zero values can be used. Mean body weight or total population size or numbers in a cohort can be used as a constraint in the harvesting subroutine. The harvest can be taken over a variable time base or as a discrete operation. The judicious use of this subroutine can determine an optimum harvesting strategy for the island which is readily understood by the manager.

Discussion

The model described has many failings which may be ascribed to a failure of logic or an insufficiency of knowledge of important functions. However, it produces output for the compartments which vary realistically. A typical output for total Soay sheep population is given in Table 2 as an example. This

Table 2. Total numbers of Soay sheep in May

Year	Actual	Model
1961	910	910
1962	1065	990
1963	1590	1340
1964	1006	1575
1965	1470	1066
1966	1598	660
1967	956	955
1968	1104	1010
1969	1197	325
1970	1062	930

shows that the numbers fluctuate in a realistic manner but the phasing of heavy mortality is not correct. It could be inferred that the model is structurally incorrect, although this is unlikely, or that one or many of the equations has the wrong form or coefficients whose values are too far from reality. This can be corrected as the model evolves. However I suggest that the model is important not only or mainly as a predictive artefact but as a stimulus to the collection of the right data determined by an intuitive use of the model and by testing the sensitivity of the output to changes in variables.

The present model for example is particularly sensitive to small changes in daily mean temperatures owing to their effect on plant growth, but particularly in the cumulative effect on sheep body weight change and survival. The inaccurate fluctuations in population size shown by the model which had

Benbecula data as input suggested that there may be larger differences between Benbecula and the St Kilda sheep environment than had been considered, and this has indicated the importance of climatological measurement within the environment of the sheep, i.e. in the sheltered Village area and among the talus of the cliffs. Preliminary measurements, using portable hygrometers, has indicated that consistent differences in ambient temperatures occur between open sites on Hirta and the environment within the areas containing *cleitean* and talus. However, if the effect of wind is examined, the differences are much greater, and, although the instrument used for these preliminary measurements is crude, it has nevertheless indicated the necessity for more relevant environmental measurements as the difference in the two wind velocities has been considerable. The model has not, however, incorporated wind speed as a factor increasing heat output, but empirical relationships are available, and, when accurate data are collected, the function can be incorporated. The model has also assumed that the environment of a mean sheep is that of a single animal. However, the Soay sheep show considerable clustering behaviour and the environment within such a group particularly within a *cleitean* is most certainly different. This environment should be recorded, if a good fit to the temperature relations of the model is to be obtained.

If the model is run for several years, it becomes apparent that the primary production model is insufficiently damped and fluctuations in total green production become extreme (as indeed do the associated sheep population variations). The model therefore reinforces the knowledge that complex feedback loops exist in the sheep/vegetation system and their study should have high priority with a major input of research required to examine this portion of the model. Experimental work on defoliation and its effects on carbohydrate reserves of the species concerned is of a high priority. This work would not of necessity be at the physiological level with sharp resolution over a short time scale but at the ecosystem level with responses of several years.

Mention has already been made of a feedback mechanism built into the original model structure, i.e. the increasing return of dung and urine to plant communities as sheep numbers increase. This mechanism would result in increased plant growth rates at higher sheep numbers and damp the oscillations caused by the exponential nature of the sub model. Work to investigate this is planned as a direct result of the experiments on the model.

The use of the model has therefore provided the resource managers and associated research staff with powerful intellectual advantages, not all of which are concrete. However, five of them can be formalized.

(1) The structuring of the model has removed ambiguities and illogicalities from our view of the island ecosystem.

(2) The search for appropriate functions has stimulated a literature search

which would have been doubtfully attainable without the stimulus of model building.

(3) The model has provided a reference framework for the condensation of the available data and has indicated the information desirable to extract from the considerable available data.

(4) The failure of the model to explain observed data and its improvement by more careful consideration of the structure and detailed mathematics of the equation set has provided a powerful stimulus to efficient data collection for a specific purpose. It has resulted in the design of appropriate experiments and the effective utilization of their results particularly in deciding the scale, response time, and accuracy desired from the experiments.

(5) It has provided the manager of the reserve with a predictive model to guide operations, and, although in this particular case, prediction would not be sufficiently exact, it allows both the reserve manager and other staff the luxury of management without the potential consequences, and at considerably lower cost than in the real world. This training on a model results in immeasurably increased confidence when faced with complexity in the real world, and adds an extra dimension to intuition.

These very concrete advantages must be considered with other less tangible effects.

Data collection for a specific purpose is clearly better than the collection of 'available' data. Technical staff quickly become aware of this, and the model can be shown to be updated and improved by a data set provided by them. This raises morale and improves the quality of data collected.

Management agencies conducting research in complex ecosystems with interest in several compartments and processes have major integration problems owing to the wide range of resolution and time scale involved in applied ecological research, and communication between groups operating within these different scales is not easy. However, a model such as the one described, in which the main program requires values for variables from subroutines, has particular relevance in improving communication. The specialists producing the subroutines are able to work at a level appropriate to the sophistication of the subject, providing that the variables requested by the main program are provided by the subroutine in an acceptable form. In this sense, then, the main program directs the team research of a number of specialist groups. Such direction is frequently more acceptable than human direction, since all members of the specialist group will have been involved in the development of the main program and have affected the levels of accuracy and generality chosen for other subroutines, whose output to the main program they will use in their particular specialist subroutine. Compromise will be necessary but the modelling sessions themselves will be more rewarding than the general discussion approach frequently mistaken for research integration. The common goal which is predictive ability at the ecosystem level and the discipline forced on the modelling groups makes the integration much closer and prevents misunderstanding of aims since the mathematical/logical flow chart or Fortran program is immediately comprehensible to all participants in the exercise. The fear of many research workers that they are abrogating their initiative to a machine or a set of equations is as unacceptable as similar sentiments expressed about any other technical aid used in modern ecology.

Mathematical modelling is a powerful strategy, not without the dangers to which all research strategies are prone. There are however four particular dangers of modelling which being stated lose much of their danger.

(1) Any mathematical model is only of value as a representation of the real (or observed) system. It has no relevance beyond that, and, if it does not represent a valid view of the real system, it should be given no other status in the mind of the modeller. (The same warning restated, using 'hypothesis' for model, is part of accepted scientific method.)

(2) The model should be as simple a view of the system as is compatible with our use of the model. Complexity in a model is of no value *per see* and may obscure important management and ecological principles. The level of complexity to be incorporated is perhaps the most important decision to be made by a modeller and there are few guides but intuition and experience. However, as a general rule simplicity is to be preferred to complexity.

(3) If, however, in seeking simplicity the important feedback processes are not included, the model may be very unstable and non-robust to errors in equation parameters. The question of instability is one of considerable importance which I am not qualified to discuss, but in practice this seems not to be the problem it has been in other fields. However, simplicity must not be sought at the expense of realism.

(4) In a complex, or fairly complex, model i.e. one with more than a few compartments, it is potentially easy to force the output into a curve which departs very little from the observed trace of population size or production for example. With increase in the number of transfer processes, expressed as non linear equations with several coefficients, the possibility of modification of a coefficient to provide an acceptable agreement increases. When it is considered that the observed variation is subject to a stochastic element and error, it becomes easier to relate the observed curve and model output by changing a small number of parameters. However, this crude curve fitting is of little value, and results in a model of limited predictive value, one that is only a demonstration of the superficial variations without giving any insight into the actual processes. Used in this way, a simulation model is only moderately better than curve fitting or multiple linear regression.

The model described has been of value in the management of an ecosystem in upland Britain. However, a conservation organization has three major responsibilities which includes the management of systems as only a part. These are:

(1) The identification of the scientific or conservation value of specific areas in upland Britain.

(2) The management of systems to retain or reconstruct an identified level of conservation value.

(3) To advise other land agencies with less general responsibilities in the management of specific areas to retain or reconstruct an identified level of scientific or conservation value.

It is considered that mathematical models of various types are capable of assisting all three of these, although the model described would be of value mainly in the last two of the functions.

The identification of conservation or scientific value is a difficult problem, but I suggest that multivariate models are more likely to assist in providing an objective base from which the inevitably subjective decisions can be made than other techniques. Once conservation value is assigned to a region, differential or difference equation models are invoked both for their predictive content and the disciplined thinking involved, as discussed earlier.

The future of models of the type described can be clearly stated. They should be constructed at whatever level of resolution is possible and necessary for all the management situations arising in ecosystem management. Ideally, indeed it will be managers, not research staff, that produce them for the specific area to be managed. The relative ease with which such simulation models can be written, particularly with the special purpose simulation languages (Chapas 1970, Radford 1970), suggests that this general tool should be a part of every scientific land manager's repertoire and replace the intuitive land agency aspects of his management strategy.

The advisory aspects of conservation are clearly concerned with both of the first two functions and, as such, empirical multivariate models are of value, as are differential difference equation models. Advice is often of limited value if it is not available when requested, and, in both decision and management, the urgency of the situation frequently requires immediate answers. Modelling provides a framework within which immediate advice can be given on the best available data. Having given this advice, if the problem warrants it, the model can be improved by experiment and more detailed analysis, resulting in a more precise and accurate predictive model. This iterative process can be continued in a changing real world situation and the advice produced is, as a consequence, of greater value. There is evidence that the output from a simulation model of the type described could be treated as the input to an optimizing linear programming model (Radford this volume) which could considerably enhance its value in this broader responsibility of conservation. Such models are more readily understood by government than are appeals to emotion, and, in these circumstances, the advice given will be given greater credibility than more intuitive techniques. Again the model increases communication.

Conservation in upland Britain, therefore, is resource management of considerable complexity since, although the resource is not easily defined, it certainly contains many elements of the system and their changes of state as part of a valuation. In such a situation a model of the type described, which has many affinities with those of Goodall (1969), provides a logical framework for management in which the effects of management practice can be explored before implementation in the field. Simulation is a powerful weapon being adopted by conservation scientists in the certain knowledge that, despite its deficiencies, a model can be readily improved upon and communicated in a way intuition and experience can not.

References

BLEDSOE L.J. & JAMESON D.A. (1969) Model structure for a grassland ecosystem. In Dix R.L. & Beidleman R.G. (eds.) The grassland ecosystem: A preliminary synthesis. *Range Science Series No. 2* Colorado State University.

BOYD J.M., DONEY J., GUNN R.G. & JEWELL P.A. (1964) The Soay Sheep of the island of Hirta, St Kilda. A study of a feral population. *Proc. Zool. Soc. Lond.* **142**, 1–129.

BOYD J.M., JEWELL P.A. & MILNER C. (eds). (in press.) *The Ecology of the Soay Sheep of St Kilda.* Nature Conservancy Monograph.

BLAXTER K.L. (1962) *The Energy metabolism of Ruminants.* London.

CHAPAS L.C. (1970) in Jones J.G.W. (ed.) *The use of models in Agricultural and Biological Research.* Grassland Research Institute, Hurley.

CLAPPERTON J.L. (1961) The energy expenditure of sheep in walking on the level and on gradients. *Proc. Nutr. Soc.* **20**, xxxi.

DARLING F.FRASER (1955) *West Highland Survey*, Oxford.

ELTON C.S. (1958) *The Ecology of Invasions by Animals and Plants.* London.

GOODALL D.W. (1969) Simulating the Grazing Situation. In 'Concepts & Models of Biomathematics: Simulation Techniques & Methods'. *Biomathematics Vol. I.* New York.

GOODALL D.W. (1967) Computer simulation of changes in vegetation subject to grazing. *J. Ind. Bot. Soc.* 46, 356.

GORE A.J.P. & OLSON J.S. (1967) Preliminary models for accumulation of organic matter in an Eriophorum/Calluna ecosystem. *Aquilo. Ser. Botanica* 6, 297–313.

HOLDGATE M.W. (1970) *Conservation in the Antartic in Antarctic Ecology Vol. 2.* ed. M.W.Holdgate, London, New York.

KELLEY J.M., OPSTRUP P.A., OLSON J.S., AUERBACH S.I. & VAN DYNE G.M. (1969) Models of seasonal primary productivity in Eastern Tennessee *Festuca* and *Andropogon* ecosystems. *Oak Ridge National Laboratory Report* 4310.

KING J. & NICHOLSON I.A. (1964) Grasslands of the Forest & Sub-alpine Zones. In Burnett J.H. (ed). *The Vegetation of Scotland*, Edinburgh.

* Copies of the print-out of a current version of the model can be obtained from the author.

MACGREGOR D.R. (1960) The island of St Kilda: a survey of its character and occupance *Scott. Stud.* 4, 1–48.

MARGALEF R. (1968) *Perspectives in Ecological Theory*. Chicago, London.

McVEAN D.N. (1961) Flora and Vegetation of the islands of St Kilda and North Rona in 1958. *J. Ecol.* 49, 1–39.

OLSON J.S. (1964) Gross and net production of terrestrial vegetation. *J. Ecol.* **52**, (Suppl.) 203–11.

PETCH C.P. (1933) The Vegetation of St Kilda. *J. Ecol.* **21**, 92.

POORE M.E.D. & ROBERTSON V.C. (1949) The Vegetation of St Kilda in 1948. *J. Ecol.* 37, 82–99.

RADFORD P.J. (1970) in Jones J.G.W. (ed.) *The use of models in Agricultural and Biological Research*, Grassland Research Institute, Hurley.

SCHULTZ A. (1969) In Van Dyne, G.M. (ed.) *The ecosystem concept in natural resource management*. Academic Press, New York.

TURRILL W.B. (1927) The Flora of St Kilda. *B.E.C. Report* 8, 428.

VAN DYNE G.M. (1969) Grasslands Management, Research, and Training viewed in a Systems Context. *Range Science Series No. 3*. Colorado State University.

WILLIAMSON K. & BOYD J.M. (1960) *St Kilda Summer*. London.

The simulation language as an aid to ecological modelling

P.J.RADFORD *Water Resources Board, Reading, Berkshire*

Summary

Examples of simulation programs written in a number of different simulation languages, illustrate the value of such problem oriented packages in the formulation of ecological models. The best of such languages are shown to assist the novice model builder without restricting that future program sophistication which experience might consider necessary.

Within the mind of almost every ecologist is a qualitative model trying to break out and become quantitative. Such models are very individualistic and they are conceived in a variety of different forms. Among the most common types are:

(i) The hydraulic analogue in which water, passing through a series of pipelines and reservoirs, represents, for example, the flow of nutrient through soil and plant. The valves and taps in the model control the rates at which the various processes proceed (Radford 1970).

(ii) The heat conduction analogue, where the dispersion pattern of a population of insects might be represented by the flow of heat through a conductor. In this example the laws governing insect movement are considered to be similar to the physical laws which govern heat dissipation.

(iii) The electric circuit analogue, such as has been used to model water movement through soil, plants and atmosphere (Cowan 1965) which envisages these processes as a network of resistors and capacitors. The phenomena which occur at the boundaries of soil-plant-atmosphere are assumed to have their proper analogy in the thermo-electric and Peltier effects which occur at junctions between dissimilar metalic conductors.

(iv) The analogue computer model, rather different, in that it is a physical representation of the mathematical relationships which themselves describe some aspect of the ecological problem under consideration. An example of this type of analogue is given by Mr J.K.Denmead in this volume, where the dynamic interactions of a two species, predator-prey system were modelled, using electronic components.

(v) The ecological flow diagram, which is by far the most common analogue. Here 'boxes', joined by a network of lines, demonstrate the interrelationships within an ecological system. Many different forms of representations are used with boxes of all shapes and sizes linked by lines of assorted thicknesses, with arrows indicating directions of information or material flow.

The actual form of the model used in a given situation is more a comment upon the thought processes of the individual ecologist than the nature of the system modelled. Each of these conceptual models does, however, indicate something of the essential dynamic nature of the system it represents. The majority of ecologists would understand that such systems, incorporating time dependent interactions, may properly be described in terms of differential equations. Should the interactions be dependent upon continuous space as well as time, then it is generally recognized that partial differential equations would be required. What would not be so readily understood is the manner in which these equations could be formulated and solved. A survey of the literature would not prove too encouraging, for it would reveal that very simple systems lead to quite advanced mathematics even when an analytic solution is sought. When no such solution is available and approximate numerical solutions are needed, the mathematical notation becomes even more obscure (Passioura & Frere 1967); the necessary algebraic manipulation would seem to be the prerogative of the professional mathematician alone. Faced with such a situation, it has been suggested that co-operation between the ecologist and the mathematician might yield the desired result. However the dangers of this approach are that (*a*) the mathematician might well misinterpret the ecology and thereby misrepresent it in the model, and (*b*) the ecologist might misinterpret the solutions from the model and thereby reach wrong conclusions regarding the ecology of the system. These dangers are of special concern when the model becomes part of the overall iterative technique of science, whereby hypothesis leads to a model, model performance leads to experimentation, and experimental results lead to model improvement. Furthermore, when each scan of this cycle involves fundamental changes in the structure of the model rather than simple parameter changes, it is of vital importance that the ecologist understands the mathematical implications and assumptions of such modifications.

These difficulties associated with model formulation and solution can lead to one of two courses of action, neither of which is particularly fruitful:

(*a*) The ecologist abandons a dynamic formulation of his problem and resorts to the more conventional statistical analyses. Such action can produce worthwhile results, but, more often, inappropriate analyses cut right across the basic concepts which research has placed in the experimenter's mind.

(*b*) The ecologist enrols on a FORTRAN computer programming course in the hope that computers will aid his problem formulation and solution. Often he returns without the necessary knowledge to transfer his problem into a computer simulation model. The reason behind this failure is that the process of modelling requires not only ecological expertise but also a modicum of knowledge of mathematics, computing and model building. The computer simulation language is able to greatly assist, by providing that modicum of knowledge without submerging the uninitiated in the generalities of differential calculus, the overall scope of general purpose computer languages, and the wide horizon of different forms of simulation techniques. This is made possible because the fundamental features of even the most complex of systems, implied by the various types of analogues previously discussed, can be presented within a very simple mathematical framework. The simulation language DYNAMO (Pugh 1963) takes advantage of this fact and has been written so that its syntax reflects the philosophy upon which the model building exercise is based. In this sense, DYNAMO may be described as an application orientated package designed to guide novice programmers/model builders to a correct basis for continuous system model formulation.

To aid the transition from the conceptual model to a computer simulation program, the DYNAMO user is encouraged to produce a system flow diagram which will indicate the broad features of the model without introducing the precise and detailed relationships which will eventually be in the program. Each of the different types of analogues discussed earlier can be presented in this form of flow diagram quite simply. Specific examples of how this can be done have been published in the proceedings of a symposium *The Use of Models in Agricultural and Biological Research* (P.J.Radford 1970). For the purposes of this paper, the techniques are illustrated by an example of a problem taken from the realm of fisheries management.

The simulation of the upstream migration of salmon using dynamo conventions

A superficial glance at data from contemporaneous records of daily river water levels at a weir, and numbers of salmon migrating past that weir, suggests a high degree of association between these variables (Fig. 1). Actual analysis of the figures reveals very different degrees of correlation from month to month during the year, and an overall correlation not significantly different from zero. A typical reaction to such a result might be to consider other variables which could be affecting the salmon migration at the weir. In particular,

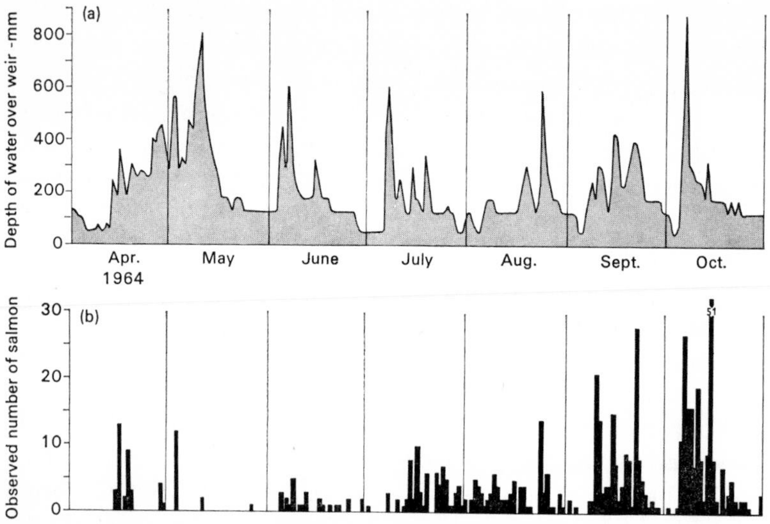

Figure 1. Daily river levels and contemporaneous observed upstream salmon migration (Stewart 1968).

variables such as water temperature, air temperature, light intensity and humidity might be considered as valid. With modern data logging equipment, it is possible to take measurements of such variables at 15 minute intervals throughout the 24 hours of each day, and records could be kept for a period of five or six years. This form of data is most amenable to multiple regression analysis, but possibly the most significant factor to emerge when the calculations have been completed might be the size of the bill which has been incurred for the use of computer time and services. This rather cynical characature of ecological research is not a true reflection of the ecologists' holistic view of his system, but rather a comment upon the rather piecemeal manner in which his data are so often analysed. Had a systems analysis approach been made in formulating this problem, a model would have emerged which could incorporate not only those data which fitted into a multiple regression framework but also other relevant factors quite unsuitable to that particular form of statistical treatment. The steps involved in this approach are illustrated using the simulation language DYNAMO and the system of flow diagrams recommended in Industrial Dynamics (Forrester 1961). In Fig. 2, the upper continuous line indicates the passage of salmon from the sea to the head of the river, via the estuary and weir. The lower line indicates the passage of water from the head of the river to the sea. The main point of interest is the weir, at which the rate of salmon migration and the water level are measured. In the flow diagram, rates are indicated by valve-like symbols, which by definition must control the flow of material (Salmon or Water) from one point to

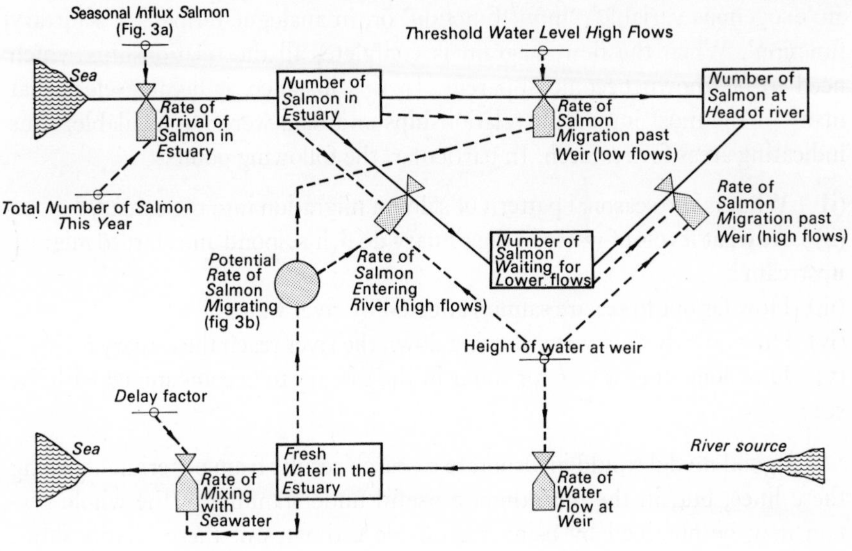

Figure 2. System flow diagram representing salmon migration.

another within the system. Rates are the fundamental components of any dynamic system. During this symposium, they have variously been described as 'rates', 'process control variables', 'decision variables' or, in analogue computing terms 'inputs to integrators'. Consideration of a rate must inevitably raise questions concerning where the material (salmon) is coming from and what is its destination. By the very nature of the system, material must come from some physical compartment, shown in the flow diagram as a rectangle. In this particular example, the salmon are known to come from the estuary where we postulate that there must be a certain number of salmon waiting for suitable conditions to migrate. Such compartments are named 'levels' by Forrester, but have variously been referred to in this symposium as 'component or compartment variables', 'state variables' or in analogue terms 'outputs of integrators'. The construction of a flow diagram serves to present the fundamental components of the system without providing the detailed mathematical relationships between them. The causal relationships are, however, indicated by the broken lines. By considering this diagram, it becomes clear that migration is not directly affected by conditions at the weir but by some effect in the estuary where the fish happen to be. In the formulation of this model, therefore, it was postulated that salmon migration depended upon the quantity of fresh water in the estuary rather than the water level at the weir. However, the latter variable was still the 'driving variable' which determined the basic environment of the salmon. Such a variable, which affects the system without itself being altered by the system, has alternatively been called an 'exogenous variable', 'input function' or, in analogue terms, an 'arbitrary function'. When the flow diagram is complete, all the relationships which need to be known become apparent. In this instance, it became clear that many of the most important relationships and data were not available, thus indicating areas for research. In particular, the following points:

(i) What is the seasonal pattern of salmon migration into the estuary?
(ii) At what levels of water in the estuary do fish respond and start to migrate upstream?
(iii) How far out to sea are salmon affected by river water?
(iv) How quickly does water flowing down the river reach the estuary?
(v) How long does it take for water in the estuary to become mixed with the sea?

A good model would incorporate research results from experiments along these lines, but, in the meantime, a useful understanding of the whole system may be obtained by using reasonable estimates of these relationships. Figure 3a shows the posulated curve of the seasonal influx of salmon into the

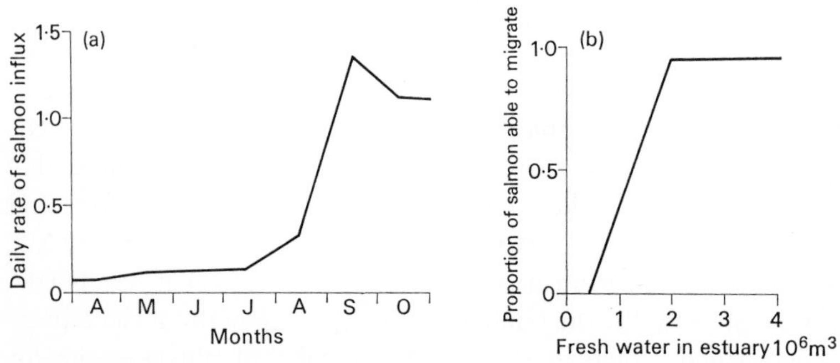

Figure 3a. Graph of seasonal influx of salmon. The graph gives the daily rate of salmon influx as a percentage of the total number of salmon migrating in the year.

Figure 3b. Potential rate of salmon migration past weir. The proportion of the total number of salmon in the estuary is assumed to be dependent upon the quantity of fresh water in the estuary.

estuary. The data for this table were based upon the average number of salmon that did in fact pass the weir during the three year period 1964–6. The data, obtained by taking the mean monthly migration as a proportion of the annual migration, are expressed as the mean daily rate of influx of salmon into the estuary. The proportion of the salmon in the estuary able to migrate upstream is given in Fig. 3b. The data for this relationship were obtained by careful inspection of the sort of data presented in Fig. 1. By taking note of the maximum water levels at which no salmon migrated, when obviously salmon must be available (i.e. just before peaks of migration), it was possible to estimate the lower extreme on this curve, and, by observing the minimum water level at which a very large number of salmon migrate, it was possible to fix the higher extreme on this curve. Linear interpolation was used to evaluate the percentage of salmon migrating at intermediate points. Further inspection of the data, indicated a reluctance of fish to migrate at very high flows of water at the weir. A mechanism was therefore introduced into the model which allowed for this.

Once the flow diagram has been drawn, and the functional relationships between the variables have been established, it is a relatively simple matter to systematically write down the relevant equations using DYNAMO. In general, there will be one line of coding for each symbol in the flow chart and the order in which these lines are written is irrelevant since the DYNAMO system is able to sort the equations into the proper computational sequence. Part of the salmon migration program is given in Table 1.

Table 1.

```
NOTE   THE UPSTREAM MIGRATION OF SALMON FORMULATED IN DYNAMO II

NOTE                     PROGRAMMED FOR AN IBM S/360 COMPUTER

L  NSE.K=NSE.J+DT*(RASE.JK-RSMW.JK-RSER.JK) NUMBER SALMON IN ESTUARY
R  RMS.KL=FWE.K/DELAY                        RATE OF MIXING SEAWATER
R  RASE.KL=TNSTY*SIS.K/100.                  RATE ARRIVAL SALMON EST
A  SIS.K=TABHL(TDRSI,TIME,15,201,31)         SEASONAL INFLUX SALMON
T  TDRSI=.083/.131/.129/.133/.323/1.370/1.108 TABLE SIS FIG 3A
A  PRSM.K=TABHL(TPRSM,FWE.K,.6,1.6,1.)       POTENTIAL RATE MIGRATION
T  TPRSM=0.0/0.95                            TABLE PRSM FIG 3B
L  FWE.K=FWE.J+DT*(RWFW.JK-RMS.JK)           FRESH WATER IN ESTUARY

.
.   OTHER STRUCTURE STATEMENTS
.

PRINT 1)NSH/2)NSE/3)RSMW/4)WLW               PRINTED OUTPUT TABULAR
PLOT   NSH=N,RSMW=R                          PRINTED OUTPUT GRAPHICAL
SPEC   DT=1/LENGTH=215/PRTPER=1/PLTPER=1     RUN SPECIFICATION
ENDJOB
```

The first line is a NOTE card which enables titles and comments to be freely intermingled with the program. In general, the letter in column one of each line is a shorthand description of the type of equation which follows. *L* indicates that a *L*evel (or state) variable is being defined, *R* a *R*ate, *N* an i*N*itial value is being set, etc. etc.

In particular, consider the state variable–Number of Salmon in Estuary (*NSE*). In DYNAMO, the number of Salmon in the Estuary at a given time (*NSE.K*) is defined in terms of the numbers that were present (*NSE.J*) one small interval of time (*DT*) before and the rates of arrival (*RASE.JK*) and departure (*RSMW.JK*, *RSER.JK*) that operated over that interval. It is assumed that the value of *DT* chosen will make it reasonable to assume that these rates are constant over that short period of time. (This form of equation is equivalent to the integration of the rates by rectangular integration otherwise known as the Euler point slope method.) A similar form of equation is written for each of the state variables; in each case, the flow diagram indicates clearly the relevant rates which must be included. The rate equations can also be written down. In general, these will be functions of the state variables and exogenous data. For example, the *R*ate of *M*ixing of the estuary water with the *S*ea over a small interval of time (*RMS.KL*) depends upon the quantity of fresh water in the estuary at time *K* (*FWE.K*) divided by the delay constant (*DELAY*). This particular form of equation has the effect of applying an exponential decay to the *FWE* with a half life of 3 days when DELAY=3. The *R*ate of *A*rrival of *S*almon in the *E*stuary depends upon the Seasonal Influx of Salmon into the estuary which is obtained from the graph given in Figure 3a. In DYNAMO, this figure may be entered into the program as a table (*T*) which is referenced by using the function *TABHL* (*TDRSI*, *TIME*, 15, 201, 31). This indicates that the data depends upon *TIME* according to some arbitrary function tabulated at points 31 days apart, starting at day 15 and finishing at day 201 (i.e. monthly values). Linear interpolation is assumed between points and the first value is taken to apply if time is less than 15, and the last value if time is greater than 201. This effectively reproduces the form of Fig. 3a.

The whole structure of the model can in this way be written down. The only remaining step is to define the output which is required and the run specifications. These are given in the PRINT, PLOT and SPEC lines of the program. The PRINT instruction is simply a list of the variable names for which output values are required, and this particular plot card states that the *N*umber of *S*almon at the *H*ead of the River (*NSH*) is to be plotted against *TIME* on the same graph as *R*ate of *S*almon *M*igrating *P*ast *W*eir using the characters *N* and *R* respectively to denote the points. The specification card (SPEC) gives:

(i) the size of the small time increments DT, by which the program will advance, i.e. 1 day;
(ii) the LENGTH of the simulation run, i.e. 215 days;
(iii) the frequency with which results are required to be printed (PRTPER= 1 day);

(iv) the frequency with which results are required to be plotted (PLTPER= 1 day).

When the program is presented to the computer, the model simulates the development of the system through time from the initial conditions until the specified end point (215 days later). The results are printed in a standard tabular form and also in a graphical representation produced on the computer's line printer. The advantages of DYNAMO as a means of formulating and solving this type of problem are:

(i) The language is oriented towards the problem rather than towards the computer.

(ii) It uses only the simplest concepts of differential calculus i.e. levels, rates and rectangular integration.

(iii) It gives excellent tabular and graphical output with very simple output instructions. This is essential for quick and easy assessment of results.

The very first simulation runs of this model gave a correlation between observed salmon number and predicted salmon number which was significantly different from zero. Results showing the predicted salmon runs relating to the observed data in Fig. 1 are shown in Fig. 4. Although the degree of

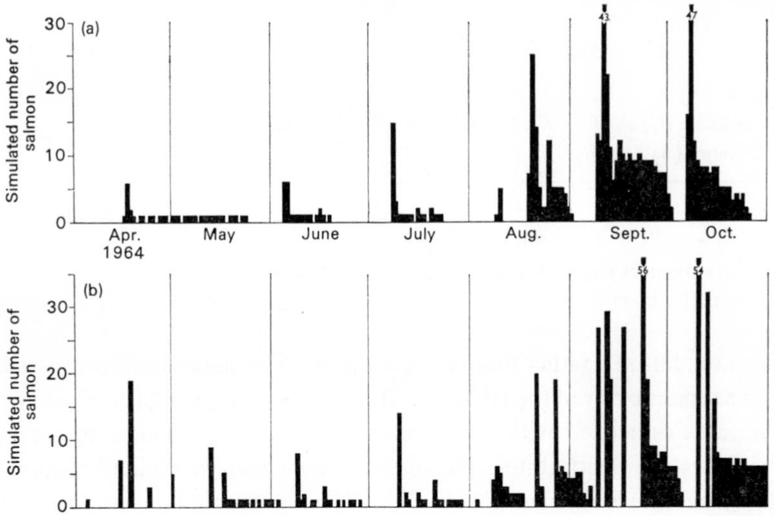

Figure 4. The expected numbers of salmon migrating obtained from simulation runs using (*a*) the simple model (*b*) a model including a factor to represent the reluctance of fish to migrate on high flows.

correlation was not large in these instances, a study of the results gave new insight into the problem suggesting fruitful lines of model development. Subsequent runs of these more refined models yield relationships between observed and expected salmon migration with correlations of up to ·82 (100 df). The reason for the comparative success of this type of model, as opposed to a multiple regression model, is probably due to the synthesis in the former of the state variable 'Number of Salmon in the Estuary'. No measure of this 'availability' had been made and in fact it would probably be impossible to measure it without substantially interfering with the migration process itself. The simulation model is able to circumvent this problem by postulating a smooth seasonal influx of salmon and an erratic migration from the estuary. In this way, the value of a key variable may be deduced and so may be used effectively in the model.

The simulation of an aquatic ecosystem using DSL 1130

Not all model builders like to consider their problem in terms of a system flow diagram such as given in Fig. 2. An alternative method of presentation is given in the following example of the simulation of an aquatic ecosystem (Parker 1968). Table 2 contains the same kind of information as is found in a

Table 2. Growth and death rates used in the model

		Death rate		
	Growth rate	Natural	Predation	Emigration
Algae (A)	$A_1\, e(T)\, P\, F$	$A_2\, T$	$A_3\, C\, T$	
Cladocera (C)	$C_1 f(T)\, A\, F g(P)$	$C_2\, T g(P)$	$C_3\, B\, T$	
Kokanee (B)	$B_1\, C\, T$	$B_2\, T$		$B_3\, h(W)$

W=time (weeks) e, f, h are normal distributions
T=temperature (°C) g involves sin P
P=photoperiod (hours/day) Subscripted symbols are constants
F=phosphate (μg/liter)

flow diagram, but in a rather more compact form. The state variables are each listed on a separate row of the table, i.e. (i) Algae, (ii) Cladocera, (iii) Kokanee, and the rates associated with each are listed in columns alongside, i.e. (i) Growth Rates, (ii) Death Rates. The inter-relationships, which in the equivalent flow diagram would have appeared as broken lines, are indicated here by the code letter of the relevant exogenous variables and state variables. Some indication is also given as to the actual form of the functional relationships which apply i.e. (i) normal distributions, (ii) sines etc. It is purely a matter of personal preference as to which mode of presentation of this information is adopted. I personally find the flow diagram easier to follow, whereas some of my colleagues prefer this tabular form. The particular mode adopted is not related in any way to the simulation language used. Both the previous example and this current one have been written in both DYNAMO and DSL 1130 (Wyman & Syn 1968). The whole of the DSL program relating to the aquatic ecosystem is given in Table 3.

Table 3.

```
TITLE    SIMULATION OF AN AQUATIC ECOSYSTEM—RICHARD A PARKER
TITLE       FROM 'BIOMETRICS' 1968. 24,803–821
TITLE         PROGRAMMED IN DSL FOR AN IBM 1130 COMPUTER
*
         W=TIME
*
*        STATE VARIABLES
*        ALGAE
         A=INTGR(IA,DADW)
*        CLADOCERA
         C=INTGR(IC,DCDW)
*        KOKANEE
         B=INTGR(IB,DBDW)
*        PHOSPHATE
         F=INTGR(IF,F1—F2*DCDW—F3*DADW)
*
*        NET RATES OF GROWTH OF ALGAE CLADOCERA AND KOKANEE
         DADW=A*(A1*EXP(—0.5*((T—18.)/3.)**2.)*P*F—A2*T—A3*C*T)
         DCDW=C*(C1*EXP(—0.5*((T—13.)/8.)**2.)*A*F*GP—C2*T*GP—C3*B*T)
         GP=0.820+0.343*SIN(((P—7.2)/10.4)*2.*3.1416)
         DBDW=B*(B1*C*T—B2*T—B3*EXP(—0.5*((W—40.)/3.)**2))
*
*
*        EXOGENOUS VARIABLES
*        PHOTOPERIOD
         P=12.2+4.1*SIN(((38.—W)/52.)*2.*3.1416)
*        TEMPERATURE
         T=4.0+14.0*EXP(—0.5*((WT—34.)/9.)**2.)
         WT=FCNSW(W—8.,W+52.,W+52.,W)
*        RATE OF FLOW OF RIVER
         RF=604.8E3*1.E—3*28.32*(4500.+52000.*EXP(—0.5*((W—22.)/4.5)**2.))
         F1=RF*(FCR—F)/VOL
*        PHOSPHORUS CONTENT OF RIVER
         FCR=1275.—1020.*EXP(—0.5*((W—24.)/7.5)**2.)
*
*        PARAMETERS—CONSTANTS FOR A GIVEN RUN
*
PARAM    A1=.00014,A2=.0020,A3=,0060,...
         C1=.01,C2=.03,C3=.0010,...
         B1=.00048,B2=.0010,B3=.0500,...
         IA=0.2,IC=0.1,IB=5.0,IF=150.,...
         F2=.6,F3=20.,...
         VOL=26529.E6
         BM3=B—3
*
INTEG RKFIX
*
```

```
CONTRL DELT=1.0,FINTI=50.
IPRINT 1.0,B,C,A,F
GRAPH 1.0,10,8,TIME,BM3
LABEL KOKANEE(KG.HA) MINUS 3 KG.HA.
SCALE .2,2.0
RANGE B,C,A,F
END
STOP
```

The TITLE cards enable a suitable heading to be inserted which will head each page of computer output. Any comments may be interspersed within the text of the program by the use of a line with an asterisk (*) in column 1. Richard Parker had worked in *w*eekly time increment and so had named the time variable *W*, hence in the first line of the program *W* is set equal to the system variable called *TIME*. In DSL, the equivalent to the *L*evel (*L*) equations of DYNAMO are the INTEGRAL equations; one is needed for each state variable. The quantity of each organism present (e.g. *A*) is equated to the integral (*INTGR*) of the net rate of change of that organism dA/dW (*DADW*), starting with the initial quantity present (*IA*). The particular rate equations which were used in this model are given in Table 4. These may be

Table 4. Differential equations used in the model

Component	Equation
Algae	$dA/dW = A\{A_1 \exp[-\tfrac{1}{2}((T-18)/3)^2]\, P\, F - A_2\, T - A_2\, C\, T\}$
Cladocera	$dC/dW = C\{C_1 \exp[-\tfrac{1}{2}((T-13)/8,^2]\, A\, F\, g(\mathrm{P}) -$
	$C_2\, T\, g(P) - C_3\, B\, T\}$,
	where $g(P) = .820 + .343 \sin\{((P-7.2)/10.4)2\pi\}$
Kokanee	$dB/dW = B\{B_1\, C\, T - B_2\, T - B_3 \exp[-\tfrac{1}{2}((W-40)/3)^2]\}$
Phosphate	$dF/dW = F_1 - F_2\, dC/dW - F_3\, dA/dW$,
	where F_1 is a function of phosphate input and output from the south arm of the lake estimated by the proportion of total volume displaced by incoming Kootenay River water multiplied by the difference between the phosphate concentration in the river and that in the lake during the previous unit of time.

transposed almost character by character into the DSL program (cf. Table 3). Similarly the exogenous variables were computed by fitting appropriate time dependent functions to historic data—these were presented by Parker in the manner illustrated in Table 5. Again these equations may be transposed almost character for character into the DSL program, except that the discontinuity expressed as an inequality in Table 5 appears in Table 3 as a special CSMP switch function (*FCNSW* ($W-8$, $W+52.$, $W+52.$, W)). The particular values of the constants in these equations were obtained by Parker initially from laboratory experimentation. These parameters are listed in the

Table 5. Equations used to model the physical components as functions of time in weeks (W)

Component	Equation
Kootenay Lake	
Photoperiod (hrs/day)	$12.2+4.1\sin[((38-W)/52)2\pi]$
Temperature (°C)	$4.0+14.0\exp[-\frac{1}{2}((W-34)/9)^2]$, $W=W+52$ for $W\leq 8$
Kootenay River	
Flow (L/sec)	$28.32\{4500+52000\exp[-\frac{1}{2}((W-22)/4.5)^2]\}$
Phosphate content (μg/L)	$1275-1020\exp[-\frac{1}{2}((W-24)/7.5)^2]$

PARAM section of the program. DSL not only allows the rectangular method of integration given in DYNAMO but also allows the user the choice of several of the well known approximate integration formulae. In this program, the method chosen was Runge-Kutta with fixed time increments (*RKFIX*) in order to conform with the method used by Parker. All that remains to complete the program is a definition of the output variables required to be printed and plotted and the run CONTROL instruction (equivalent to the DYNAMO SPECification instruction). DSL provides a general history of output variables by stating on the PRINT line the frequency with which output is to be printed (1.0 = each week) and listing the names of the required variables (B, C, A, F). If the particular IBM 1130 computer which is used for the simulation is fitted with an $X-Y$ plotter, a very useful feature of DSL is its provision for the plotting of results. Simply by requesting on a GRAPH instruction the names of variables to be plotted on the X and Y axes (*TIME*, *BM*3) and the size of the plot required, it is possible to obtain an elegant graph of these variables headed with the title specified on the LABEL instruction, and plotted on the scale specified on the SCALE instruction. It is possible to superimpose outputs of several runs of the model for comparative purposes. Figure 5 is a copy of output obtained in this way from the simulation of the model given in Table 3. The rather sophisticated cathode ray tube output which Parker shows in his paper could also be obtained simply by replacing the directive GRAPH with the directive SCOPE (provided of course that the computer is fitted with a cathode ray display!).

I believe that this example has illustrated that a simulation language such as DSL 1130 enables novice programmers/model builders to quickly formulate and solve their continuous system simulation problems. Of particular value are those features of DSL which:

(i) enable equations to be written in a parallel, analogue manner rather than in the serial form which general purpose computer languages require;
(ii) the choice of various approximate methods for integration;

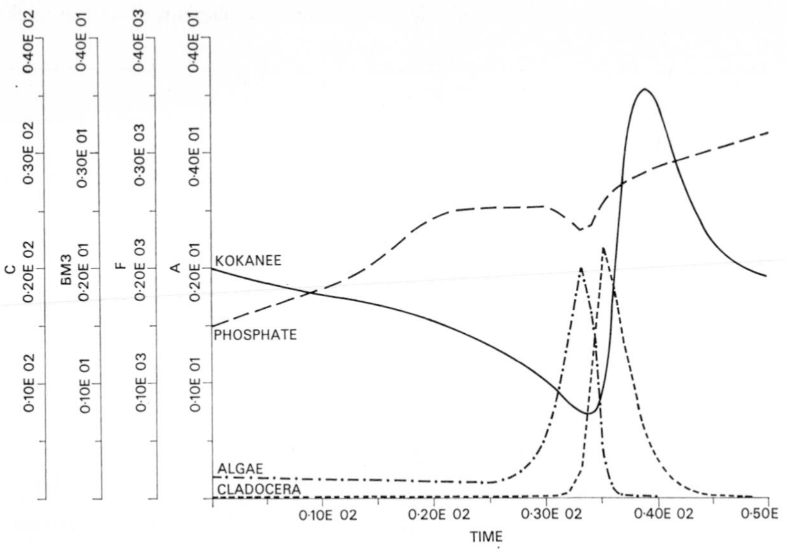

Figure 5. Simulation of an aquatic ecosystem.

(iii) the powerful output facilities which allow the model builder to quickly see and appreciate the simulation results.

Simulation languages such as DYNAMO and DSL 1130 have been shown to facilitate model formulation. This, however, is only the first stage in the whole strategy of simulation modelling. It is important that the model, once constructed, can be used to obtain insight into overall system performance. This is achieved by examining a whole series of specific simulation runs of the model with variations in parameter values, exogenous data, model structure etc., according to the requirements and interests of the model builder. DYNAMO provides excellent facilities for such manipulation, enabling a whole series of runs to be performed without human intervention. The limitation imposed by this system is that all the parameter changes etc., must be specified within the program beforehand. In this respect, DSL 1130 is more flexible, in that it is possible to program search techniques which automatically choose future runs of a model upon the basis of earlier runs. This feature is particularly valuable when it is desired to find the parameter values which minimize the sums of squares of deviations of simulated and actual results. Another application is concerned with finding the management strategy which will yield maximum output from a given system. An example to illustrate how this may be achieved follows.

Crop production model using CSMP 360

Mr Brockington of the Grassland Research Institute produced a model which simulated the growth of a grass sward under various cutting regimes. The rate of regrowth of the grass depended upon the time of year and the time since the sward was last cut; the digestibility of the herbage varied according to these and other factors. The model was originally written in DYNAMO, and, in that language, was readily used to investigate the effects of a range of different cutting regimes on the total yield of digestible dry matter. At this stage, it was desired to find the maximum harvestable yield obtainable under the whole range of feasible management plans. It was decided that the pattern chosen should conform to normal agricultural practice in so far as there should be an initial 'spring' harvest followed by a series of equally spaced successive cuts. It was not possible, within the framework of DYNAMO I, to arrange for an automatic search procedure to find the maximum yield. The program was therefore re-written in S/360 CSMP (IBM Application Program 1967) in order to make use of the rather more flexible features of this language. The syntax of CSMP is very similar to that of DSL 1130, but the user of CSMP is encouraged to subdivide his program into three distinct sections (INITIAL, DYNAMIC, TERMINAL). The main features of the program are given in Table 6. After the TITLE of the program has been given, the INITIAL section of the program is presented. This contains all the parameters of the model, the initial conditions and any calculations which must be carried out before a simulation run can proceed. In this particular model the first cutting interval (*FCX*) had to be constrained to lie between the values 3, (*MINFI*) and 221, (*MAXFI*). This was achieved by optimizing on the unconstrained variable *FCIG* and obtaining the constrained variable *FCX* using the transformation $FCX = MINFI + (MAXFI - MINFI) * \text{Sin}^2 (FCIG)$. Hence, the inclusion of this line in the INITIAL section of the program and a similar transformation for subsequent cutting intervals (*SCX*).

The DYNAMIC section contains all the structure statements of the model, including the main output variable *HDDM* (*H*arvestable *D*igestible *D*ry *M*atter). It is in the TERMINAL section that instructions can be given to enable automatic search procedures to be included. Equations in this section are executed at the end of each simulation run. In this instance, it was required to find those values of *FCX* and *SCX* which gave the maximum value of *HDDM*. The technique chosen to search for this maximum was that developed by Powell 1964. The details of the technique are unimportant, but sufficient to mention that the algorithm used was obtainable in the form of a FORTRAN sub-routine. One advantage of DSL and CSMP over DYNAMO is that FORTRAN subroutines may be included in the program and may be called in various ways from within the main structure of any model. In this case the subroutine call name was VA04A (X, E, N, F, ESCALE, I PRINT, I CONV, MAXIT, LABEL). When called, the subroutine inspects the values of the vector of parameters *X* and the resultant simulation

Table 6.

```
TITLE       HARVESTING FREQUENCY FOR MAXIMUM DIGESTIBLE
TITLE       DRY MATTER PROGRAMMED IN S/360 CSMP FOR IBM 360

INITIAL
.
.
CONSTANT  N=2, ESCALE=50., IPRINT=1, ICONV=1, MAXIT=20
CONSTANT  MINFI=3., MAXFI=221., MINSI=3., MAXSI=221.
      FCX=MINFI+(MAXFI-MINFI)*SIN(FCIG)*SIN(FCIG)
      SCX=MINSI+(MAXSI-MINSI)*SIN(SCIG)*SIN(SCIG)
.
.
DYNAMIC
.
.           (STRUCTURE STATEMENTS)
.
            HDDM=INTGRL(HDDMN, ROHD)
.
TIMER DELT=1., FINTIM=301.
METHOD RECTANGULAR

TERMINAL
            IF (LABEL.EQ.203) GO TO 4
            CALL VA04A (X, E. N, F, ESCALE, IPRINT, ICONV, MAXIT,
              LABEL)
            IF(LABEL.EQ.203) GO TO 4
            FCIG=X(1)
            SCIG=X(2)
            CALL RERUN
4           CONTINUE
END
TIMER DELT=1., FINTIM=301., PRDEL=1., OUTDEL=1.
PRINT       ADM, DMD, ADDM, HDM, HDDM, HI, ARP, HIX
END
STOP
END JOB
```

output value of F, and, on the basis of these, decides upon the next values of vector X with a view to approaching the parameters which minimize F. When the minimum value is found, the LABEL is set to equal the value 203. In this TERMINAL section, F is set equal to $-HDDM$ (To maximize $HDDM$ it is necessary to minimize $-HDDM$) and a test is made to see if the maximum value has been found. If not then VA04A is called. This will return with either the maximum value, or the values of the parameters for the next simulation run. Again, a test is made to see if the maximum has been found, and, if not, the unconstrained parameters are equated to the elements of vector *X*. The inclusion of the CSMP statement CALL RERUN causes the model to be reinstated as at time zero, but with the parameter changes as dictated in the terminal section. In this manner, the simulation can be made to proceed initially with any sensible value for *FCX* and *SCX*. At the end of the first simulation run, VA04A will decide the values of *FCX* and *SCX* which are to be inspected next and a RERUN will be called. This procedure will be repeated until the maximum value of *HDDM* is found when LABEL, having been set to 203, will cause control to jump over the CALL RERUN directive, and hence terminate the complete sequence of simulation runs. If a detailed printout is required of only the final rerun, then the addition of a TIMER and PRINT card after the END card will cause a final run of the model with the parameters set to give the maximum output of harvestable digestible dry matter (*HDDM*). The STOP directive will then cause the whole computer program to come to a halt. Results of running this model are shown in Fig. 6. The yield (*HDDM*) is shown as a contour map on the two dimensional plane, First Cutting Interval v. Subsequent Cutting Interval.

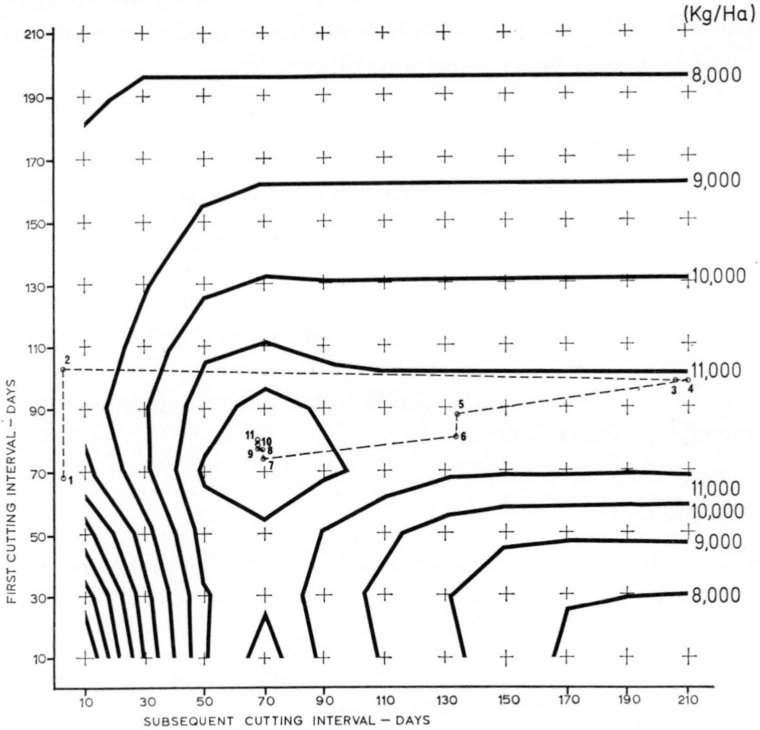

Figure 6. Computer simulation results from a model of crop production.

The discontinuous line, superimposed upon the contour map of yield, shows the path to the maximum chosen by the automatic search algorithm.

The numbers 1 to 11 joined with a discontinuous line indicate the main search path to the maximum. Each of these points represent the closest approximation to the maximum based on the best information obtained at that stage in the search procedure. For each of these decision points, a number of different simulation results were evaluated by the search algorithm. It must be remembered that the optimization technique has no other information about the form of the response surface. The contours were obtained by an entirely different procedure and are shown simply to facilitate understanding of the search method.

This example has illustrated the way in which the basic facilities of CSMP 360 may be extended by the inclusion of whole FORTRAN subroutines, as well as the normal branching facilities of that language, within the TERMINAL section of the program. In fact it is possible to include any FORTRAN instruction, and call any FORTRAN subroutine from any part of model. However if these facilities are used within the DYNAMIC section they necessarily interfere with the sorting algorithms of CSMP and so they must be included within what is called a NOSORT or a PROCEDURAL section. All such statements must therefore be written in the correct computational order and it must be noted that statements appearing before and after such a section will only be sorted within themselves.

Conclusion

The fundamental framework of CSMP enables the novice model builder, with little or no programming or modelling experience, to readily formulate his problem and obtain meaningful results. The special value of CSMP is that, as programming experience is gained, and as more ambitious models are attempted, extensions to the language are available to cope with them. In the ultimate analysis, the language can define any continuous system simulation which could be defined in FORTRAN but requiring substantially less programming expertise and effort.

Acknowledgements

I would like to thank the following for drawing my attention to the problems which have been used as examples in this paper.

Mr N.R.Brockington	Grassland Research Institute
Mr R.A.Parker	Washington State University
Mr J.C.Peters	Water Resources Board
Dr L.Stewart	Lancashire River Authority

This work was carried out whilst the author was employed by the Grassland Research Institute and the Water Resources Board. Tables 2, 4 and 5 were originally published in *Biometrics* and appear by kind permission of the Editor.

References

COWAN I.R. (1965) Transport of water in the soil–plant–atmosphere system. *J. appl. Ecol.* 2, 221–39.

FORRESTER J.W. (1961) Industrial Dynamics. Cambridge Massachusetts, USA, MIT Press.

IBM, Application Program (1967) System /360. Continuous System Modelling Program (360A—CX—16X). *Users Manual* IBM.

PARKER R.A. (1968) Simulation of an aquatic ecosystem. *Biometrics* 24, 4, 803–21.

PASSIOURA J.B. & FRERE M.H. (1967) Numerical analysis of the convection and diffusion of solutes to roots. *Aust. J. Soil Res.* 5, 149–59.

POWELL M.J.D. (1964) An efficient method for finding the minimum of a function of several variables without calculating derivations. *Computer Journal* 7, 2, 155–61.

PUGH A.L. (1963) *DYNAMO Users Manual*, Cambridge Massachusetts, USA, MIT Press.

RADFORD P.J. (1970) Some considerations governing the choice of a suitable simulation language. In Proc. ARC symposium on *The use of models in agricultural and biological research.* Ed. J.G.W.Jones, Grassland Research Institute, Hurley, Berkshire, pp. 87–106.

STEWART L. (1968) The Water Requirements of Salmon in the River Lune, Lancaster. Lancashire River Authority Fisheries Department.

WYMAN D.G. & SYN W.M. (1968) Digital Simulation Language, IBM, 1130/1806 *User's Guide* IBM.

Stochastic model fitting by evolutionary operation

G.J.S.ROSS *Statistics Department, Rothamsted Experimental Station, Harpenden, Herts.*

Summary

When an observed distribution or curve is assumed to arise from a stochastic model, it is often not possible to obtain explicit expressions for the predicted values corresponding to given parameters. However, simulation experiments, using pseudo-random numbers, can be used to obtain approximations to the likelihood function, the variance of the estimated likelihood depending on the number of trials used. Hence, approximate maximum likelihood estimates and confidence regions for the parameters may be obtained by applying evolutionary operation to the approximate likelihood function.

Definition of terms

For the purposes of this discussion, the following definitions are assumed:

Model. A set of rules for computing predicted values to which data values can be compared.

Parameters. The quantities that are varied to obtain different sets of predicted values. Quantities that take fixed values are known as *constants*.

Deterministic models. Models for which the predicted values may be computed exactly.

Stochastic models. Models for which the predicted values depend on probability distributions.

Model fitting. The selection of parameter values that generate predicted values acceptably similar to the observed values.

Likelihood function. The probability that the parameters give rise to the observed data, treated as a function of the parameters. The function is a measure of agreement between model and data, and the parameter values for which the likelihood is a maximum are known as *maximum likelihood estimates.*

The likelihood function is closely related to the chi-squared function used to test the goodness of fit of theoretical distributions to observed frequency data, and to the sum of squared residuals used in curve fitting.

Evolutionary operation. A strategy of repeated experimentation to find the optimum value of a function.

Difficulties of model fitting

I shall now classify models by the ease with which they can be fitted to data. *Analytic models* are those for which explicit formulae are derived for predicted values or distributions: examples include regression models, experimental designs and the standard theoretical distributions such as the Normal and Poisson distributions. *Simulation models* are those that can be specified by a programme of arithmetical operations, such as the solution of differential equations, the repeated application of a transition matrix, or use of pseudo-random numbers. There is no logical difference between analytic and simulation models, only that we have chosen not to, or been unable to, formulate them in algebraic terms. In practice, however, there are two important differences: simulation models are very much easier for the non-mathematician to construct, especially with the aid of languages such as CSMP, but they are much more difficult to fit to data than are analytical models.

Deterministic models can be fitted by the method of maximum likelihood (which is equivalent to the method of least squares when the observations are assumed to be independent and distributed normally about their predicted values). For analytic models, this is often a simple procedure and a computer can process as many sets of data as exist. The validity of the model is then checked by the consistency of the fitted parameters and goodness of fit in different sets of data. Fitting simulation models tends to take more time because we must use numerical optimization techniques that do not require explicit derivatives, and the models are usually non-linear, which adds to their complexity. Even so, there is no reason for not attempting to fit the model to as many sets of data as possible, unless the computer requirements are excessive. It is unfortunate that computing expense often prevents an adequate study of simulation models and that some workers are content to accept the first set of parameter values that seem to fit acceptably.

Stochastic models are characterized by the possibility of obtaining widely differing realizations from the same parameters, as in the competition model for two animal populations in which one or other species may become extinct, depending on the direction taken by the random walk representing the change of population sizes with time. The method of assessing goodness of fit depends on the experimental circumstances. If the observations are independent and are expressed as a frequency distribution, expected frequencies can be calculated either analytically or by repeated simulation, and, for the former, maximum likelihood estimation is possible. Some multiplicative processes give similar growth curves on different trial runs, and the mean of all these curves may be taken to be the predicted curve corresponding to the parameter values. This method is inappropriate when large oscillations are predicted, as in an epidemic or prey-predator cycle, because the oscillations will differ in phase and wave length, producing a mean curve with no resemblance to individual realizations. It is then necessary to study the distribution of transitions from one state to the next (Bartlett 1955).

I shall show how, where maximum likelihood estimations would be possible if an analytical model existed, stochastic models can be fitted using repeated simulation.

The interpretation of likelihood functions

The likelihood function refers to one particular set of data and one particular model. If there is only one parameter, it can be represented as a curve, and, if there are two parameters, contours of equal likelihood may be drawn. In higher dimensions, representation is less easy, but, for three dimensions, a series of parallel cuts gives a diagram resembling a sliced veal, ham and egg pie. Two-dimensional contour diagrams can be produced on a line printer as in Trend Surface Analysis.

For linear models, contours of equal likelihood are ellipses or ellipsoids, but non-linear models may give elongated banana-shaped contours, and may contain features such as saddle points and local optima. For this reason, it is inadequate to rely on the maximum likelihood estimates and their standard errors as an accurate description of the range of possible sets of parameter values in non-linear models. An approximate confidence region is obtained by finding the contour corresponding to a fit just significantly worse than the maximum likelihood fit, usually by a simple F test or χ^2 test, (Draper & Smith 1966). I suggested (Ross 1970) that parameters can often be transformed to make the model nearly linear in the region of the optimum. This makes the solution much simpler and makes the standard error of the parameters more appropriate, in that the corresponding confidence regions are much closer to the contours of the transformed likelihood. Banana-shaped contours in the original parameter space can be explained as the result of applying the inverse transformation to regular elliptical contours.

The study of likelihood functions reveals the danger of accepting the first set of parameters that seem to fit the data, or even of accepting maximum likelihood estimates without further considering the range of alternative solutions.

A deterministic model and its likelihood

As an example of a typical likelihood surface in three dimensions, I shall briefly describe a model studied by simulation (Jones *et al.* 1967). This model illustrates how applying a control measure to a pest population selects in favour of offspring resistant to that measure. New hybrids of potato resistant to the cyst nematode, *Heterodera rostochiensis*, are potentially important for growing on infested land in eastern England. When susceptible potatoes are grown continuously on infested land, the pest multiplies until an equilibrium density is reached, and crop losses are large long before this. Resistant potatoes cause cysts to hatch, but the larvae fail to form new cysts, thus decreasing the cyst population. In some soils, however, the cyst count ceases to fall after a few years and starts to rise again, indicating that a substantial fraction of the nematodes can now multiply in the previously resistant variety.

Table 1. Population density of potato cyst-nematodes on plots where resistant potatoes are grown continuously.

Year	1	2	3	4	5	6	7
Density (% of equilibrium)	15·6	6·8	3·2	1·4	1·4	5·0	13·0

The model fitted postulated the existence in the nematode of a recessive gene with initial frequency q, a multiplication rate a that was a parameter in a logistic population growth law standardized to the equilibrium density, and a carry-over parameter, c, the proportion of cysts unaffected by potato root exudates and capable of hatching in the following year.

The likelihood function (a weighted sum of squares of residuals) is represented as a contour in each plane in Fig. 1, which shows that, for given values

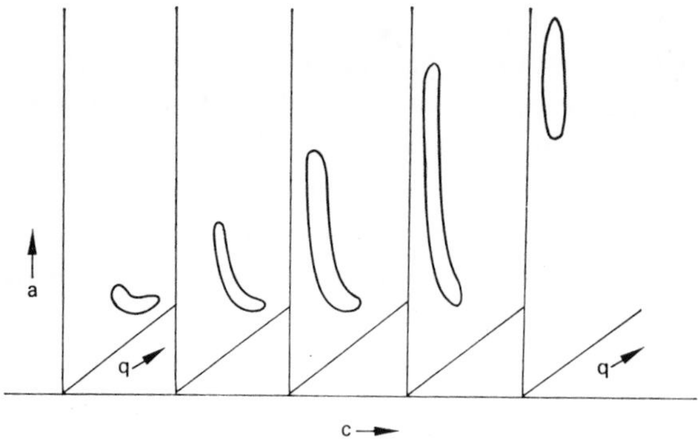

Figure 1. Contours of equal likelihood for three parameter nematode population model. c=carry-over fraction, a=multiplication parameter, q=gene frequency.

of *c*, the fit is best for some particular combinations of *a* and *q*, and that for small *c* we must have large *a* and small *q*, and for large *c* we must have small *a* and large *q*. In fact, all the combinations (*a*, *q*, *c*) within the contour gave predicted curves very close to the observed data. This is not surprising with only six points and three parameters.

Another experiment showed a continuing decline, and there was no evidence for any influence of *a* and *q*, only a best estimate of *c*. This population may not have possessed the appropriate resistance-breaking gene. The likelihood function did not possess an acceptable maximum (negative gene frequencies being a little hard to interpret).

A stochastic model fitted by simulation

Trudgill and Ross (1969) discuss how the observed relationship between sex ratio and population density of Heterodera species on the roots of tomatoes and potatoes can be explained in terms of a simple stochastic model. When a larva attacks the root and has space enough at this point, it becomes female and forms an enlarged cell or cyst. With too little space, it will become male. This model is an intractable generalization of the classical occupancy problem, or coupon collectors' problem (Feller 1950), and has analogies with the problem of finding space in a car park, or of placing beer mugs at random on a small table. The problem was easily programmed for simulation, and plotting mean curves of success rate against number of trials for different sizes of enlarged cell generated a fan of predicted curves (Fig. 2). The appropriate size

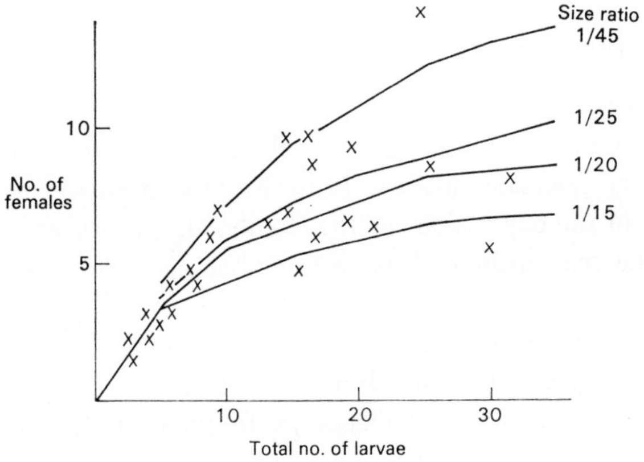

Figure 2. Observed and simulated results in nematode sex-ratio model. Lines are means of fifty trials for a range of cell sizes.

of enlarged cell was clearly shown when the data were plotted. This example suggested that, were the curves exact theoretical curves, a least squares estimation procedure would be justified, and a standard error for the estimate could be calculated.

A plant-spacing model

Unpublished data from the National Institute of Agricultural Engineering, Silsoe, on the intervals between successive seedlings in a drilled row revealed a multi-modal distribution with modes near to integral multiples of the distance between drill holes. The study was intended to provide a basis for comparing different makes of drill.

An obvious model for the distribution involves the following assumptions:

(1) We cannot distinguish between failure of the equipment to place a seed and failure of the seed to germinate.
(2) The equipment may place two seeds in one hole with probability p_1.
(3) The equipment may fail to place a seed or place an infertile seed with probability p_2. These events are assumed to be independent, although in practice seeds may become stuck in the drill, causing regular omissions.
(4) The point where the seedling emerges is displaced from the point where it is placed, the displacement being a normally distributed random variable with zero mean and variance σ^2.

The simulation program produced random samples from the distribution with parameters p_1, p_2 and σ, and it was not difficult to find values generating samples resembling the data.

Approximate likelihood functions

Repeated trials of a stochastic simulation model give approximations to theoretical frequencies and mean values. If the analytical formulation is known, the likelihood can be calculated exactly. If, instead, the approximate predicted values are used in the expression for the likelihood, a function is obtained that tends to the true likelihood as the number of replications increases.

For example, consider a simple binomial model for the phenotypic frequency of some character in a population of fish in a lake. A random sample of 10 fish reveals 4 with the character. If the phenotypic frequency is p, then the logarithm of the likelihood function for p is

$$L = 4 \log(p) + 6 \log(1-p) \tag{1}$$

which takes its maximum value when $p = 0{\cdot}4$.

If, however, the formal binomial distribution had not been worked out, but for a given value of p a series of n simulation trials had been conducted in which n successes were counted, then the approximate likelihood function would be

$$L^* = 4 \log\left(\frac{r}{n}\right) + 6 \log\left(\frac{n-r}{n}\right) \tag{2}$$

Table 2 shows a particular series with $n = 50$, and in this example it

Table 2. Approximate likelihood function for 4 successes out of 10 fitted to results of 50 binomial trials.

Parameter p	·20	·25	·30	·35	·40	·45	·50	·55	·60
Number of successes, r	6	11	18	16	24	23	24	25	35
Likelihood, $-L^*$	9·25	7·55	6·76	6·87	6·86	6·80	6·86	6·93	8·65

accidentally happens that $p = 0·3$ gives a better fit than $p = 0·4$ because the numbers of successes are 18 and 24 respectively, instead of the theoretical 15 and 20. However, a second degree polynomial fitted to the series gives an optimum at $p = 0·45$. It is clear therefore that $n = 50$ is not sufficient replication for this problem, but that the results may still be useful.

Likelihood function of a stochastic birth-and-death process

We simulated a two parameter density-dependent birth-and-death process, which deterministically gives a logistic growth curve, and fitted Gause's experimental data on *Paramecia aurelia* (Gause 1934, Leslie 1958) by calculating chi-square values for the agreement between the data and the means of 100 Monte Carlo trials. Although the suitability of the model and the lack of independence can be questioned, the general shape of the likelihood (chi-squared) function is clearly shown in Table 3. Orthogonal polynomials fitted

Table 3. Empirical χ^2 values fitting Gause's data to the mean of 100 trials. $1-\alpha$=net birth rate, β=change in death rate per unit of population.

β \ α	·80	·81	·82	·83	·84	·85	·86	·87
·0017	193	145	124	91	76	69	64	59
·0016	127	110	75	68	54	49	49	59
·0015	93	66	57	48	45	47	57	73
·0014	76	50	47	52	56	70	94	105
·0013	61	61	66	81	96	115	150	182

in the neighbourhood of the minimum revealed no significant cubic bias and a residual standard error of about 2 in the chi-squared values. The interdependence of the two parameters reflects the need to predict the right equilibrium level, α/β.

Distribution of the approximate likelihood

We shall require some general results on the effect of replication on the distribution of L^*. The argument is expounded in one dimension, but the general results apply to many dimensions if cross-sections of the likelihood function are considered.

If the trial value of the parameter is x, but the series of trials gives predicted values corresponding to a value $x + h$, where h is normally distributed with mean O and variance σ^2/n, then approximately

$$L^*(x) = L(x + h) = L(x) + h.L'(x) + \tfrac{1}{2}h^2.L'(x) + O(h^3) \qquad (3)$$

by Taylor's theorem. Then, by obtaining the characteristic function of $L^*(x)$, we find that its distribution has the following approximate moments:

$$\text{Mean of } L^*(x) = L(x) + \tfrac{1}{2}\, L'(x)\,.\frac{\sigma^2}{n} \qquad (4)$$

$$\text{Variance of } L^*(x) = (L'(x))^2\,\frac{\sigma^2}{n} + \tfrac{1}{2}\, L'(x)\,\frac{\sigma^4}{n^2} \qquad (5)$$

This argument shows that the bias and variance are proportional to n^{-1}, but that near the optimum, where $L'(x)$ tends to zero, the variance is proportional to n^{-2}, and the bias is nearly constant.

Even if the likelihood is not approximately quadratic, there are transformations to make it more nearly so, and the results will still apply.

Finding the maximum by evolutionary operation

Box and Wilson (1951) introduced evolutionary operation as an efficient technique for finding the optimum value of a function, such as the cost or purity of a chemical substance under controllable conditions such as temperature, humidity and concentration of a reagent. Initial experiments indicate the direction most likely to improve the function, and new experiments are designed in the light of previous experience. The principles are discussed by Cochran and Cox (1957) and by Davies (1956). Experimental error limits the precision of the optimization, but, when the function is cost, the resulting losses will be very small.

Evolutionary operation is successfully applied to analytical likelihood functions where it is known as optimization (Box *et al.* 1969). The problem here is simpler because the function is not subject to experimental error.

The basic procedure is to find a sequence of increasing function values until the function becomes too flat for further progress, and then to fit quadratic approximations to the function to locate the maximum. Two statistical results are useful in the context of a function with variance σ^2. If the function is calculated at two points d apart, and the average slope along the line joining them is L', then probability that one is correctly identified as being greater than the other is the normal percentage point corresponding to

$$z = \frac{L'd}{\sigma\sqrt{2}} \tag{6}$$

which means that the greater the variance the larger the step required to ensure progress towards the optimum. The second result is that, if the optimum is estimated from the parabola fitted through three equally spaced points, then, if the true optimum is at distance kd from the middle point, the variance of the estimated optimum is

$$\left(\frac{1 + 12k^2}{8}\right) \qquad \frac{\sigma^2}{d^2(L')^2} \tag{7}$$

which means again that large steps are necessary, and that the precision of the estimate declines rapidly as trial values get further from the true optimum.

Programming strategies

We wish to find the approximate solution with the minimum number of replications of Monte Carlo trials, although it is clear that the number will be large. Yet the method is quite feasible for many real problems, and solutions can be obtained in a single run on the computer.

Formulae (4) to (7) suggest that an appropriate strategy is to begin with a small value of n, assuming that the gradient L' is large and that the direction of progress will easily be found at first. Then, when progress becomes slower as L' decreases, we should double n, but keep the step length constant, else further progress cannot be expected. When several doublings have taken place a quadratic fit can be applied. The efficiency of this method can be greatly improved by the methods suggested elsewhere (Ross 1970), especially that of fitting fewer parameters by imposing constraints.

Evolutionary techniques include the following:

(1) *Random search*: Starting from the best point so far, a new point is sought in a random direction (obtained by multiplying each step length by a random normal deviate). When there is no improvement after 2^{p+1} trials (where p is the number of parameters being fitted), the current phase is ended.

(2) *Simplex search*: Starting with a simplex of $(p + 1)$ points not all in a 'plane' of fewer than p dimensions, the worst point is discarded and a new

point is taken by reflexion with respect to the remaining points of the simplex. This process is continued until progress stops.

(3) *Steepest ascent*: The gradients in each direction are calculated from the main effects of a 2^p factorial design, possibly with fractional replication if $p \geqslant 4$, then trial steps are taken along the line of steepest ascent until progress stops. The process is then repeated with a new design. The phase ends when the gradient is less than a certain value, e.g. that required to ensure a 95 per cent chance of progress by a single step.

(4) *Quadratic fitting*: Either the 3^p designs or the rotatable designs of Box may be suitable. However, when past experience suggests serious curvature, a complete print out of grid points is desirable and a transformation may be needed for future runs.

A practical example

To test the method in practice, the random search technique was applied to a four parameter multinomial model with known solution. Each phase ended when the parameter point accidentally produced a very good fit, making further improvements improbable. Doubling the number of trials gave an improved estimate of the likelihood and further progress could be made. After four phases, the true solution had nearly been reached, although the random method is not one to be recommended for its efficiency. The progress of the search is given in Table 4.

Table 4. Fitting a four-parameter model by random steps

Phase	Sample size, n	Number of steps taken	Parameter values P_1	P_2	P_3	P_4	χ^2 Empirical	Theoretical
Initial	—	—	·20	·20	·10	·10	4·59	6·75
1	50	45+32	·18	·19	·14	·20	0·14	1·71
2	100	27+32	·13	·26	·15	·16	0·48	2·32
3	200	51+32	·09	·20	·22	·17	0·35	0·98
4	400	50+32	·10	·16	·24	·21	0·04	0·28
True	—	—	·10	·15	·20	·25	—	0·00
Standard errors			·07	·08	·09	·10		

Total number of steps=311
Total number of replications=63150

The first attempt to apply this method failed because the step length was halved each time the replication was doubled, and no progress was made after the first phase, as predicted by equation (6). This model is relatively easy to fit because parameters are not strongly correlated.

Other studies showed that standard optimization routines cannot be applied to such erratic functions because they assume continuity and smoothness. Therefore, any general routine for applying this method of fitting would have to include instructions to increase the amount of replication, and would have to lower its standards of accuracy in complicated cases where computing time would otherwise become excessive.

Conclusions

Most stochastic simulation models are not studied in sufficient depth. The models are not compared with enough independent sets of data, and the likelihood functions are not studied. Instead, models are produced with many more parameters than are justified by the data, and the first set of parameter values to produce a good fit is accepted, both as proof that the model is appropriate and that the parameter values are suitable estimates. The only valid excuse for failure to study a model adequately is the amount of computing time required, and even that need not be excessive. Simulation models are often closer to nature than are the comparatively few models that admit analytical formulation. We need to know when they can be simplified and when they need extension. We need much data to show which parameters are always relevant and which only sometimes. The speed of third generation computers should do much to remove the obstacles to an adequate treatment of simulation models.

Acknowledgements

I thank Mr F.B.Lauckner and Mr R.A.Kempton for their assistance in studying some of the applications discussed.

References

BARTLETT M.S. (1955) *Stochastic processes.* Cambridge University Press.

BOX G.E.P. & WILSON K.B. (1951) On the experimental attainment of optimum conditions. *J. R. Statist. Soc., B*, 13, 1–45.

BOX M.J., DAVIES D. & SWANN W.H. (1969) Non-linear optimization techniques. *I.C.I. Monograph No. 5.* Oliver & Boyd, Edinburgh.

COCHRAN W.G. & COX G.M. (1957) *Experimental Designs. 2nd Ed.* Wiley, New York.

DAVIES O.L. (Ed.) (1956) *The design and analysis of industrial experiments. 2nd Ed.* Oliver & Boyd, Edinburgh.

DRAPER N.R. & SMITH H. (1966) *Applied regression analysis.* Wiley, New York.

FELLER W. (1950) *Probability theory and its applications.* 1, Wiley, New York.

GAUSE G.F. (1934) *The struggle for existence.* Williams & Williams, Baltimore.

JONES F.G.W., PARROTT D.M. & ROSS G.J.S. (1967) The population genetics of the potato cyst nematode, *H. rostochiensis. Ann. appl. Biol.*, **60**, 151–71.

LESLIE P.H. (1958) A stochastic model for studying the properties of certain biological systems by numerical methods. *Biometrika*, **45**, 16–31.

ROSS G.J.S. (1970) The efficient use of function minimization in non-linear maximum-likelihood estimation. *Appl. Statist.*, **19**, No. 3, 205–21.

TRUDGILL D.L. & ROSS G.J.S. (1969) The effect of population density on the sex ratio of *Heterodera rostochiensis*: a two dimensional model. *Nematologica*, **15**, 601–7.

A field experiment, a small computer and model simulation

A.J.P.GORE *Nature Conservancy*, *Merlewood Research Station*, *Grange-over-Sands*, *Lancs*.

Summary

Simulation methods have been used with simple mathematical models to explore the results of a long-term experiment designed to examine the effects of stress by clipping. The special characteristics of ombrogenous bog on which this experiment was conducted provide useful field conditions for the study of the mineral nutrient economy.

Introduction

When heather is burnt on British moorlands for sheep grazing or for grouse management, or when trees are cut down for forestry or agriculture, sequences or successions of plants follow which have generally similar forms. The pioneer vegetation, whether natural or introduced, vigorously exploits the resources of light and free nutrients. Certain practices of management permanently eliminate the woody species like *Calluna* and leave the grassy species. This has occurred widely in the Pennines resulting in the Eriophoreta described by Tansley (1949).

It is one aspect of the pioneer growth of an experimentally produced Eriophoretum with which I intend to deal today, using it as the basis for some ideas on modelling and simulation. In the context of the peat covered moorlands of the northern Pennines, it is appropriate to consider the role of mineral nutrients as being important in determining the amount of growth. To explain this briefly, ombrogenous or climatic peat at the Moor House Nature Reserve is generally a fairly deep deposit of organic matter, in this case between 60 and 100 cm deep. It has a low mineral content with phosphorus and potassium in greatest concentration near the surface.

Figure 1 illustrates the total amounts of the main elements per unit depth and brings out in particular the small increases in certain elements, notably

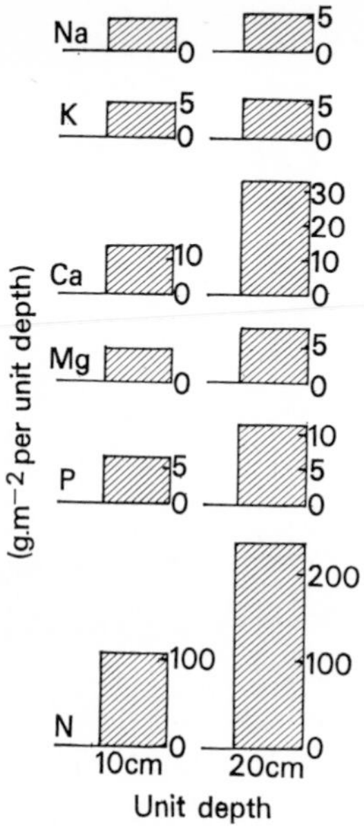

Figure 1. Total amounts of nitrogen, phosphorus and the major cations to 10 cm and to 20 cm peat depth.

potassium, with a doubling of the depth. Most of the living plant roots are confined to the upper 10 cm as shown in Fig. 2.

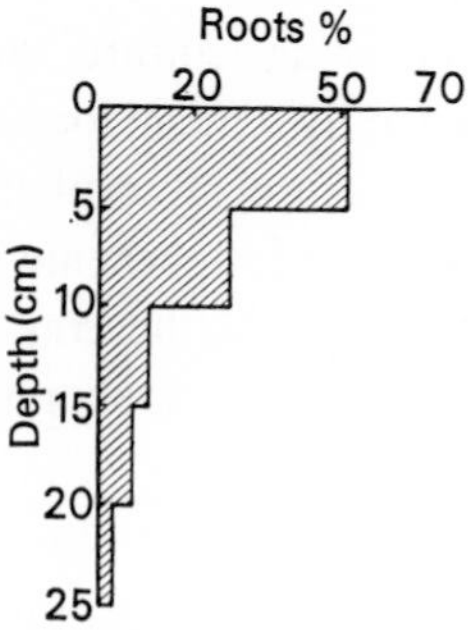

Figure 2. Mean root distribution with depth on experimental plots, mainly *Eriophorum vaginatum* (100 per cent=all roots to 25 cm).

This histogram does not include the rhizomes which also occur in the upper 10 cm.

On this soil, growth of the native plants is thought to be limited predominantly by nutrient supply in the sense that if a *full* range (all three major elements appear to be necessary) of nutrients are supplied, annual growth increases substantially. This suggests that neither the factors of climate directly nor water are primarily deficient.

The field experiment

An experiment was set up in 1958 with the idea that it should be possible to exhaust this nutrient-poor system by cropping in a fairly short time. A range of cropping intensities were imposed whose effects could be used as standards against which to assess the effects of the actual practices involving removal of nutrients by grazing and burning.

The experiment consisted of two blocks, each block containing a number of differing clipping treatments varying from annual to five yearly cycles in which all the above-ground material was removed.

Modelling ideas

1. *A simple system treating all species as behaving similarly and not distinguishing nutrients in the roots and rhizomes from those in the soil*

A suitable model concept of the kind required by this study was proposed by Jenny (1941). He examined the work of Salter & Green (1933) who used the first of the following equations:

$$\frac{dX}{dt} = \frac{-k_1X}{1} \qquad (1)$$

$$\frac{dX}{dt} = k_2 - k_1X \qquad (2)$$

$$\frac{dX}{dt} = -k_1 \qquad (3)$$

where X = amount of one element in rooting zone,
k_2 = annual external contribution of each element,
and k_1 = the fractional loss coefficients,

to calculate the rates of loss of nitrogen and carbon from soils in Ohio under different cropping systems. However, Jenny considered that equation (2) was a better approximation to the real circumstances because a decline of soil nitrogen to zero as predicted by equation (1) was unlikely to happen. Equation (2) leads to an equilibrium condition consonant with the idea of nitrogen fixation.

These equations have been used in the present context to estimate the rates of removal of each of six elements by annual clipping, the most frequent of the clipping treatments. Because of differences between the two blocks of the experiment, associated with differences in peat depth, angle of slope and position on the slope, results from one block only are used here. Results from the second block were similar and will be discussed in a detailed consideration of the experiment elsewhere. In Jenny's example, corn yields were approximately proportional to soil nitrogen content and so the equations could be used to estimate yields also. The relationships between the nutrients in the peat and production are not definitely known but it is assumed that production will be controlled by the element depleted most rapidly. The following assumptions were made:

k_2 is considered to be the rainwater input even in the case of nitrogen. The evidence for or against fixation of nitrogen either symbiotically or by free-living organisms is not yet available, but the general absence of plants likely to act in a symbiotic partnership and the low pH of the peat (3·2) suggest that fixation is low. Anaerobic nitrogen fixation is a possibility, however, although such fixation would presumably occur in horizons below the main mass of living roots.

X is taken to be the amount of each nutrient in the peat/plant system within the upper 10 cm.

k_1 is the fraction of X removed by the crop each year. Evidence for losses other than by the crop is not available. Denitrifying organisms occur, but it is not known whether they release significant amounts of nitrogen from the peat under field conditions. Crisp (1966) found that there were few net losses of nitrogen and phosphorus in solution from a peat covered catchment at Moor House.

It was not possible to estimate the losses of the major cations from the peat in his study, but such losses by run-off, principally of rainwater, must occur.

In order to allow for the errors and other variation involved in the estimates of k_2, X and k_1, the calculations were made on repeated samplings of their frequency distributions. Means and standard deviations of the results were then obtained. In this procedure, which is really an application of the Monte Carlo method in its most elementary form, it was assumed that:

(i) The crop was independent of nutrients in the rain and of nutrients in the peat/plant system over the period of observations.
(ii) k_1 remained constant with time.

The crops removed annually over 10 years did not show any marked trends (Fig. 3), but the variation experienced included an experimental error due to under-sampling in 1962. The values for 1959 were considered too transient to be representative and were omitted.

The elementary constitution of the rainwater for Moor House has been

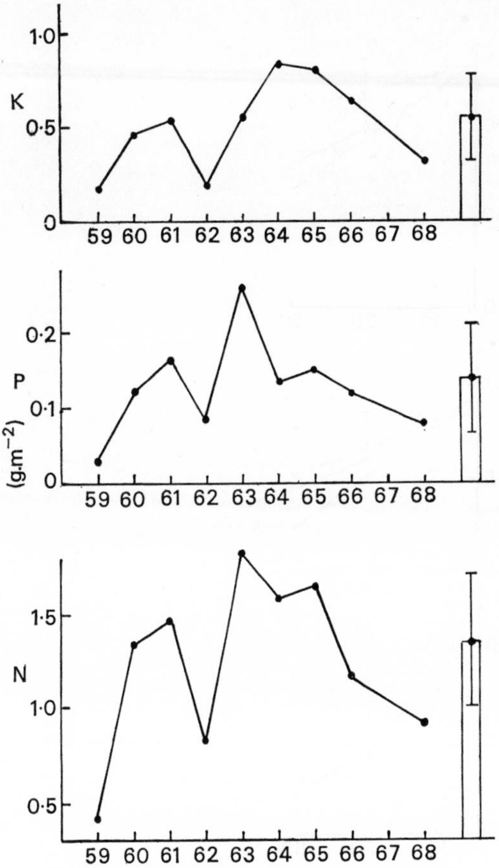

Figure 3. Amounts of nitrogen, phosphorus and potassium in *Eriophorum vaginatum* removed by cropping annually. Histograms show means of years 1960 to 1968 with one standard deviation.

published (Gore 1968) and the estimates of variation between years used here. The elements in the peat/plant system were measured directly on the plots in 1967/68.

The results shown in Fig. 4 are interpreted to mean that sodium, for example, is not controlling growth. Most of the sodium supplied by the rain presumably runs off: it obviously does not build up as the output of model (2) implies. A second loss term might be inserted to account for this, but, in the absence of data, would only serve a theoretical role. The rate of removal of sodium by cropping is low, with a mean 'half life' of about 120 years.

In the case of magnesium, removal by cropping, assuming no rain input, is high, with a mean 'half life' of 20 years. However, the supply is potentially in excess of demand and leads one to doubt that magnesium supply is likely to

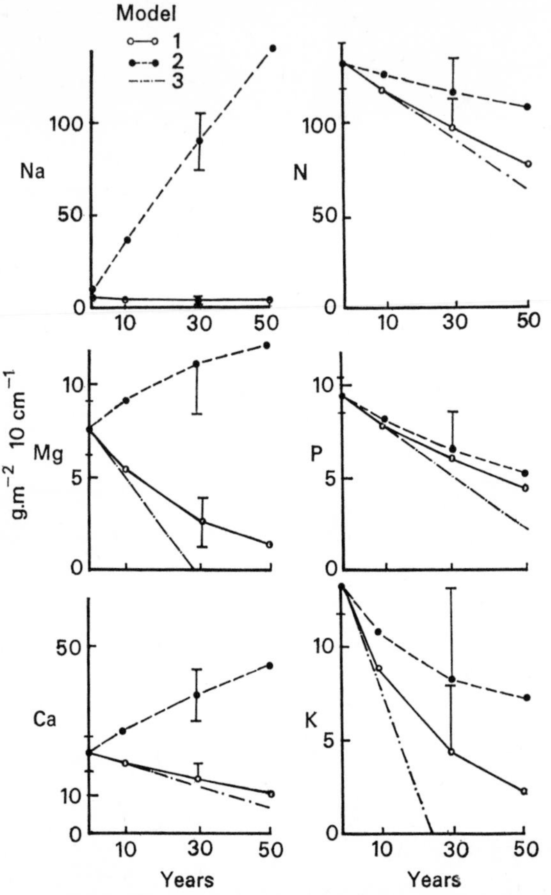

Figure 4. Results of simulations using three models. 95 per cent confidence intervals for $n=5$ are given in certain cases to illustrate the computed variation.

become a limiting factor for growth. Similar arguments apply to calcium. In this element, the additional possibility of deep rooting contributions makes a deficiency even less probable. In the case of nitrogen, loss due to cropping is such that, with no effective rainwater input, the 'half life' average is 69 years. If all the rainwater were effective in supplying nitrogen, on the other hand, supply would almost equal demand, with an equilibrium level of over 60 per cent of the initial state of the nitrogen in the peat/plant system. Again, the lower horizons could represent an appreciable input source of this element.

The demands made by clipping on potassium are the highest of all, with a 'half life' of 17 years. If the maximum possible supply from rain is allowed, a fairly rapid decline still takes place to the equilibrium level of 44 per cent of the initial level in the system. Obviously, extra sources of this element from the deeper horizons are extremely limited and it seems a strong candidate as limiting factor for growth.

Phosphorus is of interest, in that the supply from rain is so low that the theoretical rate of depletion of the peat/plant system is similar in models (1) and (2). If the actual *potassium* supply were appreciable, then the demand for phosphorus would tend to make it the most quickly deficient and therefore likely to control growth.

Model (3) results could apply to all elements before deficiency levels are reached in any one or more of them and assuming no effective rainwater supply.

These forecasts are based on crude assumptions. In fact, the assumption that k_1 would remain constant is questionable. Root masses are generally proportional to the crop, suggesting that exploitation of a given volume of soil is less effective as the crop gets smaller. The point is arguable, however, because root mass is not the same as an effective root surface. In any case, the elements might be expected to change their availability with time in respect to their form, and this would modify the predictions of change in elementary amounts owing to the existence of growth control not visualized by these models. For all these possibilities, the findings at Moor House and elsewhere confirm the attention drawn to potassium and phosphorus. Carlisle and his colleagues (Brown & White 1970) have found that phosphorus and potassium in particular are deficient for *Pinus contorta* growing on experimental plots adjacent to the ones reported on here. Nitrogen does not appear to be primarily limiting and the effects of added phosphorus are to reduce the potassium concentrations in the needles. Greatest height growth has been associated with treatments combining both phosphorus and potassium. The steady decline in the concentrations of these elements suggests that a tree crop is a form of continuous nutrient removal essentially similar to the clipping treatments. Primary deficiencies of phosphorus (Tamm 1954) and of potassium (Goodman & Perkins 1968) are well known for *Eriophorum vaginatum* growing on peat elsewhere.

2. *A model derived for* Eriophorum vaginatum (*the induced dominant species*)

The above approach to the question of the effects of clipping has many defects. It regards all species as behaving similarly and lumps the peat/plant system into one. A compartment model is being developed from the work I did with Dr J.S.Olson at Oak Ridge. In this, species behaviour is accounted for separately, and the storage organs of the plant are separated from the peat soil. A model having the minimum requirements for these purposes is shown in Fig. 5 and equations (4) to (7).

$$\frac{d(N)}{dt} = A4 - A1\,(N)\,(LA) \qquad (4)$$

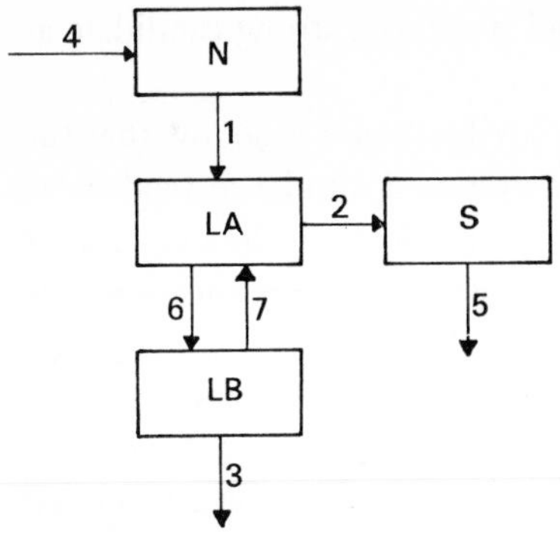

Figure 5. Box diagram of a growth model controlled by available mineral nutrients. Variable names and numbered parameters are identified in the text.

$$\frac{d(LA)}{dt} = A1\,(N)\,(LA) + A7\,(LB) - (LA)\,(A6 + A2) \qquad (5)$$

$$\frac{d(LB)}{dt} = A6\,(LA) - (LB)\,(A7 + A3) \qquad (6)$$

$$\frac{d(S)}{dt} = A2\,(LA) - A5\,(S) \qquad (7)$$

where (N) = 'available' amount of one element in rooting zone
(LA) = amount of the same element in the living above-ground material
(LB) = amount in living below-ground material
(S) = amount in above-ground dead leaf
$A1-A7$ = rate coefficients

The essential point here is that growth of the above-ground parts (leaves) is assumed proportional to themselves and also to the amount of available mineral nutrients in the soil. It is to be noted that this applies only to annual events and ignores seasonal effects at this stage. It is also to be noted that leaf dry matter is proportional to leaf nutrient mass, because leaf nutrient concentrations vary little with time. The implication of this central part of the model is not that there is a flow of nutrients directly into the leaves, but that leaf growth is controlled by nutrient availability. The box diagram in Fig. 5 is limited in its ability to convey this idea, although all the other arrows, 2 to 7, do represent physical flows.

In the earlier model (Gore & Olson 1967), lag was introduced to account for the sigmoid character of the observed growth curves. This was essentially a forcing function consisting of the interpolation of a NET PRIMARY PRODUCTION compartment between the SOLAR ENERGY INPUT and LIVE BIOMASS compartments. This was an empirical procedure which was not fully satisfactory either in biological content or as an effective description of the lag. In that paper, we suggested that the lag in growth reflected the time required to fully re-occupy the bared area with foliage. Indeed, at first sight, the growth rate of the leaf might be considered to depend on the available light, but the maximum leaf area index does not exceed 2·5, which seems well below the optimum possible value and, more convincingly, the leaf area index reached at the end of the first five year cycle was greater than that at the end of the second cycle, in spite of the fact that a substantial amount of *Calluna* was also present in the first, but absent in the second. In this way, it seems unlikely that the light climate is fully utilized, and therefore plays a minor part in the growth control in this ecosystem. Results based on a model of *Calluna* developed by Grace (1970) for the same locality showed a sensitivity to temperature but relatively little to light.

Having adopted the model, it was decided to work in terms of phosphorus, although potassium or nitrogen would probably have been suitable, since all three appear to be near the threshold of growth limitation. To have dealt in terms of dry matter or energy would have meant using translation formulae. Smith (1969) also avoided using these and in our context of a small computer they seemed an unnecessary complication when dealing with annual changes. Dry matter estimates can be made if required, by using knowledge of the phosphorus concentrations of the plant parts concerned.

It was now required to find out if the model was appropriate to the data, and, if so, what the parameter values were. The data consisted of the phosphorus equivalents of the successional changes already published (Gore & Olson 1967) plus further information which had to be derived in retrospect from the below-ground changes during the same period. Information on the 10 year stage of the experiment has also been used. The retrospective estimates of below-ground parts were based on a knowledge of the *Eriophorum* tussock areas and regressions of these against weights of the below-ground organs. The method used is due to Mountford and Bunce (in press). The help of Mr Sykes who programmed the method is acknowledged with this.

The data were first plotted and smoothed curves drawn by eye as shown in Fig. 6.

As can be seen from the wide confidence intervals, a large number of smoothed curves could be drawn in this way, but, as a first step, the smoothed curves were drawn by eye to meet the requirements of the data as closely as possible. The relative growth rates of the smoothed curves were then calculated using the usual formula:

$$\text{R.G.R.} = \frac{\log_e W_2 - \log_e W_1}{t_2 - t_1}$$

(where W_1 and W_2 are the masses of phosphorus at times t^1 and t^2 respectively).

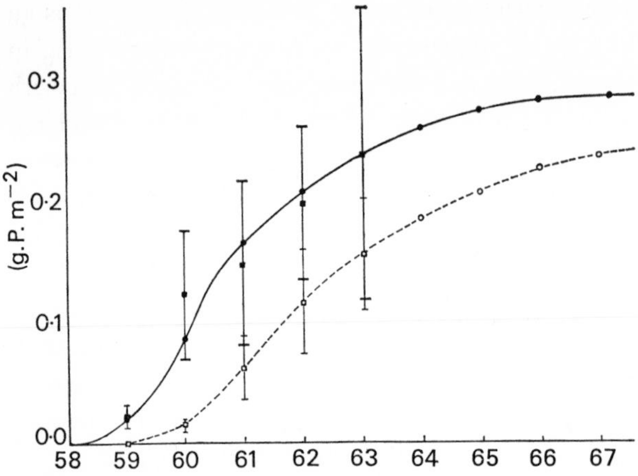

Figure 6. Smoothed curves drawn manually through the observed phosphorus values of living above-ground (continuous line with solid squares and circles) and standing dead (dashed line with open squares and circles). Variation is expressed as 95 per cent confidence intervals for $n=5$.

The next step consisted of plotting the calculated relative growth rates according to the linear equations obtained by re-arrangement of the model equations, for example:

$$\frac{1}{(LB)} \cdot \frac{d(LB)}{dt} = A6 \frac{(LA)}{(LB)} - (A7 + A3)$$

$$\frac{1}{(S)} \cdot \frac{d(S)}{dt} = A2 \frac{(LA)}{(S)} - A5$$

The points so plotted did not necessarily fall on a straight line, but could be made to do so with minor adjustments of the smoothed curve. The results of this procedure for the smoothed curves finally adopted in Fig. 6 are shown in Fig. 7. Figure 8 is a similar application for the (*LB*) and (*LA*) compartments. In this way, it was possible to estimate five of the parameters including the sum ($A3 + A7$).

The procedure is subjective, but it is still a substantial improvement on a completely trial and error fitting procedure, since it involves far less computer time and is in any case only a first approximation.

At this stage, a trial simulation was made on the computer. The value of $A1$ (*N*) (*LA*) and of $A4$ at equilibrium could be found from the parameters estimated so far. Setting the initial conditions of $A1$ (*N*) ($A1$ was set at unity at this point for want of better information) and the value of $A4$ to their equilibrium values resulted in the continuous line of Fig. 9.

The dashed line in Fig. 9 represents the values of $A1$ (*N*) for each year,

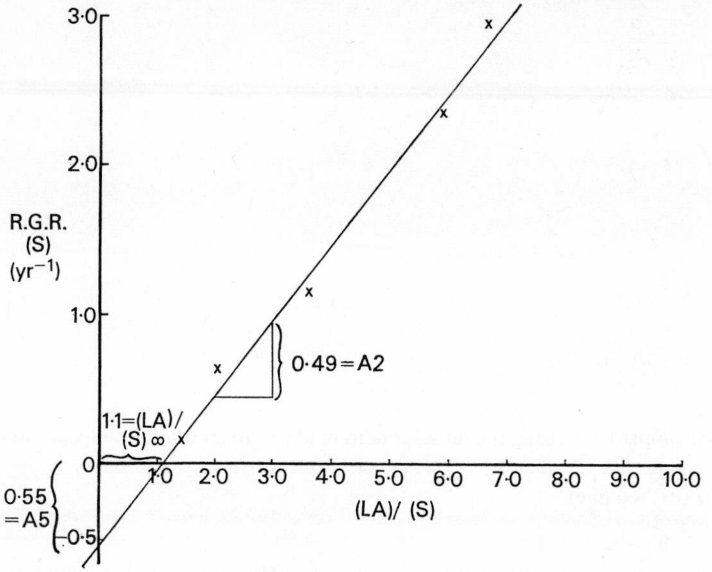

Figure 7. Graphical estimation of parameters *A2* and *A5* from smoothed curves. ∞ = steady state.

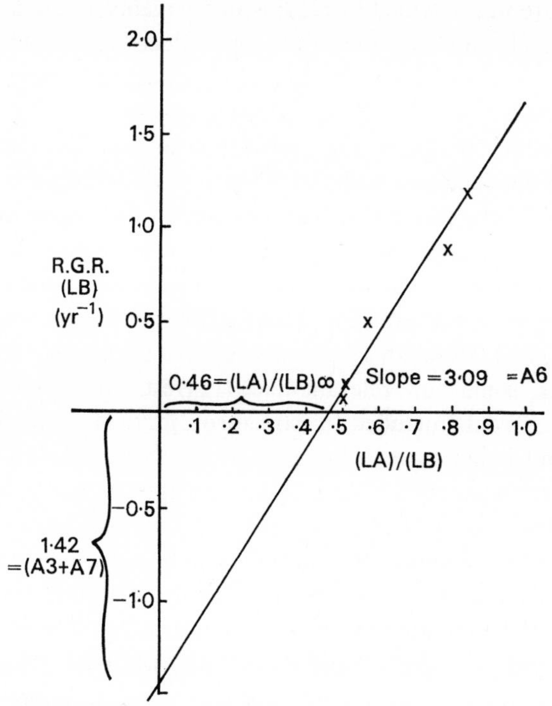

Figure 8. Graphical estimation of parameters *A6* and $(A3 + A7)$ from smoothed curves.

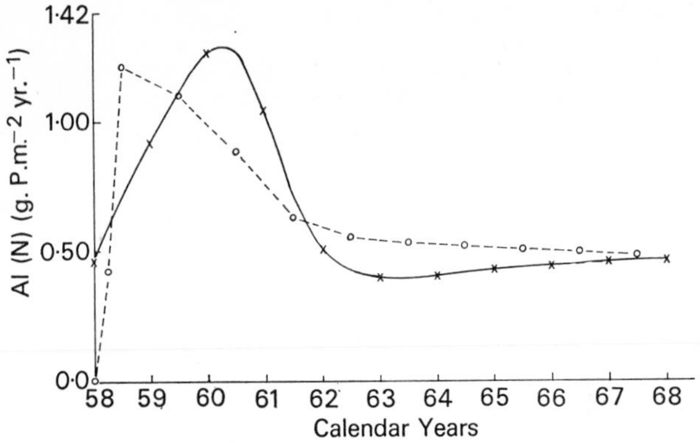

Figure 9. First simulation, setting the initial condition of (*N*) to its equilibrium value and *A*1 = 1·0 (continuous line). Values of *A*1 (*N*) for each time step calculated from smoothed data (dashed line).

calculated manually by solving equation (5) for the smoothed curves. An initial condition of zero for *A*1 (*N*) was found to be necessary.

Figure 10 shows how the first trial produced an unacceptable overshoot. The dashed line in this figure was obtained by reading in the manually calculated values of *A*1 (*N*) at each time step and of course produced a close fit. The data for the second cycle of five years are shown on this figure by open squares. The values of *A*4 were now calculated manually from equation (4), using the *A*1 (*N*) values which produced the dashed line in Fig. 10. The change of *A*4 with time so obtained was such that a sharp pulse occurred in the first year, falling to zero by the end of that year, followed by a slow rise to a low equilibrium condition thereafter. Such a pulse could be interpreted as a release of nutrients, from the roots of the mature heather plants cut initially. However, an alternative input of nutrients could have come from the below-ground compartment (*LB*) and, although I had assumed this to be equal to the calculated 1959 values, some new relevant data (Forrest, 1971) indicated a higher figure. Forrest's data included an estimate of *A*3, the fractional loss from the below-ground living parts. This estimate of *A*3 was also incorporated at this time, previously an arbitrary value, having been adopted based on the sum (*A*3 + *A*7).

Figure 11 shows the effect of these modifications on the values of *A*1 (*N*). The dashed line is derived from the manually calculated values, the continuous line represents the computed values with *A*1 set to 2·5. The increase in the value of *A*1 from 1·0 (dashed and dotted line) was made on a trial and error basis to make the computed values coincide as nearly as possible with the manually calculated ones.

Figure 12 shows the output for the three compartments corresponding to

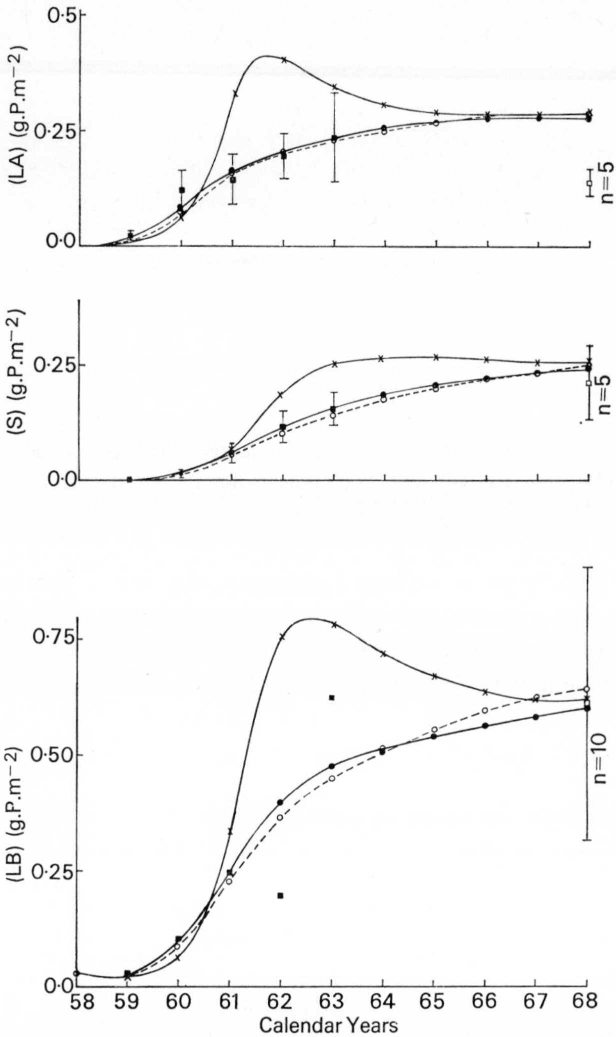

Figure 10. First simulation results corresponding to *A*1 (*N*) values shown in Fig. 9. The solid circles are the smoothed data the solid squares are the observed results used in estimating the smoothed curves. The open squares are observed results for the second five year cycle of clipping, 1968 with 95 per cent C.I.

these three sets of values of *A*1 (*N*) and can be compared with the smoothed curves and the observed values which are also shown. Improvements in the agreement between simulated and observed could be made, but, until more information on the validity of the model is obtained, such adjustments have little value. The original forcing function has, however, been replaced by change of form of the equations having a more appropriate biological interpretation.

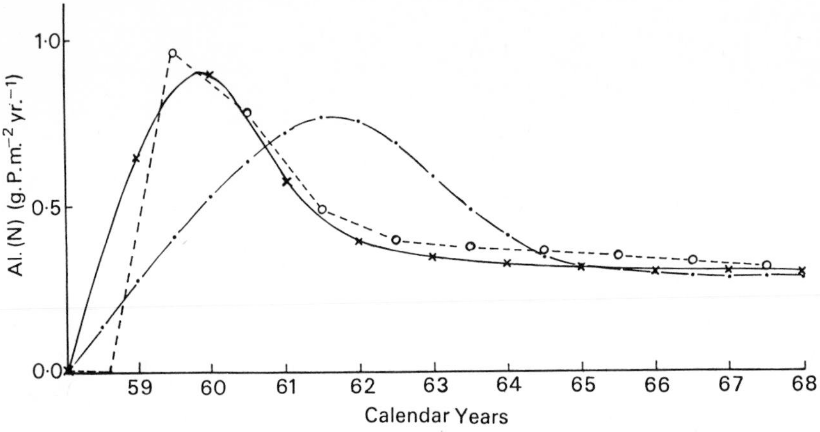

Figure 11. Final simulation, the values of *A1* (*N*) for each time step calculated manually shown by dashed line. The computed output of *A1* (*N*) shown for *A1*=2·5 and *A1*=1·0 by the continuous line and the dashed and dotted line respectively.

It was Dr J.H.Williamson of Culham Laboratory who first suggested using the non-linear model form to avoid the artificiality of the forcing function. He recently kindly carried out a check on the graphical methods I have described, by using a computer program involving a least squares fitting routine. I gave him the smoothed values for the years 1959 to 1963. The initial conditions of the living below-ground compartment were assumed, at that

Table 1. Two different methods of estimating the model parameters.

	Parameter (yr^{-1})	Least squares fitting method 1	2	Graphical method
	A1	0·950	2·017	2·50
	A2	0·388	0·545	0·490
	A3	0	0·014	0·246
	A4	0·340	0·586	1·028
	A5	0·370	0·611	0·550
	A6	2·223	3·240	3·090
	A7	0·893	1·476	1·174
	(*A3*+*A7*)	0·893	1·490	1·420
Initial conditions ($g.P.m^{-2}$)				
	(*N*)	0·7202	0·2772	0·0
	(*LA*)	0·0001	0·0001	0·0001
	(*LB*)	0·032	0·032	1·00
	(*S*)	0·0	0·0	0·0

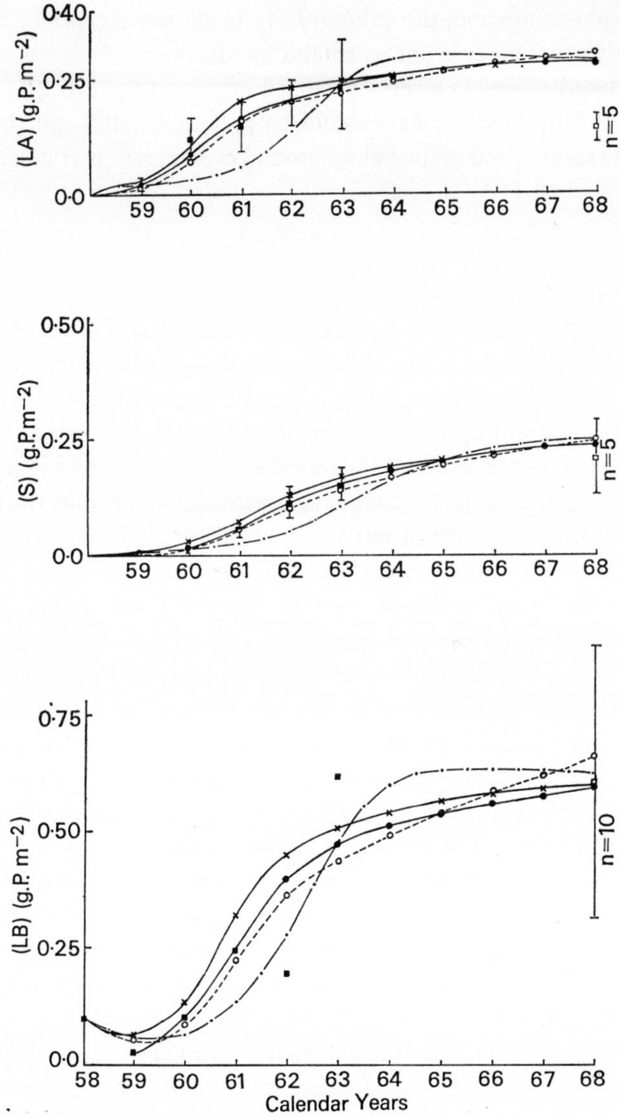

Figure 12. Final simulation results corresponding to *A*1 (*N*) values shown in Fig. 11. Other symbols as in Fig. 10.

stage, to be those used in Fig. 10 and the initial conditions of the available nutrient compartment were varied over a wide range in five separate trials. Convergence was obtained in a short time in two of these trials. From Table 1, a comparison is possible between the parameter values obtained from those two trials and the parameter values finally adopted for the output of Fig. 12 derived primarily from the graphical method.

The values in set 2 of the least squares method are similar to those obtained by the graphical method, the values of $A3$ being the exception. A near zero value for this parameter is not acceptable and the value of $A4$ has to be correspondingly larger when $A3$ is set at 0·246. Clearly, least squares application to the original (unsmoothed) data would be preferable, although the poor definition of the results, and of the below-ground estimates in particular, raises difficulties.

Conclusion

A fit, in the sense used here, does not have any special biological significance. The fairly wide limits of the observations make the range of possible parameter values wide also. It needs no emphasis that a successful application of a least squares approach depends on improved data quality, although application of the graphical method to independent sets of data could be a useful preliminary step. A test by 'cropping' the model for comparison with clipping results from the field is a more satisfactory form of test. The model has, however, to be developed to the point where it includes seasonal variation if it is to be 'cropped' and this will require information on the seasonal translocation of nutrients which is at present indirect and incomplete.

The mean annual crop of phosphorus in *Eriophorum* removed by clipping is 0·10 g.P.m^{-2} on the basis of field results over ten years, while the steady state annual input to the model is presently estimated at 0·29 g.P.m^{-2}. This may be an over-estimate, but, allowing for differences between the steady-state *Eriophoretum* and the cropped state and for below-ground turnover, the values are in the right order. The extractable phosphorus based on 0·002 N H_2SO_4 (Truog's reagent) for the climax *Calluna* community was found to be approximately 0·4 g.P.m^{-2}·10 cm^{-1} and Allen (1966) made a similar measurement on the same soil type and found a value of 0·15 g.P.m^{-2}·10 cm^{-1}. The value calculated for (N) in this model for the steady-state *Eriophoretum* was 0·12 g.P.m^{-2}.

The first modelling approach makes few demands on a knowledge of how the system 'works' and may well provide adequate answers for some purposes. The second approach attempts to incorporate more detail of the biological mechanisms and has most concern for the estimation of certain rates of transfer. The main task, and a most difficult one, is finding a form of model which incorporates only the central biological processes necessary to the problem in hand. A fine balance must be sought between empiricism and reality as we currently see it.

Having said this and identified what I take to be the heart of modelling, the question of computer size necessarily takes a subordinate place. A large computer may have an important role when it comes to the point of assembling several species sub-models interacting in an ecosystem context, but a small, fast computer has many advantages, not least of which is the need to confine attention to essentials.

References

ALLEN S.E. (1966) Chemical aspects of heather burning. *J. appl. Ecol.* 1, 347–67.

BROWN A.H.F. & WHITE E.J. (1970) Establishment and Growth of Trees at High Elevation. *Triennial Report of Merlewood Research Station* 1967–9.

CRISP D.T. (1966) Input and Output of Minerals for an Area of Pennine Moorland. The Importance of Precipitation, Drainage, Peat Erosion and Animals. *J. appl. Ecol.* 3, 327–48.

FORREST G.I. (1971) Structure and Production of North Pennine Blanket Bog Vegetation. *J. Ecol.* 59, 453–79.

GOODMAN G.T. & PERKINS D.F. (1968) The role of mineral nutrients in *Eriophorum* communities v Potassium supply as a limiting factor in an *E. vaginatum* community. *J. Ecol.* 56, 685–96.

GORE A.J.P. (1968) The Supply of Six Elements by Rain to an Upland Peat Area. *J. Ecol.* 56, 483–95.

GORE A.J.P. & OLSON J.S. (1967) Preliminary Models for accumulation of organic matter in an *Eriophorum/Calluna* Ecosystem. *Aquilo Ser. Botanica, Tom* 6, 297–313.

GRACE J. (1970) The Growth-Physiology of Moorland Plants in Relation to their Aerial Environment. Thesis for Ph.D. in the University of Sheffield.

JENNY HANS (1941) *Factors of Soil Formation—A System of Quantitative Pedology.* McGraw-Hill Book Co., New York.

MOUNTFORD M.D. & BUNCE R.G.H. (in press). Regression Sampling with Allometrically Related Variables.

SALTER R.M. & GREEN G.C. (1933) Factors Affecting the Accumulation and Loss of Nitrogen and Organic Carbon in Cropped Soils. *J. Am. Soc. Agron.* 25, 622–30.

SMITH F.E. (1969) Effects of Enrichment in Mathematical Models in the Proceedings of a Symposium entitled *Eutrophication, causes, consequences, correctives.* National Academy of Sciences, Washington, D.C.

TAMM C.O. (1954) Some observations on the nutrient turn-over in a bog community dominated by *Eriophorum vaginatum* L. *Oikos*, 5, 189–94.

TANSLEY A.G. (1949) *The British Islands and their Vegetation.* Cambridge University Press.

Data management with ASCOP

B.E.COOPER *ICL Dataskil Ltd, Reading Bridge House, Reading Bridge Approach, Reading RG1 8PN, Berkshire*

Summary

Experience in using the statistical and data management system ASCOP (Cooper 1969) has shown that the provision of a wide range of data management operations in a statistical system is at least as important as the provision of a range of statistical analyses. Data management requirements will be discussed and illustrated using ASCOP.

1. Introduction

It is not possible in the time available to give a detailed description of the ASCOP system. The paper therefore begins with a brief description of the data structures recognized and the statistical analyses that it can perform. The facility referred to as 'feedback' which allows the user to give names to, and subsequently refer to, values computed as part of analyses is described and stressed as a vital facility. Data management facilities available are then described in a little more detail. Finally, the feedback facility and the data management facilities are illustrated by detailed discussion of four typical examples. The reader is thereby introduced to the system 'by example', in an attempt to give him a feel for the data management facilities that the system provides and to highlight the need for such facilities in the analysis of data. A detailed description of version 2 of the system can be found in the ASCOP User Manual (Cooper 1969). The Manual for version 3 of the system (the version currently available and described here) has been written (Cooper, in press) and is being published by the National Computing Centre in Manchester.

2. Data structures recognized by the system

The system recognizes three basic data structures, the data matrix, a parameter, and a set of coefficients. Data matrices may be saved and retrieved (fetched) and only one data matrix is normally available for analysis at any one time. This is referred to as the present data matrix. Parameters and sets of coefficients on the other hand are always available irrespective of which data matrix is currently the present data matrix. The three structures are briefly described in the next sections.

2.1 *A data matrix*

The data matrix is a rectangular arrangement of values with columns representing variables and rows referred to as points. Variables are named by the user for example HEIGHT, WEIGHT, A17, XY294 and Y, and each point is given a positive integer identifier referred to as its label. Point labels may be specified by the user or set by the system, and the column of these labels formed as part of the matrix may be referred to as a variable with the name LABEL. A further column referred to as a special variable PMISS is computed by the system and its value for a given point is the number of missing values in that point. The user may reset the values of LABEL but not PMISS. A data matrix containing four user variables A B C D and ten points labelled 1(1)5(5)30 has the following form.

PMISS	LABEL	A	B	C	D
0	1	X	X	X	X
1	2	X	X	M	X
0	3	X	X	X	X
0	4	X	X	X	X
2	5	M	M	X	X
0	10	X	X	X	X
0	15	X	X	X	X
0	20	X	X	X	X
1	25	X	X	M	X
0	30	X	X	X	X

An M in the data matrix denotes a missing value. Data matrices may be read as original data, increased in size or amended in a variety of ways specified by the user, or created as parts of, or amalgamations of, other data matrices. Some of these ways are illustrated in examples discussed later.

2.2. *A parameter*

A parameter is a single value declared as such using the declaration

PARAMETER PA

for example. Several parameters may be declared in one declaration; for example

PARAMETERS PA PB PC

Parameters may be read as original data set explicitly by the user, or set to certain values computed during an analysis such as, for example, the residual mean square in a regression analysis.

2.3. *A set of coefficients*

A set of coefficients is an array of values indexed from zero. A set may be declared using the declaration

COEFFICIENTS CA 10

for example. This defines a set of 11 coefficients indexed from CA(0) to CA(10). They may be given values by reading them as original data, by explicit setting by the user, for example

CA(7) = 49·2

or by setting them to coefficients of functions of variables such as are computed in regression, components and other analyses. This latter method provides the user with access to the results of analyses and the ability to use them in later instructions. This facility is referred to as feedback.

3. Statistical instructions

In this section I give a rapid survey of the statistical instructions available in ASCOP. Two types of instruction may be distinguished.

1. A declaration declaring information which may be used later by analyses.
2. Instructions specifying analyses to be performed.

Both types are described here in an order dictated by the need to be brief. Not all currently available instructions have been included. For a complete and more detailed description the reader is referred to the Manual.

Many instructions require the specification of a list of variable names. Such lists may be presented in a variety of shorthand ways. A complete description of the rules is not given, but a wide variety of ways of presenting a list are used in the examples.

The analysis instructions and related declarations are now briefly described. The description is not a formal one nor is each instruction necessarily completely described. The description consists largely of a series of example instructions. Items included in lower case in quotes are user specified items which may take a number of different forms as described with each instruction in which they appear. Items in upper case are also user specified items but they are examples of the forms that the user may use. Items in upper case and underlined are system words. Finally words in upper case italics are optional words that the user may include or omit as he wishes.

3.1. *Output special forms*

<u>OUTPUT</u> 'list' <u>FOR</u> *VARIABLES* A, B, C1–10, *AND* D

The 'list' specifies what is to be output and it may include any selection from:

BRIEF SUMMARIES	—means, standard deviation, minimum and maximum values.
FULL SUMMARIES	—Brief summaries plus the correlation matrix.
COVARIANCE MATRIX CORRELATION MATRIX	—alternatives
HISTOGRAMS	—sample histograms with normal fit.
SERIAL CORRELATIONS *UP TO* LAG 40	
MOMENTS	—moments and moment ratios to degree 4.
RUNS DISTRIBUTIONS	—Distribution of runs up and down.

The word STREAM followed by a stream number may be added to the instruction if a stream other than the normal output stream is intended.

3.2. *Plotting*

<u>PLOT</u> 'Scale' <u>AGAINST</u> 'Scale'

The 'Scale' may be
1. a variable
2. a function of a variable such as LOG(X)
3. the word NORMAL or ORDER.

In the third case the word NORMAL specifies a normal probability plot and ORDER is a generated scale taking values 1, 2, 3 . . . . The instruction

PLOT LOG(Y) AGAINST NORMAL

produces a plot with variable Y organized on a Logarithmic scale against a normal probability scale and it shows the log-normality of variable Y.

The plots are produced on a SC4020 micro film plotter offering a choice of

4 possible forms including 16mm and 35mm film, 8 × 8 inch transparencies for overhead projection or 8 × 8 inch hard copy. It is possible to superimpose plots on the same frame.

3.3. *Tabulation*

TABULATE A BY B *AND* C

This specifies a tabulation of A against B and another of A against C. The cells of the tabulation may be defined in terms of boundaries or central values using a declaration such as:

BOUNDARIES A,B (1(1)10(2)20,25) C (1·0,2·5,3·7,5·1,7·2)

Boundaries or central values for each variable need only be declared once but they must be specified before the TABULATE instructions that use them. If none are specified for a particular variable the system will compute boundaries based on the sample size and the standard deviation. A wide range of lists of tables may be specified in a tabulate instruction.

3.4. *Diallel table analysis*

DIALLEL *TABLE ANALYSIS* OF *VARIABLES* X1–3 PARENTS 6 $
BLOCKS 4

Full and half-diallel table analyses may be performed on replicated or unreplicated data.

3.5. *Comparison of groups*

COMPARE GROUPS 2, 4, 7, PARTITION *VARIABLE* GRPVALS $
FOR *VARIABLES* X1–3, 5, 7–10

The present data matrix is divided into three groups so that group *i* contains those points for which variable GRPVALS is equal to V_i where V_1,V_2,V_3 are 2, 4 and 7 respectively, the values specified by the user. Points with GRPVALS equal to some other value are not allocated to a group. The means, standard deviations and other details and the correlation matrix for each group and the specified variables are output.

3.6. *Regression*

REGRESSION OF Y ON X1–10
REGRESSION OF Y ON BEST 2 *VARIABLES*

These are two examples of the regression statements available. The first specifies a simple regression while the second specifies the selection of the best 2 variables for describing variable Y. Any regression statement may be followed by a statement such as

NAME RESIDUALS YRES FITTED *VALUES* YFIT COEFFICIENTS $
CY RMS YRMS RMSDF YRMSDF *AND* RATIO YR

to define new structures and 'feedback' values computed during the analysis. This instruction has the following effect.

1. Residuals and fitted values are set up as new variables in the data matrix.

2. A set of coefficients CY are defined (if not already defined) and given values equal to the coefficients of the fitted regression function. The constant term is placed in CY(0) and the coefficient of the Ith variable is placed in CY(I).

3. Parameters are defined and set as follows:

YRMS	—residual mean square
YRMSDF	—residual mean square degrees of freedom
YR	—ratio of regression sum of squares to total sum of squares (i.e. the square of the multiple correlation coefficient).

3.7. *Components and factor analysis*

COMPONENTS *ANALYSIS* OF X1–12 B4–7

FACTOR *ANALYSIS* OF X1–21 WITH 3 *FACTORS*

These specify the obvious analyses. Feedback of analysis values may be achieved by a NAME instruction immediately following the analysis instruction. For example the statements:

NAME COMPONENTS CA CB COEFFICIENTS BA BB

NAME FACTORS FA FB COEFFICENTS BA BB

define new variables CA CB or FA FB equal to the component or factor scores for the first two components or factors and sets of coefficients of the first two components or factors.

3.8. *Discriminant analysis*

DISCRIMINATE BETWEEN *GROUPS* 2, 4, 7 PARTITION $
VARIABLE GRPVALS

Points are allocated to groups as described in the comparison instruction and a discriminant analysis is performed. Feedback of values may be specified using a NAME instruction. For example

NAME ALLOCATION A123 FUNCTIONS F1 F2 F3 COEFFICIENTS $
C1–3

defines

1. Variable A123 with values 2, 4 or 7 defining the group each point would be allocated to if the discrimination was applied to the data matrix.
2. Variables F1, F2 and F3 taking the values of the three discriminant functions respectively.
3. Sets of coefficients C1, C2 and C3 taking as values the coefficients of the three discriminant functions respectively.

3.9. *Analysis of variance*

This is restricted to complete factorial experiments but polynomial partitioning is provided. The instruction

ANOVA OF Y DESIGN FA*FB*FC

specifies an analysis of variable Y as a three way factorial experiment with factors FA, FB and FC. The levels of each factor and the degree of partitioning required are declared before the ANOVA instruction using declarations such as

LEVELS FA FB 1–4, 6 FC 14·7, 15·8, 19·2
PARTITIONS FA 1 FA FB 2

If the factors specified in the ANOVA instruction are also variables in the data matrix, the values of these variables specify the points order. Points with values of FA, FB or FC not included in the levels listed are ignored. If the factors are not variables, the points are assumed to be in standard order within the data matrix.

An ANOVA statement may be followed by a NAME statement defining a new variable in terms of the analysis of variance model. For example

NAME YF = FA + FB + FA*FB + FB*FC VARIABLE Y

specifies new variable YF as the values fitted by the model including main effects FA and FB and interactions FA*FB and FB*FC.

4. Data management instructions

In this section I describe the data management instructions available in the system including the reading of data matrices, saving them and retrieving them from magnetic tape or disc, forming new data matrices as subsets or amalgamations of other data matrices, the definition of new variables and the sorting of points. It is stressed in Section 6 and illustrated by many examples in Section 5 that much of the power of the system is rooted in the availability of a wide range of data management operations and their interaction with the statistical instructions.

4.1. *Reading a data matrix*

The basic instruction is of the form:

READ DATA MATRIX BEC 1 VARIABLES Y X1–10 Z AA AB

This causes a data matrix named BEC 1 with 14 variables Y, X1, X2, . . X10, Z, AA, AB to be read from the normal input stream. That is starting immediately after the READ instruction itself. The basic instruction may be qualified with suitable qualifying phrases to read data in a number of different arrangements in both free-field and fixed-field form. Data may be read in certain binary forms also.

It is also possible to read the correlation matrix or the covariance matrix instead of the data matrix. The relevant instructions are the same as that above with the word DATA replaced by CORRELATION or COVARIANCE as appropriate.

4.2. *Saving and fetching of matrices*

The relevant instructions include:

SAVE DATA MATRIX BEC 1 STREAM 9

FETCH DATA MATRIX BEC 2

The stream to receive data matrices is specified using:

SET UP DISC R029 AS OUTPUT STREAM 9

and this may be presented as an input stream in a later run using:

SET UP DISC R029 AS INPUT STREAM 8

Adding to an existing disc area or magnetic tape may be achieved using:

SET UP TAPE N1079 AS INPUT *AND* OUTPUT STREAM 8

4.3. *Data matrix formation*

The basic statement for the formation of a new data matrix from one or more other data matrices has the form:

FORM *DATA MATRIX* BEC4 FROM *DATA MATRICES* BEC1–3 $

VARIABLES A B C D1–7 9–12

This instruction creates a new matrix BEC4 by the concatenation of matrices BEC1, BEC2 and BEC3 and containing 14 variables. All points in all three matrices will be included in the new matrix but the points may be restricted to those that satisfy certain criteria by including statements of the form:

IF(A–B*C) CONTINUE, OMIT,OMIT,OMIT

after the FORM statement. Such statements are obeyed once for every point included in all three matrices and they control which points are to be transferred to the new matrix. In this case points for which (A–B*C) is negative will be transferred to the matrix and others will not. The IF statement is similar to that of Fortran, but it includes a fourth successor which is taken when the expression is undefined because of missing values. The acceptance criterion may be complex involving several IF statements and new variables may be created by arithmetic expression such as

$$D = A + B + C$$

included after the FORM statement. New variables so created may also be involved in IF statements. Thus variables to be included are specified in a VARIABLES phrase included with the instruction and points to be included are specified using IF statements and possibly other arithmetic statements.

4.4. *Data matrix build-up*

Data matrices may be built up to contain points from a variety of sources using statements such as:

START DATA MATRIX NEW1 VARIABLES Y X1–10

ADD POINTS FROM STREAM 2 TO DATA MATRIX NEW 1
ADD DATA MATRIX OLD TO DATA MATRIX NEW 1
ADD 'values for one point' TO DATA MATRIX NEW 1

COMPLETE DATA MATRIX NEW 1

The first of these instructions defines the framework of a data matrix, the next three add points to the matrix from a variety of sources and the last completes the matrix making it a bonefide matrix which may then be analysed by other instructions.

4.5. *Arithmetic and branching instruction*

These are of a similar form to those of Fortran and may be used to create new variables and to change existing variables. If they follow data management instructions described above, they are obeyed once for each point involved. Alternatively, they may be used to qualify the present data matrix when they are obeyed for each point contained in it.

Similar statements involving scalar quantities only (parameters, a single coefficient, or a constant) may be used to create new values for scalar quantities and to control program flow.

4.6. *Sorting*

Points in the current data matrix may be sorted into a new order using an instruction such as:

SORT ON *VARIABLES* A B *AND* C

Ascending order is assumed unless another order is specified. The points are sorted into ascending order on variable A. Points with equal values of A are sorted on B and those with equal values of A and B are sorted on C.

5. Discussion of examples

The brief description of the system given in previous sections is supplemented in this section by the discussion of 4 examples chosen to illustrate the data management aspects and to illustrate the interactive advantages of a system including both statistical and data management instructions.

5.1. *Example 1*

It is assumed that the present data matrix contains the variables X1–11. The following instructions operate on the present data matrix.

REGRESSION OF X11 ON X1
NAME FITTED VALUES X11FIT AND RESIDUALS X11RES
PLOT X11 AGAINST X1
PLOT X11FIT AGAINST X1   JOIN   SAME GRAPH
PLOT X11RES AGAINST NORMAL.

The fitted values and residuals from the regression of X11 on X1 are set up as new variables with names X11FIT and X11RES. The third and fourth instructions produce a single plot showing the scatter of original data and the regression line fitted to it. The last instruction produces a normal probability plot of the residuals and provides a useful pictorial check on the assumption made in tests applied in the regression analysis.

5.2. *Example 2*

This example assumes that the present data matrix is named BEC1 and that it contains variables X1–10.

```
SETUP DISC COMMON AS OUTPUT STREAM 9
COMPONENTS ANALYSIS USING VARIABLES X1–10
NAME COMPONENTS CA CB
PLOT CA AGAINST CB
SAVE DATA MATRIX BEC1 STREAM 9
FORM DATA MATRIX QUAD1 FROM DATA MATRIX BEC1
  IF(CA) OMIT,CONTINUE,CONTINUE,OMIT
  IF(CB) OMIT,CONTINUE,CONTINUE,OMIT
  analyses on matrix QUAD1
FORM DATA MATRIX QUAD2 FROM DATA MATRIX BEC1
  IF(CA) OMIT,CONTINUE,CONTINUE,OMIT
  IF(CB) CONTINUE,CONTINUE,OMIT,OMIT
  analyses on matrix QUAD2
```

After setting up a common disc area as output stream 9 to eventually receive data matrix BEC1 a components analysis on 10 variables is performed. The first two components are set up as new variables with names CA and CB and these are plotted by the fourth instruction. The fifth instruction saves the data matrix including new variables CA and CB on the common disc area. This is necessary because the next instruction replaces matrix BEC1 as the present data matrix and the second FORM instruction requires access to matrix BEC1. The first FORM statement and the two qualifying IF instructions create a matrix QUAD1 containing all variables in matrix BEC1 and only those points that fall in the first quadrant of the plot of the first two components CA and CB. Instructions may be placed after this and before the next FORM instruction to specify analyses on the new matrix QUAD1. The second FORM instruction and the two following IF instructions create a new matrix called QUAD2 containing those points in the second quadrant of the plot of CA and CB. Other instructions may be placed after this to specify analyses on matrix QUAD2.

This example well illustrates the interaction between instructions to produce a very useful overall analysis. The first two components are computed as new variables, they are plotted against each other and new matrices are formed for separate analysis and representing the first two quandrants in the components plot.

5.3. *Example 3*

Many users of ASCOP have used the system for its data management facilities and this example illustrates the use of the system to establish, update and interrogate a bank of experimental data. The example also illustrates the definition and call of subroutines written in the ASCOP language. The example is in three parts. The first describing the subroutine used in the other two parts. The second part describes the establishment of the bank with the first months' data; and the third part describes the typical monthly updating process.

5.3.1. Part 1 of Example 3

The subroutine described here reads one month's data and adds the month's average values for each of the six variables to a matrix named MONTHAVE. Monthly averages will accumulate in this matrix. The month's data is contained in a separate matrix appropriately named.

```
SUBROUTINE RDMONDAT(MONNAM,MONNO,YEARNO)
FORMAT FMT1(12H CHECK POINT,F5·0,4X,8HX4-6 ARE,3F10·4)
PARAMETERS P1,P2,P3,P4,P5 AND P6
READ DATA MATRIX MONNAM YEARNO VARIABLES X1-6 $
   LABEL IN POSITION 7   STREAM 4
   Y1 = ANGLE(X4)
   Y2 = X1*X2 + X3/43·2
   IF(X6—X5*X4) 1,1,2,OMIT
 2 PRINT FMT1,LABEL,X4,X5,X6
   LABEL = LABEL+1000
 1 MONTH = MONNO
   YEAR = YEARNO
PRINT FULL SUMMARIES FOR VARIABLES X1-6

   instructions for analyses on current months data
   P1 = MEAN(X1)
```

```
  P2 = MEAN(X2)
  P3 = MEAN(X3)
  P4 = MEAN(X4)
  P5 = MEAN(X5)
  P6 = MEAN(X6)
ADD P1,P2,P3,P4,P5,P6,MONNO,YEARNO TO DATA MATRIX $
  MONTHAVE
  RETURN
  END
```

This subroutine reads one month's data from Stream 4 into a data matrix given a name and number specified as arguments. The data matrix contains six variables X1–6 and for one point seven values are read from Stream 4 (probably seven values per card). The first six are the values of the variables X1–6 and the seventh is specified as the point label. The arithmetic statements and the IF statement are obeyed once for each point as they are read and four new variables Y1,Y2,MONTH and YEAR are created and given values as specified. The IF statement compares variables X6,X5 and X4 and if (X6–X5*X4) is positive, a line is printed with format FMT1 containing the point label and the values of the three variables. In addition, points with the value of X4,X5 or X6 missing are not accepted as part of the data matrix. After the data have been read, a summary containing the means, standard deviations, variances, minimum and maximum values for all 10 variables is printed followed by the correlation matrix. The remaining instructions add one point to the data matrix MONTHAVE containing the mean values of the six variables X1–6, the month number, and the year number. This matrix must, of course, be defined to the system before the subroutine is called. This is achieved by the START instruction in part 2 of this example .

5.3.2. Part 2 of Example 3

In the second part I describe the program to establish the data bank and to read the first month's data.

```
SETUP TAPE 1 AS OUTPUT STREAM 9
START DATA MATRIX MONTHAVE VARIABLES AX1–6, MONTH, $
YEAR
CALL RDMONDAT(JAN,1,70)
```

instruction for the monthly analysis of data for January 1970

```
FORM DATA MATRIX COMPDATA FROM DATA MATRIX $
  JAN 70
SAVE DATA MATRIX COMPDATA
COMPLETE AND SAVE DATA MATRIX MONTHAVE
```

Data for January 1970 is read into matrix JAN 70 and this is analysed according to the instruction included as indicated after the call instruction. The data are then transferred to a matrix called COMPDATA which will be used to accumulate the raw data. This matrix is saved on tape 1. Finally the matrix MONTHAVE which at this stage includes only one point including average values for January 70 is saved on tape 1 also. Thus at the end of this run, tape 1 will contain matrix COMPDATA with one month's raw data and the matrix MONTHAVE containing just the monthly averages for one month. This has established the beginning of the data bank on tape 1. Subsequent runs will add to this as described in the next section.

5.3.3. Part 3 of Example 3

The instructions in part 3 to add a further month's data to the data accumulated so far have been organized in a subroutine called MONANAL. By doing this the changes that have to be made from one month to the next are confined to the arguments in the call instruction.

```
SUBROUTINE MONANAL (MONNAM,MONNO,YEARNO)
PARAMETERS PA,PB
  PA = REMAINDER(MONNO,2)
  PB = 2—PA
  IF(PA) 1,1,2
1 SET UP TAPE 1 AS INPUT STREAM 8
  SET UP TAPE 2 AS OUTPUT STREAM 9
  GO TO 3
2 SET UP TAPE 2 AS INPUT STREAM 8
  SET UP TAPE 1 AS OUTPUT STREAM 9
3 CALL RDMONDAT(MONNAM,MONNO,YEARNO)
  ADD DATA MATRIX MONNAM YEARNO TO DATA MATRIX $
  COMPDATA
  COMPLETE AND SAVE DATA MATRIX COMPDATA
  analyses of data accumulated so far
COMPLETE AND SAVE DATA MATRIX MONTHAVE
```

```
  analyses of monthly averages accumulated so far
FORMAT FM1 (10H DATA FOR ,A8,34H HAS BEEN ADDED AND $
  STORED ON TAPE,F3·0)
PRINT FM1,MONNAM,PB
FORMAT FM2 (22H MONTHLY AVERAGES FOR ,A8,69H HAVE $
  BEEN ADDED TO DATA MATRIX MONTHAVE WHICH $
  IS ALSO STORED ON TAPE,F3·0)
PRINT FM2,MONNAM,PB
RETURN
END
CALL MONANAL (FEBRUARY,2,70)
```

To run the job for March 1970 the call instruction becomes

```
CALL MONANAL(MARCH,3,70)
```

That is, the only values to change from one month to the next are the arguments in the call instruction.

The first few instructions set up tapes 1 and 2 as input and output tapes in a way which depends on the month number and the last few instructions print messages to make it clear which tape is the latest version of the data accumulation. The call statement in subroutine MONANAL calls subroutine RDMONDAT to read the current month's data, to analyse this, and to add the monthly averages to the matrix MONTHAVE. The ADD instruction adds the month's data to the matrix COMPDATA and this matrix is completed and saved on the latest output tape. Analyses of the accumulated data may be inserted at this point. After this the matrix MONTHAVE is completed and saved and analyses on the monthly averages so far would be included here. The remaining instructions simply print messages explaining which tape is the updated tape.

In this example two tapes are used alternatively so that if a run fails before the updating is complete, the run can be repeated. The added security provided by using three or four tapes in cycle would also be possible with little change to the example given here.

5.4. *Example 4*

In this last example a number of different representations have been used for missing values and it is the purpose of the subroutine to read cards from a stream specified in the argument list of the subroutine and replace each occurrence of each representation by the systems proper missing value representation. The user prepares three lines of values such as:

6					
1	3	3	4	7	10
9	9	99	5	8	0

The number 6 on the first line specifies that there are 6 different representations to consider. The second and third lines specify that for

variable 1 the value 9 means missing
variable 3 the value 9 means missing
variable 3 the value 99 means missing
variable 4 the value 5 means missing
variable 7 the value 8 means missing
variable 10 the value 0 means missing

The subroutine is:

```
SUBROUTINE MVCHANGE(ST)
PARAMETERS COUNT,W
READ PARAMETER PA STREAM ST
        PA = PA-1
READ COEFFICIENTS VA PA STREAM ST
READ COEFFICIENTS MA PA STREAM ST
        BEGIN
        COUNT = 0
1       W = VA(COUNT)
        IF(VARIABLE(W) — MA(COUNT)) 3,2,3,3
2       VARIABLE(W) = MISSING
3       COUNT = COUNT + 1
        IF(COUNT-PA) 1,1,4
4       END
RETURN
END
```

The reduction by 1 of the value of PA is necessary because sets of coefficients are indexed from zero and the 5th and 6th instructions define and read sets of 6 coefficients indexed from 0 to 5 if the value of PA is 5. The value of the parameter COUNT starts at zero for the same reason. The pseudo-function VARIABLE allows access to variables by number and the word MISSING is a legal right hand side referring to the missing value symbol.

This subroutine has been used many times to identify missing value codes. It is common, particularly in surveys, to use the value 9 to denote missing for a variable taking one column on a card. The value 99 is often used for a variable taking 2 columns and so on. Thus a number of different values are used and a simple recognition and substitution procedure such as that described here is very useful.

6. Conclusion

The examples described above were chosen to illustrate the multi-stage nature of data analysis and management. It is vitally important that the burden of data management left on the shoulders of the user by so many packages of programs should be recognized and made as light as possible. It has always been a driving force in the development of ASCOP that the user should be able to store and retrieve data and form new structures with ease. If the user has to devote much of his attention to organizing data for himself he has less to give to the real problem of analysis. Thus it is argued that a system for the analysis of data must contain a wide and easily used range of data management facilities.

References

COOPER B.E. (1969) ASCOP *User Manual.* National Computing Centre.
COOPER B.E. (in press) ASCOP (Version 3) *User Manual.* National Computing Centre.

Discussion of the application of mathematical models to a specific ecological problem

The last morning of the Symposium was devoted to an experiment. The basic idea was that a specific ecological problem should be outlined to the participants, who would then be asked to propose possible formulations of mathematical models which might be capable of solving the problem, and to discuss the relevance of mathematical models to the solution of problems of the kind proposed. It was hoped, in this way, to obtain further insight into the application of mathematical models to ecology, and to achieve a synthesis of the basic concepts which had been advanced by the individual speakers at the Symposium. To avoid any chance of contributors to the discussion taking up entrenched points of view, the specific ecological problem to be discussed was not revealed until the start of the discussion. It was recognized that there was some risk that a useful discussion would fail to materialize, and plans were made for an alternative topic for discussion in this eventuality. The alternative topic was, however, not required.

The ecological problem for discussion was introduced by Mr J.N.R. Jeffers, Merlewood Research Station. It was based on the recently completed feasibility study of proposals to build a barrage or, alternatively, pumped storage reservoirs on Morecambe Bay, part of which was visible from the hotel in which the Symposium was held. Morecambe Bay consists of a large intertidal area of mudflats, sand, and saltings which are rich in wildlife and biologically highly productive. The various proposals to use the Bay and the rivers which flow into it as a source for water, principally for Lancashire towns such as Manchester, would result in extensive changes to a complex ecosystem, and some of these changes were outlined. If the complete barrage were built, the present marine and brackish water ecosystems would be converted into a freshwater system on the landward side of the barrage, with drastic changes in the deposition of silt on the seaward side of the barrage. The various proposals for pumped water storage would have similar, though less marked, implications. Apart from the changes in the rich invertebrate fauna of the Bay, any of the proposals would have marked effects on the large populations of waders and wildfowl which use the Bay for breeding or over-wintering. In part, these changes would result from loss of feeding grounds, but the loss of roosting and nesting sites might also be severe. The salt-marsh grazing around the shores of the Bay was an important resource for local sheep farmers, and would be lost through any full barrage scheme. It is, however, important to note that marked changes have taken place in the Bay, in the absence of any deliberate attempts to manipulate the ecosystem, and it is far from clear what future changes may take place.

The above points represent only a very brief summary of the main ecological consequences presented to the participants as an outline of the problem. From these, the direct question was posed: 'Is it possible to construct a mathematical model of the Morecambe Bay ecosystem, from which predictions could be made about future changes in the ecosystem whether or not a barrage, or some similar scheme, was constructed?' The following notes of the discussion are taken mainly from an account prepared by Dr O.W.Heal, Merlewood Research Station. The various points made are not given in the order they occurred in the discussion, but have been sorted into topics.

1. Why model and is it feasible?

What can we hope to obtain from a mathematical model of Morecambe Bay that we do not already know? It is probably simpler to continue to use the human brain which is the best computer we have available. Can any mathematical model be any better? It was pointed out that one possible reason for developing a mathematical model of the ecological situation was to provide arguments of equivalent credibility to the physical models already developed by engineers. This objective was later criticized, the feeling being that there were better ecological reasons for models and that respectability was not a valid reason for the development of the ecological model. It was also suggested that modelling of Morecambe Bay was a very appropriate problem to consider because it puts the biologists into a situation where a decision must be made, and that the proposals for a barrage provide a focal point for the model development. However, the importance of the reason for the development of a mathematical model was stressed. If the object of the exercise is to counter the proposal for a barrage, do we undertake the modelling on the engineers' terms, or on other terms? In particular, can we develop an economic model, and one which includes the effects of agriculture? Can we build a model which includes the cost and benefit of conservation of the Knot and Bar-tailed Godwit? It was felt that the latter was impossible, because it is impossible to compare geese and churches in economic values, as had been attempted in the studies on the third airport in London.

It was suggested that mathematical models offer a much wider range of ideas to those who ultimately have to make decisions, and highlight features which they would never normally take into account. An example of this was given from India, where the development of new strains of rice have led to higher yields of rice production which simply put money into landowners' pockets where it was intended to develop more efficient agriculture. In another example, irrigation had caused salt to be brought to the surface of the soil where it had subsequently killed the plants. It is this type of unexpected situation upon which the model can focus attention. A further point was made that such models can provide data with which to compare other possibilities, e.g. the biological or social changes around Morecambe Bay. Another example was mentioned of the opportunity of models to predict both the unexpected and the expected, i.e. that of the Salvinia development in the Khariba Dam, which probably arose from the considerable enrichment of the water after irrigation. Many techniques were tried to kill and remove the Salvinia but the nutrient depletion of the water eventually killed the Salvinia without any need for human intervention and this situation could have been predicted.

2. Range of models

Several speakers referred to the need to produce a series of models and it was suggested that it was necessary to produce separate chemical, physical, biological and verbal models. It was also stressed that it was necessary to produce at least two specific models, i.e. with and without a barrage. It was also essential to include the economic and social effects which would arise from various developments in the Bay, again either with or without a barrage. The barrage on Morecambe Bay would be of very little value in the development of open water, of little value in the development of harbours with locks; there are no opportunities for the development of power, and the improvement in transport from Morecambe to Barrow by car is of uncertain value, so that the barrage developments must be viewed in terms of production of water only, and this from a national context.

The need for an ecological model to be set alongside those of the engineering and economic models was emphasized by several participants and many felt this was the main aspect which should be developed, but, again, it was stressed that any ecological model must be viewed in a series of contexts, depending on the organisms involved, and that the development of local, regional, national, and even international, models concerned with wader populations was an important aspect.

3. Points of detail

Several specific points of detail were requested, for example, the size of area and depth of water if the barrage was developed, the depth of the water outside the barrage, the daily, seasonal and annual changes which occur at present in Morecambe Bay. It was mentioned that the cost of the feasibility study was in the order of £0·5 million, and that the barrage itself would cost about £84 million to construct. It was, therefore, implicit in the discussion that financial limitations were an important consideration in the development of the model or models. It was suggested that the ecological stiuation was several orders of magnitude more difficult than the engineering situation, and that the funds available for modelling the ecology should be equivalent to or greater than the funds available for physical studies, whereas at the moment they are considerably less. Strong emphasis was placed on the fact that much of the existing experience in ecological modelling has been carried out on static ecosystems, while Morecambe Bay was a dynamic ecosystem.

4. Problems of modelling

Some of the problems envisaged have already been mentioned, but it was suggested that many of the ecological changes were likely to be drastic, and there was some doubt as to whether it was possible to include such qualitative changes in an ecosystem model. In studies of other dams (e.g. the Aswan dam), the physical or engineering changes could be modelled reasonably well, but it had been less easy to predict the more subtle changes which might result from the development of a dam. Several examples of such changes were given, including the possibility of liverfluke development in sheep unable to graze on saltings. The problem of putting a financial estimate on the conservation value of birds was mentioned again in this context.

Several speakers mentioned the possibility of obtaining useful data from other examples of estuarine and low-level dams, e.g. the Aswan and Khariba dams, the studies in the Bristol Channel, and in Holland. It was suggested that valuable data could be obtained by examining other estuaries in Great Britain in which changes which might occur in Morecambe Bay could be identified. The present and future patterns of Morecambe Bay can be specified and we could look in other estuaries to find the biological relationships between these patterns. It was also mentioned that good physical and chemical data were available from the structural and water resource engineers.

At this point in the discussion, Professor Van Dyne presented an outline formulation of a model framework of the ecosystem of Morecambe Bay. Although he had had no advance information of the topic of the discussion, Professor Van Dyne had sketched out his model framework during the discussion, and it was necessarily tentative, but he advocated a confident and vigorous approach to ecological models, as opposed to the hesitant and rigorous approach that had so far been proposed.

He suggested that the estuary ecosystem is best approached as a series of sub-systems or sub-areas. The framework of this series of sub-systems is defined by the following variables:

let $i = 1 \ldots\ldots n$ areas
$j = 1 \ldots\ldots m$ time periods
x_{ij} = state variables in area i in time period j
z_{ij} = inputs in area i in period j
y_{ij} = outputs.

The variables can be further identified as

1. *State variables*

x mean channel volume
water volume (depth)
water chemical load (disolved)
silt and suspended solids
birds
plants
animals
decomposer activity

(birds, plants, animals, decomposer activity: Numbers/ biomass/ indexes, e.g. transmission of water)

(Include 'empty or null' variables to allow for invasions)

2. *Inputs*

z · to each system from 'outside'
· to each system from other systems
fresh water
salt water
pollutants
particulates
climatic parameters
temperature
wind
organism supply

3. *Outputs*

y fish
sheep
birds
evaporation
net water flow
net particulate flow

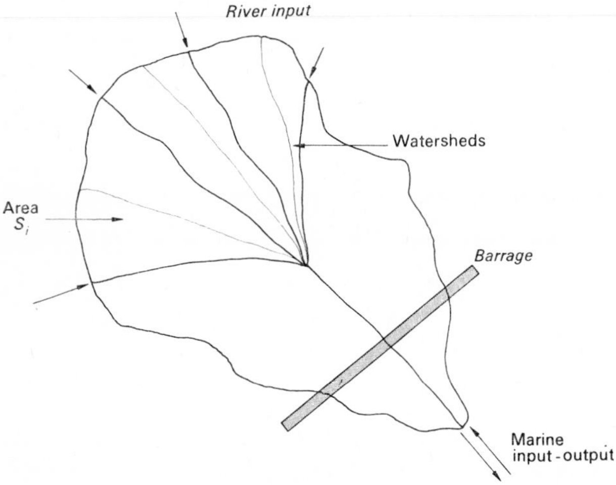

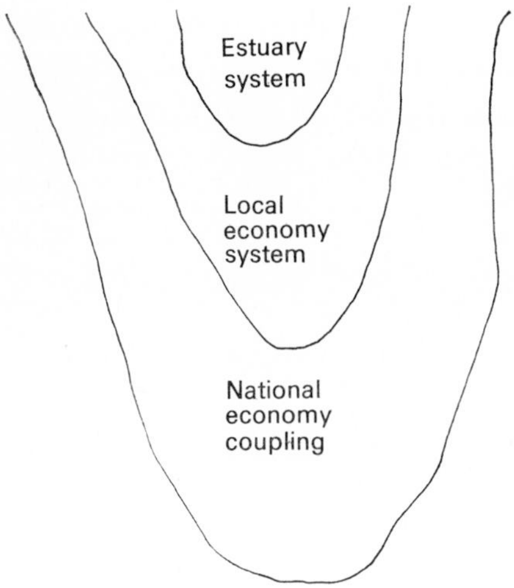

He proposed that the dynamic framework of the model should be represented by sets of differential equations, i.e.

$$x_i = f_1\, x_i + f_2\, z_i$$
$$y_i = f_3\, x_i + f_4\, z_i$$
$$Y = f_5\, y_i$$

In these equations, the functions or couplings were represented by the f_i, to indicate:

- who eats whom matrices
- time lag effects
  - population change
  - particulate/water movement
  - erosive forces
- barrier effects on f_i

Within this framework, he proposed a regional—national system optimization which would maximize economic output, i.e.

$$\underset{Q}{\text{Max}} = \sum_{i}^{\text{areas}} \sum_{j}^{\text{periods}} \frac{C_{ij}\, V_i}{(1 + p)^r} - \text{barrier}$$

or

$$\underset{Q}{\text{Max}} = \simeq \int_0^t \sum_i^a \frac{C_{ij}\, V_i}{(1 + p)^r} - \text{barrier}$$

where $C_{ij} = i \pm$ costs of each V_i
V_i = selected x_i
selected y_i

Subject to the following constraints:

$$x\ \text{min}_i \leqslant x_i \leqslant x\ \text{max}_i$$

$$x_i = f_i\,(x_j)\ \dot{C} \neq j$$
$$V_{it} = f(V_{i\,(t-1)})$$

The search for the optimum values would require the following stages:

(a) Solve $\dot{x}$ with given parameter settings

(b) Use x_i and y_i to calculate v_i at each period

(c) Find $\dfrac{\delta \text{Quax}}{\delta p_{ii}}$ at least set $p_{ij} = \begin{matrix} \text{max } p_{ij} \\ \text{min } p_{ij} \\ \text{mean } p_{ij} \end{matrix}$

$$\frac{\delta \text{Quax}}{\delta C_{ij}} \qquad \frac{\delta \text{Quax}}{\delta \text{barrier}}$$

$$\frac{\delta \text{Quax}}{\delta r}$$

It should be possible to quantify the argument for and against the barrage, from the manipulation of such a model. Further considerations were that:

(i) In the model, the variables were all commensurate with one another; it should be possible to test the effect of qualitative evaluations of certain situations.

(ii) It was essentially an optimization model, which has the advantage of presenting two solutions, i.e. one that maximizes profit and the other that minimizes loss. It would be possible to find out which variables are constraining the system; then a concentrated effort could be made to improve the evaluation of those variables which are constraining the solution.

It was agreed that Professor Van Dyne's model provided an excellent framework, but it assumed that all the variables are measurable and quantifiable and could be put into relevant units. Professor Van Dyne replied that the model would allow us to test what effect various levels of costing would have on the outcome. It was also indicated, from further discussion, that not all the parameters would have equal effects upon the outcome of the model.

The 'black box' problem was discussed by several participants. Many elements were included within the 'black box', but it was often the case that the input and output remained the same, despite the fact that the time period and frequency were important facets. Many of the variables are constant or may be regarded as constant, because they react far too quickly to affect the scale of the model (just as the lights in a room are flickering at a very rapid rate, but this flickering is unimportant in the context of normal illumination) and it is sufficient in many ecological situations to regard such variables as being held at a steady state.

One participant suggested the desirability of relating populations of birds to their nesting site availability, rather than to food supply, but it was thought that such factors could be incorporated readily into the model, including the number of birds and the critical population density which may or may not be affected by food or nesting sites.

Two kinds of uncertainty were mentioned. These included uncertainty or error in the initial measurements, and the uncertainty of a small area of bare ground being colonized by one of many types of organisms. Both would produce a series of possible outcomes, each with a low probability. Would such a wide range of outcomes from the model invalidate or make less useful the mathematical model? Professor Van Dyne answered that, where there is uncertainty in the parameters and the functional forms of the relationship, it would be useful to examine their additive and compensating characteristics through direct simulation.

Uncertainty of protection was emphasized by other participants. The example of Moran, who in 1952–3 developed a simple model simulating the life cycle of the mink, was cited, for which the predictions were poorer the further ahead they were attempted, with the result that there was a widening confidence interval for the estimates. It was felt, however, that this problem could be overcome by more recent developments in mathematical formulation. This led into discussion of meteorological models. There were uncertainty and observational errors in most of our ecological measurements, whereas the meteorologists were able to measure with great accuracy. Despite this accuracy, meteorologists' predictions were generally poor, partly, it was thought, because many small errors would accumulate to bias the predictions. It was emphasized that the meteorological situation was several orders of magnitude less difficult than the ecological situation. Professor Goodall suggested that, although he was conscious of the inadequacies of many of our attempts to model ecology and of the problems of accumulation of errors, he was optimistic, despite the high order of magnitude of difficulty. The reason for his optimism was that ecological systems are highly cybernetic, with feedback systems, and these provide much better opportunities for predicting. Other speakers agreed with Professor Goodall on this, and felt that ecosystem studies had already shown that predictions were possible after catastrophic occurrences, although intermediate situations were sometimes much more difficult to estimate.

An alternative model was suggested by Dr Hughes. This model was based on a more limited biological aspect, in which the Bay was divided into a series of spatial compartments to be characterized and classified on qualitative and quantitative attributes of the species present. They would also be characterized by a set of physical and chemical variables. A series of triggers would be set in the model to connect these variables, for example as a biological community being switched in response to chemical or physical changes. The model is limited in that experience and data need to be provided from outside, but it avoids the need to build complete 'who eats whom' matrices and it summarizes our present physiological knowledge. It was admitted that it revealed few of the scientific features but, on a limited budget, it might be a valid alternative. The most serious criticism of the model was that it did not allow any feedback, in which physical features would be altered as a result of biological changes.

In summing up the discussion, Mr Jeffers said that he was agreeably surprised at the success of the experiment. His objective had been to obtain some feeling among the participants for the modelling philosophy, and he believed that this had been achieved. One of his criticisms of much ecological research is that the assumptions made are seldom defined, and a major advantage of modelling is that it focuses attention on the definition of such assumptions. As scientists, we are working in the real world, and are in a strong position to influence decisions about that world. We must be particularly careful, therefore, to separate our scientific responsibilities from our civic responsibilities. The comparison between answers derived from modelling and from informed opinion had been stressed in the discussion, and we could not expect decision-makers to make correct value analyses without adequate quantitative information. Too much of scientific work is involved with closed systems, especially in ecology, but the methods discussed had shown the way to break out of such systems. The Symposium as a whole represented a new initiative for the British Ecological Society, and suggestions would be welcomed for continuing that initiative, within the British Ecological Society or in conjunction with any other society, e.g. the Biometric or the Royal Statistical Society, in developing the application of mathematical models to ecology.

Summary and assessment: a research director's point of view

J.N.R.JEFFERS *Nature Conservancy, Merlewood Research Station, Grange-over-Sands, Lancashire*

In listening to the presentations at this Symposium, and to the discussions, formal and informal, which followed them, I have been acutely aware of the implications of what has been said for the director of a research team, that is, for someone who is himself a scientist, but for whom at least part of his time is occupied by the management and administration of other scientists. I cannot, of course, claim that my views on the importance of mathematical models to a research director are unbiased, because many of the papers deal with topics which fall directly within my own field of research interest. For those who view almost any quantitative technique with suspicion, if not dislike, mention of this bias may be sufficient to enable them to ease their consciences by discounting anything I may say as an assessment of this Symposium!

For this reason, I shall base my assessment on what I believe to be the three fundamental divisions of the responsibilities of any research director, in the hope that we can at least start from a common point of agreement on the functional role of the director. These responsibilities may be summarized as follows:

1. his responsibilities to science, national and international, as represented by the disciplines of the scientists under his direction;
2. his responsibility to the organization which provides the resources enabling research to be undertaken;
3. his responsibility to the scientists themselves, as individuals.

It seem to me that much of what we have heard and discussed during the last few days can be viewed from the point of view of these responsibilities, and that conclusions can be drawn which touch upon the central problems of the organization, management, and administration of science.

In my introduction to the Symposium, I commented upon the inefficiency of much of our scientific activity, inefficiency arising from bad design of experiments and surveys, ineffective, and often invalid, analysis of data, and the failure to use techniques which are available for the solution of practical problems. Skellam, in his masterly and delightful consideration of the philosophical problems in the use of models in relation to meaning, has highlighted many of the difficulties at the conceptual levels of scientific research. Such difficulties are not often discussed, but they reveal the fact that science is not going to be transformed merely by the widespread introduction of mathematical models. If mathematical models are to be used, they have to be appropriate, and the scientists who use them have to be aware that such models are no different from any other form of hypothesis, and must be rejected, or at least modified, as soon as evidence is available to show that they do not fit with observations. Nevertheless, it is clear, from some of the examples presented at this Symposium alone, for example by Usher, Austin, Lambert, Conway, Plinston and Radford, that there are already families of models which extend our conceptual thinking about the problems that confront us.

The conceptual models now available to us through mathematics are essentially those of the 20th century. Our responsibility, as scientists, to advance beyond the conceptual models of the eighteenth and nineteenth centuries, and the responsibility of the director of scientific work to see that this happens, is not related solely to our wish to penetrate more deeply into the unknown areas of knowledge. As the competition for limited scientific resources becomes more intense, we owe it to the scientific community at large that we do not waste our own share of these resources, or make a bid for a larger share than is justified by the importance, tangible or intangible, of the research we undertake. As the number of scientists increases, the competition for resources also increases, and even a small group of scientists, working ineffectively or inefficiently with inadequate conceptual models, can reduce the total impact of the available resources.

In this connection, it is clear that we cannot ignore the value of mathematical models for unifying the research of large numbers of separate individuals and separate disciplines, as described by Van Dyne and Goodall. The clear definition of conceptual models, and the clarity achieved by the abstract nature of mathematics, all help to centralize effort from a multidisciplinary team towards the common ends of the research. Faced with these advantages, it is difficult to see how a research director can fail to be impressed by the opportunities provided by mathematical models in obtaining a clear and consistent strategy towards the solution of practical problems in ecology. The implications of the first revolution in mathematics were mainly in reducing the amount of time spent by scientists in work which was improperly planned and from which the consequences were invalidly drawn. The implications of the second revolution in the use of mathematics in modern science are related to the ability to concentrate enormous scientific power through the strategic use of conceptual models which have a mathematical basis.

The director's responsibility to the organization which provides the resources for his research team is similar, in some respects, to his responsibilities to science as a whole. The wastage of scientific effort through ineffective research, or through conceptual models which are not sufficiently advanced to develop the subject of the research, may be as important as wastage through more easily recognized sources, for example loss or misuse of scientific equipment, arriving late or leaving early, or even direct embezzlement of research funds. Yet the loss is just as severe and may be even more crippling than direct larceny of public funds! So, too, is the failure to adopt techniques which are available when those techniques are necessary to solve a practical problem effectively, and, as in law, ignorance is no excuse.

One of the impressive features of many of the papers presented at this Symposium is the fact that each of the models discussed belongs to a family of similar models. This is perhaps most clearly shown in the paper by Usher where a whole series of parallel developments has taken place from the formulation of a basic data matrix. Anyone who has listened to this presentation, or read this paper, will now recognize those situations for which this family of models is appropriate. There have been many other similar examples in the papers presented at the Symposium. The details of the application of a particular model may require considerable help from experts, whether mathematicians or ecologists, but the recognition of the basic family to which the model belongs is something which can be done with a minimum of mathematical training. For the research director, it is clear that one of his special tasks must be in the selection of appropriate families of models for the research which is undertaken by his organization, and in discussing the possibilities for future research with those for whom the research is being undertaken. This concept should not be too difficult for the ecologist, who is after all used to recognizing the families of flowering plants as a prerequisite to the recognition of the individual species or variety, but the concept of families of mathematical models may perhaps be new to many ecologists.

In considering the responsibilities of the research director to his organization, we must also take into account the necessary investment in expertise and in capital equipment to enable mathematical models to be introduced at a practical level in a research organization. A correct balance in the level of investment in staff, expertise and capital equipment is his direct responsibility, and will greatly affect the efficiency of the total organization. Several of the papers at this Symposium have stressed the importance of expertise in various mathematical aspects of modelling, for example the paper by Plinston has stressed the importance of parameter sensitivity and the interdependence of parameters, as well as the need to consider methods for improving the efficiency of optimization in the choice of parameters, all of which require some considerable knowledge of numerical procedures. While, as Goudriaan and de Wit have stressed, special languages have been developed for simulation of ecological systems, the use of such languages is partly dependent upon expert knowledge of mathematical relationships. Ross has demonstrated, in his paper, the importance of a correct approach to the fitting of stochastic models, which again is dependent upon expert knowledge of the underlying mathematics of such models. Radford has shown that the use of simulation languages written in DYNAMO, FORTRAN and CSMP may assist the novice model builder without restricting the future program sophistication, but has also, I think, illustrated that the use of such sophisticated languages also requires a parallel degree of sophistication in mathematical knowledge.

Much of the required expertise will be provided through the use of consultants, but, as I have emphasized in my introduction, relatively few organizations have found it possible to make effective use of consultant mathematicians and statisticians. I make again my plea for the research director to consider this aspect of the organization of the research under his control, as a direct function of his responsibility to the organization which provides the research resources. The solution may lie in the type of programme described by Van Dyne, where the mathematicians, statisticians and computer experts become part of a multidisciplinary team. I suspect that the only practical solution is for the research director himself to have some direct knowledge in one or more of these special fields. They are, after all, particularly relevant to the choice of research strategies and the assessment of the value of research tactics.

Relatively little emphasis has been given, in this Symposium, to the need for computers in the use of mathematical models, although this has at least been implicit in many of the papers presented. That data management, in terms of the actual handling, editing, and storage of data, is one of the most time-consuming parts of the use of mathematical models has been emphasized by Cooper, whose experience in this field has led him to develop a special statistical and data management system which takes into account the importance of these aspects of statistical analysis and mathematical model building. As Gore pointed out, even relatively complex simulations do not necessarily require large computers, but the computing equipment must be adequate for the research undertaken by the organization, and it is the research director's responsibility to ensure that the necessary investment is made in such equipment. He must also ensure that the management of computing services is sufficiently effective for the staff to be able to use the computer as a practical working tool. The analogue computer has perhaps been neglected in recent years as a relatively cheap and convenient method of obtaining results from mathematical models formulated as differential equations. The paper by

Denmead, and his demonstration, helped, I think, to show that the use of such computers is not as formidable as many of us have pretended, and the development of the hybrid computer may well influence future applications of mathematical models within ecology.

Finally, the responsibility of the director to the individual scientists must be related to their training to take their place in the scientific world of the future. Can we honestly say that we are fulfilling this responsibility, if we continue to encourage ecologists to pretend that they do not need mathematics as a basic discipline? There are many ecologists who do take this view and who, indeed, became ecologists because they wanted to escape the formalism of the discipline of mathematics. It will be relatively easy to allow them to continue to ignore the rapid and recent developments in the use of mathematical models, and so effectively isolate themselves from the important developments of the next ten or twenty years. I believe nevertheless that one of the important duties of every research director must be to encourage the training or retraining of the scientists for whom he is responsible towards a better understanding of numerical methods. Many scientists have never had any adequate training in mathematics since they left school, and are inhibited from further contact with the subject. Most, if not all, of these scientists respond readily to training, if this training is given with understanding of the difficulties which the non-mathematician sees in the formalization of concepts, and if the scientist concerned understands that the research director is himself aware of the need for such training. Similarly, it is the research director's responsibility to see that the experts in mathematical techniques and in computing techniques receive the necessary training in biology and ecology to enable them to communicate effectively with their colleagues. Both of these aspects require adequate and well-conceived training schemes.

The considerations, promoted by the papers and discussions at this Symposium, all confirm my view that, if the research director is to fulfil his responsibilities, he cannot ignore the developments of mathematical models in ecology.

Summary and assessment: an agricultural research scientist's point of view

N.R.BROCKINGTON *Grassland Research Institute, Hurley*

A farmer is essentially a manager of whole biological systems: he has a holistic view of biology from necessity. The agricultural ecologist adopts a similar viewpoint because he is interested in the interrelationships between plants and animals and their environment; he appreciates that any one component of a system cannot be described properly in isolation. This attitude brings both the farmer and the ecologist face to face with the same problem, the complexity of biological systems. Analytical research confronts them with more and more information on the individual components of these systems and the urgent need in applied research is to make sense and use of all these data.

This excellent and wide-ranging symposium has served to strengthen my personal belief that a range of powerful and appropriate tools is available to us for this task in the form of mathematical models. It has been well demonstrated, if that was necessary, that mathematics has the important advantages of brevity, precision and ease of manipulation, for the job of describing and studying complex ecological and agricultural systems. In his opening address, Mr Jeffers stated that the challenge to ecologists is whether they will be willing to take advantage of these tools. I wish to return to his interesting point that this challenge should be particularly related to research strategies, rather than only the tactics of research. Meanwhile, accepting the general proposition that mathematical models can provide the tools we need, there remains a number of important questions around which many of the papers and discussions have centred. Which tools are most appropriate for particular jobs? How does one use them? What, precisely, can they do and what can they not do? Finally, what are the specific risks and dangers inherent in their use?

The choice of tools

One may be sure that where two or more model-builders are gathered together a spirited discussion of alternative techniques will occur! The symposium has confirmed this general rule, but in a new and rapidly developing field of work it is both necessary and valuable to examine the tools of the trade in some detail, and I believe we have had our thoughts clarified and our horizons widened in this respect. The attention given to 'event-orientated' formulations which do not appear to have received the consideration they deserve in biology, has been especially valuable. Similarly the use of models incorporatiug stochastic variation has been rightly emphasized.

The importance of models being understandable to field biologists has been emphasized by a number of speakers, and this is particularly relevant when considering which tools to use. I hold unashamedly to the view that simplicity of operation is one of the most important criteria. The biologist is employed to solve biological problems, and the simpler the tools he has to use and to explain to his colleagues, the better. If he is obliged to spend much of his time on the mechanics of using a tool, he will be in danger of losing sight of the original problem. If he asks an 'expert' to do the whole job for him, this can lead to difficulties of communication, misunderstandings and errors. The examples which have been described of successful team projects, involving both biologists and mathematicians, are very encouraging; but simplicity of formulation must be equally, if not more, important in these circumstances.

The problem-orientated simulation languages of which we have heard are important if one is seeking for simplicity. The suggestion that the special attributes of analogue computers may be harnessed via digital machines, without being directly involved in the complexities of the electrical circuits, is also relevant.

Using the tools

The fact that models are, by definition, simplifications of reality has been a recurrent theme of the symposium. Nevertheless, it is worth reiterating that if a model were to be an accurate representation of reality, complete in every detail, then clearly it would be essentially equivalent to the real object. Because it faithfully mimicked every detail it would be no less complicated, no less difficult to understand or to manage. This truism illustrates the principle that simplification is not only usual but essential.

Apart from the practical considerations of whether it is feasible, with the technical resources that are available, to build complete models of large and complicated biological systems, there appears to be some danger of creating 'monsters'. Even the creator of one of these may not fully comprehend it, and could end up simply pushing buttons and producing results which cannot be used sensibly.

There is an equal danger, at the other extreme, of over-simplification, leading to models which are easy to understand but which are valueless or even positively misleading.

To arrive at a satisfactory compromise, the importance of a clear definition of the purpose of a model has been noted by a number of speakers. A vague definition of the objective on the lines of wanting to understand the system is not sufficient, and can only result in trying to build a model which is all things to all men. Even if it is successful in the technical sense, such a large and complicated formulation will be so cumbersome as to be virtually unusable. A limited, and clearly defined objective is essential in order to judge how much simplification can be tolerated for that particular purpose.

An integral part of the process of deciding how much detail can be sacrificed in the interests of clarity and usability is to check whether the behaviour of the model is in sufficiently close agreement with that of the real system; in other words to test or validate the model. Clearly this cannot be attempted without a closely defined objective for the modelling exercise.

A major problem in model testing which has emerged in a number of discussions is the danger of 'circularity'. In the simplest case, the trap for the unwary model builder is that he will be tempted just to confirm his arithmetic, or the computer's arithmetic, by testing the model with the identical data which have been used in its construction. De Wit's rigorous formula for avoiding this problem[1] involves constructing the model from data at a different, more detailed, level of understanding from that at which it is tested e.g. the use of data from physiological experiments under controlled conditions to model crop growth in the field. The corollary of this is that, when there is a discrepancy between predicted and measured behaviour of the system, one resorts to further experimentation at the detailed level of understanding to improve the model. In this way, the independence of the data used in construction and testing can be guaranteed. However, this is perhaps an ideal solution, which cannot be achieved invariably in practice. It has been suggested that an alternative solution may be to draw the data for construction and testing from the same level of understanding, perhaps from the same field situation, but on different occasions in time e.g. once data has been accumulated for, say, one growing season it may be used legitimately to predict model behaviour in subsequent seasons. While there may be some doubt as to the philosophical validity of this procedure, perhaps a more important question is whether it is an efficient use of available resources in practical, day-to-day model building.

Attempting to answer the more specific question, how to use models in applied agricultural research, the choice of an appropriate degree of detail in the formulation is especially relevant. Much agricultural research has relied on simple input/output experiments in the field. The main justification for this approach is that it ensures, as far as possible, that the results can be directly applied in farming practice, since the trials are carried out under field conditions which closely resemble those on commercial farms.

Alternatively, more detailed experiments can be carried out, in the laboratory, greenhouse or animal house, on specific components and relationships within an agricultural system. In this case, one has much more control of the experimental material and the treatments imposed on it, and the hope is that such work may help in elucidating the mechanisms which contribute to the overall behaviour of the system.

Both these approaches appear to have inherent limitations in applied research. In field experiments, it is seldom possible to do more than measure the treatment inputs and the gross outputs in terms like final crop yield or animal weight changes. The results can only indicate what happened in one set of ill-defined circumstances and not why or how: one has a 'black-box' situation. If the farmer is to manage the system efficiently, he needs to know the effects of the many possible treatment variations and combinations, under varying conditions of climate and weather, soil type, etc. But, with the sort of information which the field experiment produces, it is dangerous to attempt to extrapolate, or even interpolate. Consequently, a vast number of field experiments is necessary to cover all the possible treatment and site variations. This is not to belittle the very considerable progress which has been made in scientific agriculture using this technique, simply to recognize its inherent limitations.

The use of more detailed experiments under controlled conditions is equally open to criticism in an applied context. A blanket justification that it is aimed at understanding how all the bits and pieces in a system function is not good enough: again a very large programme of work will be necessary to cover them all, and there is no guarantee that one of the most vital cogs in the whole machine will not be among the last to be investigated.

Within a particular specialist discipline it is not difficult to see how mathematical models can assist in formulating hypotheses in a precise and convenient way; further, they can provide a useful framework for investigating the components in a systematic fashion. But agriculture, like ecology, is essentially multidisciplinary. Although it may be more difficult, the application of mathematical models in multidisciplinary studies, their strategic use in Mr Jeffers' words, has a much greater potential reward than within single disciplines. In agricultural research, models could form the essential link between the agronomist in the field and the specialist in the laboratory. The field man must split up his black box and find out what is going on within it: the specialist needs a more realistically defined purpose than simply understanding any or all of the mechanisms involved in isolation. Both these needs could be met by a common mathematical model, providing a common framework for the development of a viable team effort.

I believe we are just beginning to appreciate the strategic use of mathematical models in agricultural research to make multidisciplinary projects possible. As an agricultural ecologist, my definition of the challenge of mathematics is 'Can we make use of modelling in a multidisciplinary context?' Shall we be bold enough to change our ways of thought, and our organizational structures as necessary, to take advantage of this opportunity?

I am encouraged by the enthusiasm of the participants, and the valuable exchanges of ideas at this symposium with its ecological theme, which I translate as a multidisciplinary approach.

Reference

1. DE WIT C.T. (1971) Dynamic concepts in biology. *In* Proc. of A.R.C. Symp. *The Use Models in Agricultural and Biological Research*. Ed. J.G.W. Jones.

Summary and assessment: a statistician's point of view

J.A.NELDER *Rothamsted Experimental Station, Harpenden*

Introduction

This symposium has given me, as a statistician, a good perspective of present quantitative large-scale biology, and because this is entitled 'a statistician's point of view' I shall try in return to give back a perspective of statistics and its place in our efforts. Naturally I believe that its place is an important one. The headings I have used are

Model specification
Data analysis
Experimental design
Computing software

all essential components of the model-building process in 1971. Before looking at these, I must briefly outline the subject matter of statistics as I see it.

What is statistics?

Fisher's famous paper of 1922, which quantified information almost half a century ago, may be taken as the fountainhead from which developed a flow of statistical papers, soon to become a flood. This flood, as most floods, contains flotsam much of which, unfortunately, has come to rest in many text books. Everyone will have his own pet assortment of flotsam; mine include most of the theory of significance testing, including multiple-comparison tests, and non-parametric statistics. Wading into the floodwater, we can see two component streams; one derives from the methods of mathematics, in which axioms are stated, definitions set up and deductions made. The ecologist will notice that this process is self-fuelling and requires no data as input. The other stream, which largely derives from Fisher, begins with the data, and its thinking has the data always in mind. The trouble with mathematics, as Russell's aphorism tells us, is that it is '. . . the subject in which we never know what we are talking about, nor whether if what we are saying is true.' In other words, mathematics is a formalism that need not be relevant to the real world. Now in my version, statistics is a subject in which we *do* know what we are talking about, and we *do* care that what we say, if not guaranteed true, is at least well corroborated by true data. As I see it, statistics is a pattern-matching activity. We match data, with their pattern blurred and roughened by uncontrolled variation, against the clean, pure patterns from our theoretical library. We have heard a great deal about how these pattern books come to be written and about the techniques involved. Three stages are discernable in this process of model specification.

Model specification

The outline begins with a *topological* specification—important elements are identified and postulated to affect each other according to a graph; no quantitative elements enter here. Next, we add the *algebraic* specification—and postulate that this relationship has this form with these parameters; a shape emerges but no sizes. Lastly comes the *arithmetic* specification—these parameters have these values.

As we go down this sequence, the statistician becomes more and more involved. I shall say little about the topology of models, except to note a snag about the tree as a logical structure. Trees have appeared explicitly in classification problems (Lambert) and been implied in talk of hierarchies of models. Now, by definition, trees contain no closed loops and therefore seem to exclude feedback, an essential component of biological systems. Note also that, if we look for a tree in a classification analysis, we shall certainly find one, even if the real pattern is a web.

At the algebraic level of model construction, much dissatisfaction has been expressed with regression-type models, mostly on the grounds that they are purely empirical. In my view all models are empirical, from a simple linear relation to the general theory of relativity, and I suspect that what is wrong is rather that they are *static*, when a dynamic system is required. However we need not despair: the Box-Jenkins models for time series have recently appeared (1970). They extend regression analysis to dynamic systems and the estimation procedures are eminently computable. They release the analysis of time series from the crippling assumption of stationarity, a property shown by few biological series. Though developed for economic analysis, their scope is much wider.

Finally, at the arithmetic level, we reach the greatest involvement of the statistician, in making quantitative inferences from data considered in relation to one or more models (theories). I want to consider the current state of our tools for making inferences.

Data analysis

Likelihood

A central part is played in inference by the *likelihood function*, first defined and explored by Fisher. It is the pivotal function that joins the data to the model, defined as

$$\text{Probability (data given the model)}$$

and is to be thought of as a function of the parameters of the model with the data fixed.

There is no implication of repeated sampling here, of considering sets of data we might have got, but didn't. The likelihood ratio for two different sets of parameter values defines their relative plausibility (Edwards 1969).

Note that the use of likelihood implies that stochastic elements have been defined in the model. This is unavoidable. Those who make inferences using ostensibly deterministic models are really making inferences *as if* certain kinds of stochastic element were present. In general, these 'as-if' elements have the property that

$$\begin{pmatrix}\text{mean output from the model}\\ \text{for variable input}\end{pmatrix} = \begin{pmatrix}\text{output from the model}\\ \text{for mean input}\end{pmatrix}$$

As is well known, this assumption breaks down when irreversible states such as population extinction can occur in a model. The correct specification of the stochastic elements in a model can never be just a subsidiary exercise; this has been shown repeatedly during the symposium. Major changes in the channels of an estuary can be triggered by a single storm, epidemics break out in a way only partly related to population density. Such stochastic inputs affect the predictive capacity of models profoundly, and it is essential to be aware of the scope of these limitations in prediction.

At a lower level, random components in a model can produce traps for the unwary if their action is not taken into account. I have twice recently noted in published papers inferences being made from graphs when the ratio of two variates u/v is plotted against v. Impressive-looking linear relations had been found. However, it is not difficult to see that, even in the absence of any relation between u and v, the occurrence of v on both sides of the supposed relation will generate an apparent relation purely from the stochastic variation in v. The correct treatment of this situation requires a model incorporating the joint distribution of the random components of u and v. In this simple situation, the danger is clear, but, in more complex models, there is much more scope for the misleading generation of apparent order out of actual chaos.

Tools for analysing the likelihood function

(1) Graphical techniques; these include,

contour maps for two dimensions,
perspective maps for three dimensions,
moving perspective maps for four or more dimensions.

The last needs a CRT display facility; the others can be done by hand, but will usually require a computer with line printer or plotter.
(2) Transformations to produce quadraticity, i.e. changes of parameters to make the surface in the neighbourhood of the maximum as near quadratic as possible.
(3) Transformations to produce additivity, resulting in the splitting of the likelihood into components to be investigated independently. The data themselves could be made to suggest empirical transformations for these purposes.
(4) Optimization algorithms for finding the position of the maximum. Special strategies (Ross 1970) can drastically improve efficiency in common problems.
(5) Contour-following algorithms would be useful for checking quadraticity, but seem to have been little investigated.

Goodness of fit

The likelihood surface describes the information on the parameters of a model for a given set of data, assuming that the model is correct. Classical techniques use a single statistic, χ^2 or an F-ratio, to show if the residual variation after fitting the model is small enough. It is much better to look at the whole set of residuals. We write

$$\underset{\text{observed}}{y} = \underset{\text{fitted}}{Y} + \underset{\text{residual}}{e}$$

A negative rule is that a model is not adequate if pattern is discernable in the residuals. The converse may not be true. Any residual pattern is relative to present knowledge, and a new idea may throw up a variable highly correlated with residuals.

Contrast this statement of adequacy with the classical significance test, where the residual sum of squares $\Sigma\, e^2$ is compared with the noise level s^2 of background variation. If the ratio is too large, the discrepancy is statistically significant. The test takes no account of the relative sizes of ΣY^2 and Σe^2, i.e. how far we have to go in relation to how far we have come. It also takes no account of the predictive accuracy of the model, which may be entirely adequate for our purpose without accounting for all outstanding variation. As Jeffreys (1939) points out, the Newtonian theory has always given a significantly bad fit if a significance test is used as the criterion.

Gross errors

There are errors and errors; there are stochastic elements and blunders, i.e. gross errors in data. No account of data analysis is complete without considering their effects and detection. An eminent U.S. consultant is in print as saying that between 1 and 10 per cent of yields from complex factorials are wrong. The upper figure implies a level of competence below the acceptable; but even 1 per cent of gross errors can produce a false inference. The problem is one of detection; Daniel (1959) and Tukey (1962) have worked on this, chiefly with multiple classified data, and much more work is needed. Should we acknowledge human fallibility and work with a strategy of discarding the extreme 1 per cent of all data sets initially?

Internal consistency of data

A consequence of thinking of inferences in terms of repeated sampling, of assessing one's data in relation to other sets of data that might have been obtained but weren't, is to ignore powerful methods for looking at the internal consistency of analyses. For example, if a classification technique produces a tree of relationships, this tree may be meaningless, and reflect an unrepeatable stochastic process. However, if the data are divided into random parts and trees formed from each, the stable portion will be revealed. Similarly, when we are faced with intractable likelihoods (Ross), an approximation can be generated by taking random cuts and building up pseudo sub-samples. Again, much remains to be done here.

Experimental design

The crucial role of designed experiments in model-building has been repeatedly stressed. Most of the data used in the projects described at this symposium are necessarily of the survey type, generated by unknown processes, and their information content is not controllable. By contrast, experiments can act as efficient information generators, provided they are properly designed. I doubt if we make sufficient use of two important techniques. The first is that of fractional replication, the use of supersaturated designs with intentional aliasing of effects likely to be large with others likely to be small. High information per experimental unit is possible. The other technique is the calculation of expected information at the design stage. This has two uses, first to optimize the design, and secondly to see if the amount of information is of the right order. Much money and effort can be saved by detecting in advance the hopeless situation where resources are inadequate to generate sufficient information.

Good design is important in simulation also. Experiments can be done on a simulated system with many parameters whose values are ill-defined. Computing costs money and a good experimental design will generate information efficiently.

Much work is needed on the generation of effective treatment patterns that vary with time. Here we enter the territory of the control engineer, who will say that highly variable inputs are the most informative. If this is so, should we revise our ideas on the efficient use of controlled environment cabinets, and stop striving for constancy of inputs in favour of controlled variation? What will the generalization of factorial design look like with dynamic treatment factors?

Computing software

Computers are an essential tool in our activities, and statisticians have been drawn into computing science by their need to use computers effectively in data analysis (Cooper). This has required the augmentation of facilities of general purpose languages. In the archetypal program

```
GET
DO
PUT
```

the algorithmic element DO is flanked by the data-handling elements GET and PUT. Our present languages, Fortran and Algol 60, are primarily algorithmic languages, yet most of the burden of data analysis for large problems is in the data handling. This is not to say that the problem of providing good algorithms is solved. We need better algorithms; in particular, we need to ensure they work in a sequential way, i.e. can be updated by including new data or excluding subsets of existing data. We also need better standardization of language to ensure co-operation between centres (Cooper *et al.* 1968).

But the main problem concerns data handling. The large-scale interchange of data is too often more difficult than the interchange of programs. There is a lack of standards in hardware (we can't load the other man's tape), in operating systems (we can't read the header record), and in the programming languages (we can't interpret his data structures). We urgently need standards for data structures to facilitate data exchange and we should press the manufacturers about the hardware and operating systems so that the make of the other man's computer becomes no barrier.

There are other barriers facing us if we don't do serious work on standards for data description. There is a new generation of languages coming, of which Algol 68 is representative (Van Wijngaaden *et al.* 1969), in which the user can define and name instances of arbitrary data structures and use them as parameters in algorithms. Now, unless we can agree on the form of these structures, algorithms will become non-transferable and the present chaos in their exchange will become worse and not better. So far, our data structures have been defined for us by the language designer, but now he is throwing the responsibility back to the user, which is where it rightly belongs.

Conclusion

Ecology is entering a new era. I view the future with excitement, and I hope to have persuaded you that the statistician has much to offer to its successful development.

References

Box G.E.P. & Jenkins G.M. (1970) *Time-series forecasting and control.* Holden-Day: San Francisco.

Daniel C. (1959) Use of half-normal plots in interpreting factorial two-level experiments. *Technometrics* 1, 311–41.

Edwards A.W.F. (1969) Statistical methods in scientific inference. *Nature* **222**, 1233–7.

Fisher R.A. (1922) On the mathematical foundations of theoretical statistics. *Phil. Trans. R. Soc. Ser. A* **222**, 309–68.

Jeffreys H. (1939) *Theory of probability*. Clarendon Press: Oxford.

Ross G.J.S. (1970) The efficient use of function minimisation in non-linear maximum likelihood estimation. *Appl. Statist.* 19, 205–21.

Tukey J.W. (1962) The future of data analysis. *Ann. math. Statist.* 33, 1–67.

Van Wijngaaden A. (ed.), Mailloux B.M., Peck J.E.L. & Koster C.H.A. (1969) *ALGL* 68. Amsterdam: Mathematisch Centrum.

Cooper B.E., Craddock J.M., Gower J.C., Harrison P.J., Hill I.D. & Nelder J.A. (1968) The construction and description of algorithms. *Appl. Statist.* 17, 175–9.

Summary and assessment: an ecologist's point of view

A.J.RUTTER *Imperial College of Science and Technology, London*

I should like first to emphasize that this is an individual ecologist's point of view, not the output from a generalized model called *the* ecologist. This meeting represents a wide spectrum of ecologists and on the whole I identify myself not with those who have already assayed the mathematical simulation of complex systems, but rather with those who see that something new and potentially very useful is being attempted, namely the simulation of complete ecosystems, and have come here to learn more about it. Like Moliere's Bourgeois Gentilhomme, who was gratified to learn that he had been speaking prose all his life, we discover at the outset that we have been building models all our lives.

I can claim a little preliminary experience since, with colleagues at Imperial College, I have recently constructed a predictive model of one aspect of the hydrological cycle on which there is a wealth of empirical observation but little analytical treatment, namely the interception of precipitation by forest canopies. We proceeded in a somewhat untutored way to produce a model which appears to distinguish and describe the processes involved, and to be capable of making useful predictions. Anyone like ourselves who has assayed the reduction of field observations to a dynamic mathematical model, without much previous contact with other workers in this realm, will recognize the benefit to be gained from further study of the papers on mathematical techniques which have been presented or referred to in this symposium.

However, my task now is not to comment on mathematical techniques, but to assess the value of mathematical models in ecology. In the introductory papers, there were several pieces of cautionary advice given to ecologists embarking on mathematical model building. Mr Skellam warned us against being overpowered or misled by the complexity of mathematics and advised us, before making or using mathematical models, to be clear about both their assumptions and their purpose. In order to make progress, mathematicians often make assumptions about the properties of living organisms, and in my own experience I know that I have sometimes been slow to recognize the biological assumptions inherent in equations whose predictive value, given these assumptions, is readily recognizable. Another important point, which we were reminded of by Mr Jeffers, is that there is no mathematical procedure for the *formulation* of hypotheses.

The aims of the ecologist, as of any other scientist, may be listed as follows:

(1) to observe and describe natural phenomena;
(2) to systematize what he observes;
(3) to explain what he observes;
(4) to extrapolate and predict, from his observations, what is likely to occur in other, as yet uninvestigated, situations which may be provided by nature or contrived by experiment.

As Mr Skellam has pointed out, in (1) we are taking nature as our model; in (2) and (3) we are constructing our own, necessarily simpler, models of nature. I cannot draw a distinct line between (2) and (3), but in (3) I am thinking of models of mechanism rather than of structure and pattern, and finally I am adopting the point of view that the better is the understanding of mechanism, the more reliable is extrapolation and prediction.

We have heard two papers concerned mainly with the systematization of observations, in both cases the observations being the composition of stands of vegetation. To some extent, they dealt with different models of the same thing, the distribution of plants over some defined part of the earth's surface, and they provide a convenient illustration of the need to assess the value of a model by reference to its purpose. Both ordination (Dr Austin's paper) and classification by association analysis (Dr Lambert's) result in an orderly arrangement of large numbers of observations which clarifies for the mind the extent of similarity and difference between stands. Either arrangement may then be used to investigate the relation of the composition of vegetation to its environment. However Dr Austin showed that the development of techniques of ordination, which were more acceptable or convenient from a mathematical point of view, had, almost unnoticed, distorted the relation between environmental gradients and mathematical axes to such an extent as to obscure rather than illuminate the relation of vegetation to environment, which is surely one of the important concerns of ecology.

One of the advantages of association analysis is that it produces a classification from which a map can be made—a useful and practical purpose. However, Dr Lambert's table 2 appears to show that there is little correlation between the classifications of stands which result from the use of different measures of species abundance. Assuming that many ecologists who use maps of vegetation will require maps which distinguish vegetation types by the combination of species and biomass (or some measure closely related to biomass), then maps based solely on the combinations of presence of certain species will have limited value unless the two bases of description give fairly closely correlated results.

Ecologists who are interested in the relation of organisms to environment should make full use of the well developed models of environment which are available to them. Ecologists with primarily descriptive interests still tend to regard the physical environment as definable by the 'levels' of factors and do not always recognize that these levels are in part determined by exchanges of materials and energy between organisms and environment; or, as Clements said at the beginning of the century, organisms react on their environment. There are highly developed mathematical models which portray these exchanges and at the same time clarify the relations between environmental factors which have so often confused the ecologist. Although there are many aspects of environment which are of interest to ecologists, but which have not been exactly modelled, many mathematical models of environment are more advanced than those of biomes, to which they have much to contribute.

During the course of this symposium, some concern has been expressed about the extent to which the analysis of complex systems rests on regression analysis and comparable mathematical techniques. In the happy circumstance where one can fit a linear regression with low residual error, it is often possible to interpret the relation between the variables in terms of mechanism, and to have some confidence in extrapolation. But anyone who has tried to handle data by fitting multiple regressions, and has observed the effect on the size and statistical significance of partial regression coefficients when a further independent variable is introduced, will recognize that multiple regressions are mathematical patterns (I class them under 2 above) which may bear little relation to process and are to be extrapolated only with extreme caution. Ecologists long ago recognized the power of regression and correlation techniques, especially in the very uncontrolled situations in which they have to work, but they have often and justifiably been criticized for too great a readiness to interpret correlation in terms of cause and effect. As they have come to be more experimentally minded, they have recognized that, although hypotheses can be formulated from observations of correlation, these hypotheses require experimental verification, and that, through the experiments, mechanisms may be investigated. One recognizes that biological systems can be analysed at different levels, e.g. community, population, organism, organ, cell, molecule, and that what is an account of mechanism at one level may be merely empirical observation at the next lower level; also that some systems are so complex, e.g. the growth responses of plants to different combinations of mineral nutrients, or the behavioural responses of animals to environmental gradients, that they may at present defy quantitative description in terms of mechanism and can only be described in empirical mathematical terms. Nevertheless, it is more satisfying to most scientists, and more satisfactory for science, that mechanisms should be investigated. Referring to the models of environment mentioned above, it is doubtful if any of them could have been derived by fitting multiple regressions to however large a body of data.

Turning to the fourth aim of the scientist which I have distinguished, namely extrapolation and prediction of what is likely to happen, we find that ecologists are having to extrapolate their data for two distinct purposes. When hypotheses have been formulated to account for observations, they are tested by extrapolating them into as yet uninvestigated circumstances and comparing the predictions with further observations or the results of further experiments. If a hypotheses cannot account for the expanded field of information, it has to be discarded or modified. Extrapolation is therefore part of the testing of hypotheses, and part of the process of science. Increasingly, however, the ecologist is also being asked to use extrapolation not to test hypotheses but as a guide to policy. Since the result of extrapolation has often been to discard hypotheses rather than to confirm them, the ecologist is understandably hesitant that his hypotheses should be used as guides to policy. However, the threat of catastrophic changes in our environmental and biological resources has become so clear and immediate, that most ecologists recognize a responsibility to look into the future.

There is probably some divergence of view as to how the behaviour of an ecosystem or other complex system should be studied, i.e. whether from the outside inwards, by developing explanations of the behaviour of the whole system in terms of its components, or whether to start with the component subsystems and gradually build up to the more complex. Many scientists who can do what they like prefer to work on subsystems in which they can have a reasonable hope of obtaining fundamental understanding. But others, faced with a pressing need to foresee the future of large ecosystems, would rather start with an embracing model, however crude, use it as far as seems justifiable and at the same time proceed to refine it as far and as quickly as they are able. The first is a course for the individualist, the second for a directed and co-ordinated team. Dr Usher and Drs Conway and Murdie gave elegant examples of mathematical models of subsystems, in particular of the changing age-structure of particular populations, and their response to changes in other components of their ecosystems. In effect they illustrate the technique of working outwards from the subsystem, and Dr Usher showed how the analogy of flow of individuals through an age structure could be used in wider context for the flow of energy through a food chain. Professor Van Dyne's paper was of especial value in indicating the aims and problems of organizing co-operative work for the analysis of a whole ecosystem.

When a subsystem is studied, the aim is to analyse the mechanism of its response to given external circumstances. These circumstances are the output from other surrounding subsystems on which the first subsystem may itself react. The inter-relation of the subsystems is not, however, the first object of the analysis.

On the other hand, when an embracing system is studied, the complexity of the inter-relation of the subsystems within the whole may not be apparent because feed-back between them may limit the possible responses to changes in either the environment or the components of the system. Such feed-back between subsystems is one of the reasons why simple models may be made of complex systems. Dr Plinston's paper was concerned with methods of testing the sensitivity of a model to variation in its parameters and it is therefore no criticism of his paper to discuss the model which he took as an example. Part of this model of catchment behaviour involved the transfer of water from the soil through plants to the air, and Dr Plinston will be aware that the physics of water movement through this series of systems has been the subject of complex theoretical analysis. Nevertheless he was able to describe it fairly simply, employing an empirical factor to relate potential transpiration to open water evaporation, and a further empirical quantity, the root constant. This model has been extensively used by, among others, the Meteorological Office. Had he endeavoured to use the more complex models available, he would have required information about both vegetation and soil which in most circumstances would not be available. The simpler model appears to be adequate over a wide range of circumstances, and the overall behaviour of the system is simpler than a detailed study of its complex mechanism would suggest.

Conway and Murdie's paper suggested that the characteristics required of mathematical models are generality, realism and accuracy. Accuracy of prediction of output no doubt depends on the accuracy of the input, of the information given to the model, but, in the preceding paragraph, I have suggested that a good deal of realism may be sacrificed without loss of accuracy in prediction. However, ecologists are not dealing with systems like the motor car whose components remain virtually unchanged in a changing environment, but with ecosystems which maintain their identity only within a limited range of environment. Realism in the model may be less important while the system modelled is not much changed, but this forces me to the conclusion that much of what we call prediction is interpolation, and that if we wish to extrapolate far, realism in description of the mechanism becomes increasingly important and increasingly difficult to attain. To focus our ideas, Professor Goodall suggested that, in the ecosystem dominated by *Artemisia tridentata*, he is interested in predicting the effects of changing climate and grazing pressure on, e.g., vegetational productivity, ground water recharge and deer population. These predictions, though ambitious, do not appear unattainable, provided the vegetation is not changed to one with different dominants, as occurred in Mr Gore's experiment on various frequencies of clipping of ombrogenous bog. Far more difficult is the prediction of change on the scale which would result from the construction of a barrage across Morecambe Bay, which we discussed in the penultimate session. This problem baffled us while we thought about its solution by a mathematical model, until it was suggested that changes in the environment are more readily predictable, and we should then be able to say from previous knowledge what biomes should develop in the new environments. It was pointed out that steady states are predictable from experience, but the time required to attain them, and the nature of the intervening stages, present problems far more complex than are posed for instance even by weather forecasting. The main result of the discussion was perhaps to illustrate that not all the ecologist's experience can at present be reduced to mathematics, and that he is still able to make intelligent predictions without mathematical models. However, few who have attended this meeting would deny the value of mathematical models in ecology and the precision of thought, observation and analysis which their construction entails.

List of participants

Mr P.G. Adlard, Commonwealth Forestry Institute, South Parks Road, Oxford OX1 3RB.
Dr M.B.Alcock, Department of Agriculture, University College of North Wales, Bangor, Caernarvonshire.
Mr S.E.Allen, Nature Conservancy, Merlewood Research Station, Grange-over-Sands, Lancashire.
Mr J.M.Anderson, 51 Rose Street, North Lane, Edinburgh.
Mr M.Anderson, School of Plant Biology, University College of North Wales, Bangor, Caernarvonshire.
Dr F.Andersson, Department of Plant Ecology, University of Lund, O. Vallgatan 14, S-223 61, Lund, Sweden.
Dr M.V.Angel, National Institute of Oceanography, Wormley, Godalming, Surrey.
Mr J.E.Ashe, University of East Anglia, Norwich NOR 88C.
Mr T.P.Atkinson, Department of Geography, University of Birmingham, P.O. Box 363, Birmingham B15 2TT.
Mr S.Attanapola, Department of Botany, University of Southampton, Southampton SO9 5NH.
Dr M.P.Austin, CSIRO, Division of Land Research, P.O. Box 109, Canberra City, A.C.T. 2601, Australia.

Mr A.D.Bailey, Nature Conservancy, Merlewood Research Station, Grange-over-Sands, Lancashire.
Dr J.P.Barkham, School of Environmental Sciences, University of East Anglia, Norwich NOR 88C.
Mr N.Barlow, Department of Forestry and Natural Resources, University of Edinburgh, King's Buildings, Mayfield Road, Edinburgh EH9 3JU.
Mr J.Barrs, Department of Botany, University of Southampton, Southampton SO9 5NH.
Mr B.O.Bartlett, Agricultural Research Council, Letcombe Laboratory, Wantage, Berkshire.
Mr P.J.Beckett, School of Biological Sciences, King's College, 68 Half Moon Lane, London SE24.
Mr J.Beddington, Department of Forestry and Natural Resources, University of Edinburgh, King's Buildings, Mayfield Road, Edinburgh EH9 3JU.
Mr C.B.Benefield, Nature Conservancy, Merlewood Research Station, Grange-over-Sands, Lancashire.
Prof. B.Berthet, Department of Zoology, University of Louvain, 4 Rue Kraken, Louvain, Belgium.
Mr J.B.van Biezen, National Institute for Nature Management, Kemperbergerweg 11, Arnhem, Holland.
Dr H.J.B.Birks, The Botany School, University of Cambridge, Downing Street, Cambridge CB2 3EA.

Dr P.Biscoe, Department of Physiology, School of Agriculture, University of Nottingham, Sutton Bonnington, Loughborough, Leicestershire.

Dr R.E.Blackith, Department of Zoology, University of Dublin, Trinity College, Dublin 2, Eire.

Mr K.L.Bocock, Nature Conservancy, Merlewood Research Station, Grange-over-Sands, Lancashire.

Mr J.Booth, Institute for Marine Environmental Research, Oceanographic Laboratory, 78 Craighall Road, Edinburgh EH6 4RQ.

Dr J.Boss, Department of Physiology, University of Bristol, Bristol BS8 1TH.

Dr D.J.Bradley, Sir William Dunn School of Pathology, University of Oxford, South Parks Road, Oxford OX1 3RE.

Mr P.N.Bradley, Department of Geography, University of Cambridge, Downing Place, Cambridge CB2 3EN.

Dr E.Broadhead, Department of Zoology, University of Leeds, Leeds LS2 9JT.

Mr N.R.Brockington, Grassland Research Institute, Hurley, near Maidenhead, Berkshire.

Mr A.H.F.Brown, Nature Conservancy, Merlewood Research Station, Grange-over-Sands, Lancashire.

Mr M.A.Brunt, Directorate of Overseas Surveys, Land Resources Division, Tolworth Tower, Surbiton, Surrey.

Dr R.G.H.Bunce, Nature Conservancy, Merlewood Research Station, Grange-over-Sands, Lancashire.

Mr R.F.Burden, Department of Landscape Architecture, University of Sheffield, Sheffield S10 2TN.

Mr M.Cahn, School of Plant Biology, University College of North Wales, Bangor, Caernarvonshire.

Mr M.Chambers, Department of Zoology, University of Liverpool, P.O. Box 147, Liverpool L69 3BX.

Mr Champion, M.A.F.F.

Mr B.Chandler, Department of Biological Sciences, Hatfield Polytechnic, Bayfordbury Annexe, near Hertford, Hertfordshire.

Mr R.A.Cheke, University of Leeds, Leeds LS2 9JT.

Mr J.M.Christie, Forestry Commission Research Station, Alice Holt Lodge, Wrecclesham, Farnham, Surrey.

Dr R.S.Clymo, Westfield College, University of London, Kidderpore Avenue, London NW3 7ST.

Mr D.Cohen, The Hebrew University, Jerusalem, Israel.

Mrs P.D.Coker, University College, Gower Street, London WC1.

Dr J.M.Colebrook, Institute for Marine Environmental Research, Oceanographic Laboratory, 78 Craighall Road, Edinburgh EH6 4RQ.

Dr G.R.Conway, Imperial College Field Station, Silwood Park, Ascot, Berkshire.

Dr V.M.Conway, 101 Chequers Avenue, Lancaster, Lancashire.

Mr B.E.Cooper, ICL, Dataskil Limited, Reading Bridge House, Reading RG1 8PN.

Prof. R.T.Coupland, Department of Plant Ecology, University of Saskatchewan, Saskatoon, Canada.

Dr J.R.Crabtree, Grassland Research Institute, Hurley, near Maidenhead, Berkshire.

Dr C.B.Cuellor, London School of Hygiene and Tropical Medicine, Keppel Street, London WC1.

Mr I.E.Currah, National Vegetable Research Station, Wellesbourne, Warwickshire.

Dr J.W.Czerkawski, Hannah Dairy Research Institute, Ayr.

Mr J.Dancer, Department of Biological Sciences, Portsmouth Polytechnic, Hay Street, Portsmouth.

Mr F.H.Dawson, Freshwater Biological Association, River Laboratory, East Stoke, Wareham, Dorset BH20 6BB.

Mr G.J.W.Dean, Rothamsted Experimental Station, Harpenden, Hertfordshire.

Dr C.C.de Bruin, State University of Leyden, Rapenburg 67–33, Leyden, The Netherlands.

Mr J.K.Denmead, Department of Systems Engineering, University of Lancaster, Lancaster, Lancashire.

Dr A.Duncan, Royal Holloway College, Englefield Green, Surrey.

Mr W.Doubleday, Department of Probability and Statistics, University of Sheffield, Sheffield S10 2TN.

Mrs G.Ebdon, Department of Geography, University of Birmingham, P.O. Box 363, Birmingham B15 2TT.

Dr F.E.Eckardt, U.N.E.S.C.O. Paris, France.

Dr C.A.Edwards, Rothamsted Experimental Station, Harpenden, Hertfordshire.

Dr J.E.Ellis, Psychology Department, University of Bristol, Bristol BS8 1TH.

Mr G.W.Elmes, Nature Conservancy, Furzebrook Research Station, Wareham, Dorset.

Miss L.Evangelisti, National Research Council, Rome, Italy.

Mr M.J.R.Fasham, National Institute of Oceanography, Wormley, Godalming, Surrey.

Mr G.C.Fisher, School of Biological and Environmental Studies, New University of Coleraine, Northern Ireland.

Dr E.D.Ford, Institute of Tree Biology, c/o Department of Forestry and Natural Resources, University of Edinburgh, King's Buildings, Mayfield Road, Edinburgh EH9 3JU.

Dr J.Foster, Department of Biological Sciences, Hatfield Polytechnic, Bayfordbury Annexe, near Hertford, Hertfordshire.

Dr D.F.Fourt, Foresty Commission Research Station, Alice Holt Lodge, Wrecclesham, Farnham, Surrey.

Dr J.C.Frankland, Nature Conservancy, Merlewood Research Station, Grange-over-Sands, Lancashire.

Dr A.I.Frazer, Forestry Commission Research Station, Alice Holt Lodge, Wrecclesham, Farnham, Surrey.

Dr D.French, Department of Forestry and Natural Resources, University of Edinburgh, King's Buildings, Mayfield Road, Edinburgh EH9 3JU.

Mr A.S.Gardiner, Nature Conservancy, Merlewood Research Station, Grange-over-Sands, Lancashire.

Mr A.W.Ghent, University of Illinois, Urbana, Illinois 61801, U.S.A.

Dr D.Gifford, Department of Forestry and Natural Resources, University of Edinburgh, King's Buildings, Mayfield Road, Edinburgh EH9 3JU.

Dr F.B.Goldsmith, Department of Botany and Microbiology, University College, University of London, Gower Street, London WC1.

Prof. D.W.Goodall, Ecology Center, Utah State University, Logan, Utah 84321, U.S.A.

Mr R.Goodier, Nature Conservancy, 12 Hope Terrace, Edinburgh EH9 2AS.

Mr A.D.Gordon, Statistical Laboratory, University of Cambridge, Cambridge CB2 1SB.

Mr A.J.P.Gore, Nature Conservancy, Merlewood Research Station, Grange-over-Sands, Lancashire.

Ir. J.Goudriaan, State Agricultural University, Wageningen, Salverdaplein 10, The Netherlands.

Mr R.Goulder, Freshwater Biological Association, Ferry House, Ambleside, Westmorland.

Mr J.Grace, Department of Forestry and Natural Resources, University of Edinburgh, King's Buildings, Mayfield Road, Edinburgh EH9 3JU.

Mr J.E.Grainger, 7 Oakmount Avenue, Highfield, Southampton.

Dr J.S.Gray, Wellcome Marine Laboratory, Robin Hood's Bay, Yorkshire.

Miss S.M.Green, Anti-Locust Research Centre, College House, Wright's Lane, London W8.

Mr K.Gregson, School of Agriculture, Sutton Bonnington, Loughborough, Leicestershire.

Prof. P.Greig-Smith, School of Plant Biology, University College of North Wales, Bangor, Caernarvonshire.

Mr H.M.Grimshaw, Nature Conservancy, Merlewood Research Station, Grange-over-Sands, Lancashire.

Mr R.L.Gulliver, School of Agriculture, University of Nottingham, Department of Agriculture and Horticulture, Sutton Bonnington, Loughborough, Leicestershire.

Dr M.J.Hadley, Faculte de Medecine, 45 Rue des Saints-Peres, Paris 6, France.

Mr R.J.Haggar, Weed Research Organization, Begbroke Hill, Yarnton, Oxford.

Mr Hamilton, University of Strathclyde, George Street, Glasgow C1.

Mr J.Handley, Department of Botany, University of Liverpool, P.O. Box 147, Liverpool, L69 3BX.

Mr J.R.W.Harris, Department of Biology, University of York, Heslington, York YO1 5DD.

Dr A.F.Harrison, Nature Conservancy, Merlewood Research Station, Grange-over-Sands, Lancashire.

Dr M.P.Hassell, Imperial College Field Station, Silwood Park, Ascot, Berkshire.

Mr T.J.Hawes, Nature Conservancy, Monks Wood Experimental Station, Abbots Ripton, Huntingdon.

Dr O.W.Heal, Nature Conservancy, Merlewood Research Station, Grange-over-Sands, Lancashire.

Mr I.Heaney, Freshwater Biological Association, Ferry House, Ambleside, Westmorland.

Mr D.R.Helliwell, Nature Conservancy, Merlewood Research Station, Grange-over-Sands, Lancashire.

Mr J.K.Hibberd, Nature Conservancy, Merlewood Research Station, Grange-over-Sands, Lancashire.

Mr M.O.Hill, School of Plant Biology, University College of North Wales, Bangor, Caernarvonshire.

Mr I.Hodkinson, University of Lancaster, Bailrigg, Lancaster, Lancashire.

Mrs D.M.Howard, Nature Conservancy, Merlewood Research Station, Grange-over-Sands, Lancashire.

Mr P.J.A.Howard, Nature Conservancy, Merlewood Research Station, Grange-over-Sands, Lancashire.

Mr W.J.Howard, Department of Biology, University of York, Heslington, York YO1 5DD.

Miss E.Y.Howarth, Freshwater Biological Association, Ferry House, Ambleside, Westmorland.

Mr R.S.Howell, Forestry Commission Research Station, Alice Holt Lodge, Wrecclesham, Farnham, Surrey.

Mrs G.M.Howson, Nature Conservancy, Merlewood Research Station, Grange-over-Sands, Lancashire.

Dr M.K.Hughes, Department of Botany, University of Durham, South Road, Durham.

Dr R.N.Hughes, School of Plant Biology, University College of North Wales, Bangor, Caernarvonshire.

Mr C.S.Hutchinson, School of Plant Biology, University College of North Wales, Bangor, Caernarvonshire.

Miss C.H.Jackson, East Malling Research Station, Maidstone, Kent.

Dr R.James, School of Biology, University of East Anglia, Norwich NOR 88C.

Dr R.L.Jefferies, School of Biological Sciences, University of East Anglia, Norwich NOR 88C.

Mr J.N.R.Jeffers, Nature Conservancy, Merlewood Research Station, Grange-over-Sands, Lancashire.

Dr J.G.Jones, Freshwater Biological Association, Ferry House, Ambleside, Westmorland.

Dr H.E.Jones, Nature Conservancy, Merlewood Research Station, Grange-over-Sands, Lancashire.

Mr S.Jonsson, University of Uppsala, Aspvagen 10, Storvreta, S-754 60 Uppsala, Sweden.

Mr L.Kareniampi, Department of Botany, University of Turku, 20 500 Turku 50, Finland.

Dr A.J.Kayll, Faculty of Forestry, University of New Brunswick, Fredericton, New Brunswick, Canada.

Mr R.A.Kempton, Rothamsted Experimental Station, Harpenden, Hertfordshire.

Mr J.W.Kent, University of Newcastle, Newcastle upon Tyne NE1 7RU.

Mr T.J.King, The Botany School, University of Oxford, South Parks Road, Oxford OX1 3RA.

Miss C.Kipling, Freshwater Biological Association, Ferry House, Ambleside, Westmorland.

Dr J.M.Lambert, Department of Botany, University of Southampton, Southampton SO9 5NH.

Dr P.O.Larsson, Ivl, Box 4052, 400 40 Goteberg 4, Sweden.

Miss P.M.Latter, Nature Conservancy, Merlewood Research Station, Grange-over-Sands, Lancashire.

Dr R.Laughlin, Imperial College Field Station, Silwood Park, Ascot, Berkshire.

Mr R.M.Lawton, Director of Overseas Surveys, Land Resources Division, Tolworth Tower, Surbiton, Surrey.

Dr E.D.Le Cren, Freshwater Biological Association, River Laboratory, East Stoke, Wareham, Dorset BH20 6BB.

Miss Leese, Department of Applied Physics, University of Strathclyde, George Street, Glasgow C1.

Mr B.G.Lewis, Central Electricity Research Laboratory, Kelvin Avenue, Leatherhead, Surrey.

Mr M.J.Liddle, School of Plant Biology, University College of North Wales, Bangor, Caernarvonshire.

Mr D.K.Lindley, Nature Conservancy, Merlewood Research Station, Grange-over-Sands, Lancashire.

Dr T.Lindstrom, Institute of Freshwater Research, 2-17011, Drottningholm, Sweden.

Dr P.S.Lloyd, Department of Botany, University of Sheffield, Sheffield S10 2TN.

Mr B.C.Longstaff, Department of Biology, University of York, Heslington, York YO1 5DD.

Mr V.P.W.Lowe, Nature Conservancy, Merlewood Research Station, Grange-over-Sands, Lancashire.

Dr M.L.Luff, Department of Agricultural Zoology, University of Newcastle, Field Laboratory, Close House, Heddon on the Wall, Newcastle upon Tyne 5.

Mr B.H.McArdle, 1 Richmond Terrace, Bristol BS8 1AB.

Mr I.McDonald, Thames Conservancy, River Purification Department, Reading Bridge House, Reading RG1 8PR.

Dr W.W.MacDonald, Sub-Department of Entomology, School of Tropical Medicine, Liverpool L3 5QA.

Prof. A.MacFadyen, New University of Ulster, Coleraine, Northern Ireland.

Mr D.Machin, School of Plant Biology, University College of North Wales, Bangor, Caernarvonshire.

Mrs P.G.McIntyre, Department of Forestry and Natural Resources, University of Edinburgh, King's Buildings, Mayfield Road, Edinburgh EH9 3JU.

Dr S.McNeill, Imperial College Field Station, Silwood Park, Ascot, Berkshire.

Dr R.H.K.Mann, Freshwater Biological Association, River Laboratory, East Stoke, Wareham, Dorset BH20 6BB.

Mr Markov, U.S. Army Material Command, 6 Frankfurt am Main, West Germany.

Mrs S.E.Meacock, Department of Mathematics, University of Southampton, Southampton SO9 5NH.

Mr R.Mead, Department of Applied Statistics, University of Reading, Reading, RG6 2AH.

Dr H.G. Miller, Macaulay Institute for Soil Research, Craigiebuckler, Aberdeen.

Dr C.Milner, Nature Conservancy, 12 Hope Terrace, Edinburgh EH9 2AS.

Father J.J.Moore, Department of Botany, University College, Belfield, Dublin 4, Eire.

Mr P.G.Moore, Wellcome Marine Laboratory, Robin Hood's Bay, Yorkshire.

Miss S.K.Morrell, Nature Conservancy, Merlewood Research Station, Grange-over-Sands, Lancashire.

Mr D.Morris, School of Plant Biology, University College of North Wales, Bangor, Caernarvonshire.

Dr A.J.Morton, Imperial College Field Station, Silwood Park, Ascot, Berkshire.

Mr Mott, University of Dundee, Dundee DD1 4HN.

Mr M.D.Mountford, Nature Conservancy, Biometrics Section, 19 Belgrave Square, London SW1X 8PY.

Mr R.Muetzelfeldt, Department of Forestry and Natural Resources, University of Edinburgh, King's Buildings, Mayfield Road, Edinburgh EH9 3JU.

Dr J.A.Nelder, Rothamsted Experimental Station, Harpenden, Hertfordshire.

Dr G.A.Norton, Imperial College Field Station, Silwood Park, Ascot, Berkshire.

Mr I.Noy-Meir, Department of Botany, The Hebrew University of Jerusalem, Israel.

Mr Orsham, The Hebrew University, Jerusalem, Israel.

Mr Paige, Bangor.

Mr W.M.Patefield, National Vegetable Research Station, Wellesbourne, Warwickshire

Dr Pearson, Scottish Marine Biological Association, Oban, Argyll.

Mr Peters, Water Resources Board, Reading Bridge House, Reading RG1 8PS

Dr T.J.Pitcher, School of Biological and Environmental Studies, New University of Ulster, Coleraine, Northern Ireland.

Dr D.Plinston, Institute of Hydrology, Howbery Park, Wallingford, Berkshire.

Mr D.J.Pratt, Directorate of Overseas Surveys, Land Resources Division, Kingston Road, Tolworth Tower, Surbiton, Surrey.

Mr J.E.A.Procter, Nature Conservancy, 12 Hope Terrace, Edinburgh EH9 2AS.

Mr P.J.Radford, Water Resources Board, Reading Bridge House, Reading RG1 8PS.

Mr P.F.Randerson, Department of Botany and Microbiology, University College, Gower Street, London WC1.

Dr E.R.C.Reynolds, Commonwealth Forestry Institute, South Parks Road, Oxford OX1 3RB.

Dr T.B.Reynoldson, Department of Zoology, University College of North Wales, Bangor, Caernarvonshire.

Miss D.Richards, Imperial College Field Station, Silwood Park, Ascot, Berks,

Mr T.O.Robson, Weed Research Organization, Begbroke Hill, Yarnton, Oxford.

Mr J.Rodwell, Department of Botany, University of Southampton, Southampton SO9 5NH.

Mr G.J.S.Ross, Rothamsted Experimental Station, Harpenden, Herts.

Dr T.Rosswall, Department of Microbiology, Agricultural College, S-750 07, Uppsala 7, Sweden.

Prof. A.J.Rutter, Imperial College of Science and Technology, Prince Consort Road, London SW7.

Dr Samuel, Forestry Commission Research Station, Alice Holt Lodge, Wrecclesham, Farnham, Surrey.

Mr J.Sarukhan, School of Plant Biology, University College of North Wales, Bangor, Caernarvonshire.

Dr J.E.Satchell, Nature Conservancy, Merlewood Research Station, Grange-over-Sands, Lancashire.

Mr I.Sauvezon, Universite de Montpellier, 31 Rue de l'Universite, Montpellier, France.

Dr F.Schiemer, Limnol Abteilung, Berggasse 18/19, A 1090 Wien 9, Austria.

Mr M.W.Shaw, Nature Conservancy, Merlewood Research Station, Grange-over-Sands, Lancashire.

Miss W.Simmonds, Department of Agriculture, University College of North Wales, Bangor, Caernarvonshire.

Mr J.G.Skellam, Nature Conservancy, Biometrics Section, 19 Belgrave Square, London SW1X 8PY.

Mr F.M.Slater, Department of Botany, University College of Wales, Aberystwyth, Wales.

Mrs P.J.Smartt, Department of Botany, University of Southampton, Southampton SO9 5NH.

Dr V.Standen, Department of Zoology, University of Durham, Durham.

Mr D.Stewart, Ministry of Agriculture, Dandonald House, Upper Newtownards Road, Belfast, Northern Ireland.

Mr J.T.Stoakley, Forestry Commission Research Station, Alice Holt Lodge, Wrecclesham, Farnham, Surrey.

Mr Stokes, Department of Biological Sciences, Hatfield Polytechnic, Bayfordbury Annexe, near Hertford, Hertfordshire.

Mr M.D.Swaine, School of Plant Biology, University College of North Wales, Bangor, Caernarvonshire.

Mrs S.Swaine, Department of Agriculture, University College of North Wales, Bangor, Caernarvonshire.

Mr J.M.Sykes, Nature Conservancy, Merlewood Research Station, Grange-over-Sands, Lancashire.

Dr P.M.Symmons, Anti-Locust Research Centre, College House, Wrights Lane, London W8.

Mrs R.M.Taylor, Department of Botany, University of Southampton, Southampton SO9 5NH.

Dr L.R.Taylor, Rothamsted Experimental Station, Harpenden, Hertfordshire.

Dr P.Tett, Scottish Marine Biological Association, Oban.

Mr J.E.Thorpe, Freshwater Fisheries Laboratory, Pitlochry, Perthshire.

Mr H.Thomas, Agronomy Department, Welsh Plant Breeding Station, Aberystwyth, Wales.

Dr F.B.Thompson, Commonwealth Forestry Institute, South Parks Road, Oxford OX1 3RB.

Dr L.Tieszen, Augustana College, South Dakota, U.S.A.

Mr B.Tveite, Norwegian Forestry Research Institute, Vollebekk, Norway.

Dr J.W.Twidell, Department of Applied Physics, University of Strathclyde, Glasgow C1.

Mr C.Urquhart, Department of Biological Sciences, University of Aston-in-Birmingham, George Alexander Building, Birmingham 4.

Dr M.B.Usher, Department of Biology, University of York, Heslington, York YO1 5DD.

Dr Vanderaart, State University of Leyden, Rapenburg 67–73, Leyden, The Netherlands.

Prof. G.M.Van Dyne, Natural Resource Ecology Laboratory, Colorado State University, Fort Collins, Colorado 80521, U.S.A.

Dr B.H.Walker, Division of Biological Sciences, University of Rhodesia, P.O. Box MP 167, Salisbury, Rhodesia.

Prof. D.Walker, Australian National University, Canberra, Box 4, ACT 2600, Australia.

Mr S.D.Ward, Nature Conservancy, Penrhos Road, Bangor, Caernarvonshire.

Mr Waughman, Department of Biology, University of Dundee, Dundee DD1 4HN.

Dr D.F.Westlake, Freshwater Biological Association, River Laboratory, East Stoke, Wareham, Dorset BH20 6BB.

Mr E.J.White, Nature Conservancy, Merlewood Research Station, Grange-over-Sands, Lancashire.

Mr J.White, Department of Botany, University College, Belfield, Dublin 4, Ireland.

Dr J.B.Whittaker, Department of Biology, University of Lancaster, Bailrigg, Lancaster, Lancashire.

Prof. F.E.Wielgolaski, Botanical Laboratory, University of Oslo, Norway.

Prof. G.Williams, Department of Zoology, University of Reading, Reading RG6 2AN

Mr J.B.Wilson, Hartley Botanical Laboratories, University of Liverpool, P.O. Box 147, Liverpool L69 3BX.

Mr I.P.Woiwod, Rothamsted Experimental Station, Harpenden, Hertfordshire

Mr T.Wyatt, Fisheries Laboratory, Ministry of Agriculture, Fisheries & Food, Lowestoft, Suffolk.

Author Index

Bold figures refer to pages where full references appear

Subject Index